Lecture Notes in Bioengineering

Lecture Notes in Bioengineering (LNBE) publishes the latest developments in bioengineering. It covers a wide range of topics, including (but not limited to):

- Bio-inspired Technology & Biomimetics
- Biosensors
- Bionanomaterials
- Biomedical Instrumentation
- Biological Signal Processing
- Medical Robotics and Assistive Technology
- Computational Medicine, Computational Pharmacology and Computational Biology
- Personalized Medicine
- Data Analysis in Bioengineering
- Neuroengineering
- Bioengineering Ethics

Original research reported in proceedings and edited books are at the core of LNBE. Monographs presenting cutting-edge findings, new perspectives on classical fields or reviewing the state-of-the art in a certain subfield of bioengineering may exceptionally be considered for publication. Alternatively, they may be redirected to more specific book series. The series' target audience includes advanced level students, researchers, and industry professionals working at the forefront of their fields.

Indexed by SCOPUS, INSPEC, zbMATH, SCImago.

Francesca Vipiana · Lorenzo Crocco
Editors

Electromagnetic Imaging for a Novel Generation of Medical Devices

Fundamental Issues, Methodological Challenges and Practical Implementation

Springer

Editors
Francesca Vipiana
Department of Electronics
and Telecommunications
Politecnico di Torino
Turin, Italy

Lorenzo Crocco
CNR-IREA (Institute for Electromagnetic
Sensing of the Environment)
National Research Council
Naples, Italy

ISSN 2195-271X ISSN 2195-2728 (electronic)
Lecture Notes in Bioengineering
ISBN 978-3-031-28665-0 ISBN 978-3-031-28666-7 (eBook)
https://doi.org/10.1007/978-3-031-28666-7

This Springer imprint is published by the registered company Springer Nature Switzerland AG
The registered company address is: Gewerbestrasse 11, 6330 Cham, Switzerland

Preface

Nowadays, healthcare systems are facing an ever-growing number of challenges, due to aging populations, which causes a growing burden of care on medical services. In this framework, medical imaging technologies, such as X-ray, ultrasound and magnetic resonance imaging, play a key role, being the essential clinical tool to deliver accurate initial diagnosis and monitor the evolution of disease over time. Despite the excellent results obtained by means of these modalities, improvements are still required in terms of, e.g., specificity and sensitivity for some diseases, reduction of costs to allow a widespread access and non-invasiveness, considering the use of contrast agents and ionizing radiations (in X-ray).

For these reasons, a whole range of novel medical imaging modalities is currently being investigated and developed to supplement and support current modalities, such as, for instance, optical coherence tomography, hyperspectral imaging and multi-modal systems, e.g., photoacoustic imaging. Among these technologies, there is electromagnetic (EM) imaging, which involves the illumination of the portion of the body under investigation with low-power EM waves (in the GHz range of spectrum, often referred to as the microwave spectrum) and the use of the resultant backscattered signals arising from the interaction between the human tissue and the incoming EM waves to generate images of the internal structures of the body. EM imaging can bring unique contributions to a number of clinical applications overcoming the above-mentioned limitations, thanks to its inherent appealing properties. EM imaging at microwave frequencies exhibits favorable penetration depths that allow imaging tissues deeper in the human body as compared to other emerging modalities (e.g., optical techniques). Moreover, EM imaging is completely harmless, since the involved waves are non-ionizing and used in very low doses (low-power EM waves). Finally, EM technology is economically sustainable for the healthcare system. Due to the progress in mobile industry and microwave devices in the recent years, EM imaging has the potential to provide mobile, low-cost imaging platforms well suited to the above-mentioned future healthcare needs.

This book presents, for the first time, a comprehensive overview of all aspects related to the development of devices for medical imaging using electromagnetic imaging technologies. The basis of this book was laid by the Marie Skłodowska-Curie Innovative Training Network (MSCA-ITN) "EMERALD—ElectroMagnetic imaging for a novel genERation of medicAL Devices", run from 2018 to 2022 within the Horizon 2020 program of the European Union (www.MSCA-EMERALD.Eu/).

EMERALD was a coherent action of leading European engineering groups involved in EM technology for medical imaging to accelerate the translation of research in EM medical imaging into clinical prototypes. The international and inter-sectoral consortium consisted of ten beneficiaries and 17 partner institutions spread across 11 European countries, which joined their forces to implement a unique scientific and training program. In this way, EMERALD has established a group of 13 outstanding early-stage researchers capable of driving the future developments of EM imaging technology, thanks to the targeted skills they have attained and the connections they established among them and with clinicians and stakeholders.

The book collects the main achievements of the EMERALD network, and it is divided into 11 chapters. The first six chapters deal with issues and challenges which are common to the design of any microwave imaging medical device, whereas the last five chapters are focused on specific applications.

The availability of phantoms is a crucial aspect in the development of EM imaging medical devices, as it enables reliable testing before reaching the stage of pre-clinical validation on volunteers. In addition, defining shared phantom production rules and recipes is very important toward the standardization of devices. In this respect, the first chapter "Standardized Phantoms", authored by Nadine Joachimowicz et al., deals with the design and development of several anthropomorphic phantoms. The phantoms mimic physical and dielectric characteristics of the human body at microwaves and are adapted to the geometry of the EM imaging systems developed in the EMERALD project. The ongoing work relevant to this chapter is constantly updated and accessible through a devoted website.

Microwave imaging systems are essentially constituted by a hardware component dedicated to signal generation and acquisition and a software component in charge of the image formation. To bridge the gap between these two parts, the second chapter "Hardware Acceleration of Microwave Imaging Algorithms", authored by Mohammad Amir Mansoori and Mario R. Casu, identifies the computational kernels that are recurrent in EM imaging devices and presents methodologies to design specific hardware accelerators in Field-Programmable Gate Arrays (FPGAs) for these kernels. In so doing, this work paves the way to the development of systems in which dedicated hardware is exploited to perform the device-specific imaging task in real time.

The following chapter, "Metasurface Technology for Medical Imaging", is focused on the improvements of the hardware component of microwave imaging systems and investigates the use of the emerging electromagnetic metasurface technology to replace standard antennas. To this end, Eleonora Razzicchia et al. present metasurface antennas and sensors specifically designed and developed for some of the EMERALD devices and show their capability of enhancing detection by efficiently tackling the impedance mismatch that electromagnetic waves encounter when probing human tissues.

Another aspect which is common to the development of EM imaging medical devices is the need of an EM simulation environment capable of reliably predicting the behavior of the designed device before realization and experiments. This requires the capability of performing numerical simulations that properly model the electromagnetic behavior of human tissue, while being accessible to users. In addition, reliable EM solvers are also important in the development of imaging methods. Toward this goal, the contribution of Tushar Singh et al., in the fourth chapter "Numerical Modeling of Complex 3D Electromagnetic Scenarios for Medical Microwave Imaging", describes a general-purpose three-dimensional (3-D) EM software that includes editors for importing and processing files describing realistic human phantoms, triangular and quadrilateral meshers to discretize the imported geometries, libraries of human phantoms with a variety of tissues and antennas templates.

As far as the software component of microwave imaging systems is concerned, there is a vast body of literature covering inverse scattering algorithms. However, in the development of medical microwave imaging devices, it is important to design and develop algorithms that are meant to cope with the specific challenges posed by the clinical application which is targeted. For the case of brain stroke imaging, Olympia Karadima et al. illustrate, in the fifth chapter "Assessment and Validation of 2-D and 3-D DBIM-TwIST Algorithm for Brain Stroke Detection and Differentiation", how the distorted Born iterative algorithm using with two-step iterative shrinkage thresholding (DBIM-TwIST) can be incorporated in the data-processing flow aimed at generating diagnostic images from a prototype system.

Like in any field of applied science, also in microwave imaging the paradigm of machine learning is emerging as a powerful opportunity to deal with the complex challenges inherent to nonlinear inverse problems in a more effective way. The sixth chapter "Deep Learning Enhanced Medical Microwave Imaging" authored by Álvaro Yago Ruiz et al. is devoted to this growing topic and introduces the reader to the thriving field of deep learning-enhanced medical microwave imaging. In particular, the chapter describes the way in which deep learning can be exploited in microwave imaging and presents two examples showing the potential of these novel techniques.

The first of the five chapters dealing with specific by David O. Rodriguez-Duarte et al. and is entitled "Towards a Microwave Imaging Device for Cerebrovascular Diseases Monitoring: from Numerical Modeling to Experimental Testing" and presents an ad hoc EM imaging device for monitoring of cerebrovascular diseases. In particular, the design choices underneath the developed system are outlined, and their realization is shown. The performances of the system are analyzed using full-fledged numerical simulations and validated with controlled laboratory experiments involving a full-scale prototype.

The eighth chapter "The Dielectric Properties of Axillary Lymph Nodes", authored by Matteo Savazzi et al., is related to a novel application of medical microwave imaging, namely the early detection of breast tumors by inspecting axillary lymph nodes (ALN). Given the novelty of the proposed application, the main goal of the authors is to provide a reliable characterization of ALN dielectric properties in the microwave frequency range. To implement a cross-validation framework, the authors consider two approaches, one based on the coaxial probe measurement method and the other exploiting magnetic resonance. The obtained results confirm the potential of microwave imaging for ALN inspection and pave the way to the design of a microwave tomography system for breast cancer diagnosis based on this principle.

The chapter authored by Aleksandar Janjic et al. "SAFE—Microwave Imaging Device for Breast Cancer Early Screening and Diagnostics" also deals with breast cancer diagnosis and present the clinical validation of a novel EM device developed by one of the industrial partners of the EMERALD consortium. The study was conducted on 59 patients, and outcomes were evaluated based on the clinical reports provided by radiologists. Overall, the results demonstrate that the device could be employed to distinguish between healthy and affected breasts, with an accuracy of 81%, sensitivity of 81% and specificity of 83%. Interestingly, a machine learning model was used for detection, while qualitative microwave imaging algorithms were used for localization.

The contribution of Alexandra Prokhorova et al., in the tenth chapter "Microwave Ultra-Wideband Imaging for Non-invasive Temperature Monitoring During Hyperthermia Treatment", is devoted to microwave imaging for non-invasive temperature monitoring during hyperthermia therapy. Hyperthermia is a promising treatment strategy for cancer because of its safety and cost-effectiveness in which malignant tissue temperature is increased in order to improve the effect of chemo- and radiotherapic drugs. In this framework, it is important to control the temperature in the whole exposed region to ensure treatment effectiveness and avoid unwanted heating in healthy tissue. To this end, considering the case of neck cancer, the authors present a temperature estimation device based on ultra-wideband M-sequence radar technology, studying the most convenient antenna arrangement and providing an initial experimental validation of the full prototype of the device.

Finally, the eleventh chapter "An Initial Assessment of a Microwave Imaging System to Monitor Microwave Ablation Treatments", authored by Mengchu Wang et al., describes a microwave imaging system to monitor thermal ablation treatment for liver cancer. Opposite to hyperthermia, thermal ablation is a (minimally) invasive approach in which the EM energy is directly administered at the tumor site, in order to kill cancer cells. In so doing, monitoring the extent of the treated region is essential to avoid under- and over-treatment. The authors present the guidelines that have inspired the design of the device, the basic components and an initial experimental assessment of the monitoring principle.

Fig. 1 First EMERALD network event at Politecnico di Torino (Torino, Italy), February 2019

Fig. 2 EMERALD supervisory board meeting during the pandemic season, October 2021

As described above, this book presents a multi-disciplinary and multi-sectorial overview of the electromagnetic imaging technology for a novel generation of medical devices, gathering contributions by researchers from engineering, applied physics and medical communities. From a broader perspective, this book represents an important step to reinforce the multi-disciplinary scientific community that has been generated around the EMERALD network (Figs. 1–3) and will beyond it, providing a proper environment for approaching challenges of electromagnetic imaging technology once translated into clinical applications.

Fig. 3 EMERALD final conference at Sorbonne University (Paris, France), April 2022

Finally, we would like to thank all our colleagues, for the energy and efforts in preparing and writing all the chapters: without them, it would have been impossible to achieve this outcome.

Turin, Italy — Francesca Vipiana
Naples, Italy — Lorenzo Crocco

Acknowledgments This book was supported by the European Union's Horizon 2020 Research and Innovation Program under the EMERALD project, Marie Sklodowska-Curie grant agreement No. 764479.

Contents

Contributors

Soroush Abedi GeePs, Sorbonne Université, CNRS, Laboratoire de Génie Electrique et Electronique de Paris, Paris, France

Ibrahim Akduman Mitos Medical Technologies, Maslak, Istanbul, Turkey; Faculty of Electrical and Electronics Engineering, Istanbul Technical University, Maslak, Istanbul, Turkey

Mustafa Erkin Aribal Mitos Medical Technologies, Maslak, Istanbul, Turkey; Department of Radiology, Breast Health Center, Altunizade Hospital, Acibadem M.A.A. University, Atasehir, Istanbul, Turkey

Onur Bugdayci Department of Radiology, School of Medicine, Marmara University, Pendik, Istanbul, Turkey

Tiago Castela Departamento de Radiologia, Hospital da Luz Lisboa, Luz Saúde, Lisbon, Portugal

Mario R. Casu Department of Electronics and Telecommunications, Politecnico di Torino, Turin, Italy

Marta Cavagnaro Department of Information Engineering, Electronics, and Telecommunications (DIET), University of Rome "La Sapienza", Rome, Italy; IREA-CNR, Institute for Electromagnetic Sensing of the Environment, National Research Council of Italy, Naples, Italy

Mehmet Cayoren Mitos Medical Technologies, Maslak, Istanbul, Turkey; Faculty of Electrical and Electronics Engineering, Istanbul Technical University, Maslak, Istanbul, Turkey

Raquel C. Conceição Instituto de Biofísica e Engenharia Biomédica (IBEB), Faculdade de Ciências, Universidade de Lisboa, Lisbon, Portugal

Sandra Costanzo National Research Council of Italy Institute for Electromagnetic Sensing of the Environment (CNR-IREA), Napoli, Italy;
Department of Computer Engineering, Modeling, Electronics and Systems (DEIS), University of Calabria, Rende, Italy;
Inter-University National Research Center on Interactions Between Electromagnetic Fields and Biosystems (ICEmB), Genoa, Italy

Lorenzo Crocco IREA-CNR, Institute for Electromagnetic Sensing of the Environment, National Research Council of Italy, Naples, Italy

João M. Felício Instituto de Telecomunicações, Instituto Superior Técnico (IST), Universidade de Lisboa, Lisbon, Portugal;
Centro de Investigação Naval (CINAV), Escola Naval, Almada, Portugal

Carlos A. Fernandes Instituto de Telecomunicações, Instituto Superior Técnico (IST), Universidade de Lisboa, Lisbon, Portugal

Ondrej Fiser Department of Biomedical Technology, Czech Technical University in Prague, Prague, Czech Republic

Navid Ghavami Faculty of Natural and Mathematical Sciences, King's College London, Strand, London, UK

Daniela M. Godinho Instituto de Biofísica e Engenharia Biomédica (IBEB), Faculdade de Ciências, Universidade de Lisboa, Lisbon, Portugal

Marko Helbig Institute of Biomedical Engineering and Informatics, Technische Universität Ilmenau, Ilmenau, Germany

Niko Ištuk Translational Medical Device Lab, National University of Ireland Galway, Galway, Ireland

Aleksandar Janjic Mitos Medical Technologies, Maslak, Istanbul, Turkey;
Faculty of Electrical and Electronics Engineering, Istanbul Technical University, Maslak, Istanbul, Turkey

Nadine Joachimowicz GeePs, Sorbonne Université, CNRS, Laboratoire de Génie Electrique et Electronique de Paris, Paris, France;
Université Paris Cité, Paris, France

Olympia Karadima Faculty of Natural and Mathematical Sciences, King's College London, Strand, London, UK

Branko Kolundzija WIPL-D, Belgrade, Serbia;
School of Electrical Engineering, University of Belgrade, Belgrade, Serbia

Panagiotis Kosmas Faculty of Natural and Mathematical Sciences, King's College London, Strand, London, UK

Pan Lu Faculty of Natural and Mathematical Sciences, King's College London, Strand, London, WC2R 2LS, UK

Mohammad Amir Mansoori Department of Electronics and Telecommunications, Politecnico di Torino, Turin, Italy

Marija Nikolic Stevanovic School of Electrical Engineering, University of Belgrade, Belgrade, Serbia

Branislav Ninkovic WIPL-D, Belgrade, Serbia

Darko Ninković University of Belgrade, Belgrade, Serbia

Martin O'Halloran Translational Medical Device Lab, National University of Ireland Galway, Galway, Ireland

Cristina Origlia Department of Electronics and Telecommunications, Politecnico di Torino, Turin, Italy

Maria L. Orvalho Departamento de Radiologia, Hospital da Luz Lisboa, Luz Saúde, Lisbon, Portugal

Emily Porter Translational Medical Device Lab, National University of Ireland Galway, Galway, Ireland;
Department of Electrical and Computer Engineering, The University of Texas at Austin, Austin, USA

Alexandra Prokhorova Institute of Biomedical Engineering and Informatics, Technische Universität Ilmenau, Ilmenau, Germany

Eleonora Razzicchia Faculty of Natural and Mathematical Sciences, King's College London, Strand, London, UK

David O. Rodriguez-Duarte Department of Electronics and Telecommunications, Politecnico di Torino, Turin, Italy

Hélène Roussel GeePs, Sorbonne Université, CNRS, Laboratoire de Génie Electrique et Electronique de Paris, Paris, France

Matteo Savazzi Instituto de Biofísica e Engenharia Biomédica (IBEB), Faculdade de Ciências, Universidade de Lisboa, Lisbon, Portugal

Rosa Scapaticci IREA-CNR, Institute for Electromagnetic Sensing of the Environment, National Research Council of Italy, Naples, Italy

Tushar Singh WIPL-D, Belgrade, Serbia

Ioannis Sotiriou Faculty of Natural and Mathematical Sciences, King's College London, Strand, London, WC2R 2LS, UK

Mladjen Stevanetic WIPL-D, Belgrade, Serbia

Marija N. Stevanovic University of Belgrade, Belgrade, Serbia

Miodrag Tasic WIPL-D, Belgrade, Serbia;
School of Electrical Engineering, University of Belgrade, Belgrade, Serbia

Jorge A. Tobón Vasquez Department of Electronics and Telecommunications, Politecnico di Torino, Turin, Italy

Giovanna Turvani Department of Electronics and Telecommunications, Politecnico di Torino, Turin, Italy

Francesca Vipiana Department of Electronics and Telecommunications, Politecnico di Torino, Turin, Italy

Jan Vrba Department of Electromagnetic Field, Czech Technical University in Prague, Prague, Czech Republic

Mengchu Wang Department of Information Engineering, Electronics, and Telecommunications (DIET), University of Rome "La Sapienza", Rome, Italy;
National Research Council of Italy Institute for Electromagnetic Sensing of the Environment (CNR-IREA), Napoli, Italy

Álvaro Yago Ruiz DIET, Department of Information Engineering, Electronics and Telecommunications, University of Rome "La Sapienza", Rome, Italy;
IREA-CNR, Institute for Electromagnetic Sensing of the Environment, National Research Council of Italy, Naples, Italy

Standardized Phantoms

Soroush Abedi, Hélène Roussel, and Nadine Joachimowicz

Abstract Microwave imaging offers a variety of devices for patient diagnosis, monitoring, and treatment. To develop and evaluate these devices, it is necessary to design anthropomorphic reference phantoms with physical and dielectric characteristics similar to those of the human body. This chapter deals with the design and development of several anthropomorphic phantoms adapted to the geometry of the microwave imaging systems. They are composed of 3D printed anthropomorphic cavities filled with liquid mixtures of TritonX-100 and salt water. The concentrations of the latter, to mimic tissues over a wide band, are given by a Gauss–Newton algorithm combined with the use of a binary mixture law. The process is extended to the production of liquid mixtures mimicking any biological tissue, at least in the [0.5–6] GHz frequency range. This opens new avenues in the standardizing of phantoms for assessing microwave imaging systems.

Keywords Microwave imaging · Wideband anthropomorphic phantom · Dielectric characterization · Diagnosis · Monitoring · Preclinical testing

1 Introduction

Experiments on humans are extremely controlled, complex to implement, and very costly. It is therefore particularly interesting to have anthropomorphic phantoms to carry out preclinical studies. This can be useful for the design of the systems, the evaluation of the inversion algorithms as well as to understand the alteration of the pathological tissues studied, thanks to the possibility of varying the values of the dielectric permittivity in the different parts of the phantoms. In many laboratories

S. Abedi · H. Roussel · N. Joachimowicz (✉)
GeePs, Sorbonne Université, CNRS, Laboratoire de Génie Electrique et Electronique de Paris, 75252 Paris, France
e-mail: nadine.joachimowicz@paris7.jussieu.fr

N. Joachimowicz
Université Paris Cité, 75006 Paris, France

F. Vipiana and L. Crocco (eds.), *Electromagnetic Imaging for a Novel Generation of Medical Devices*, Lecture Notes in Bioengineering,
https://doi.org/10.1007/978-3-031-28666-7_1

canonical objects are used as alternatives of human body. However, these objects are neither physically nor dielectrically close to human tissues. Conversely, since a phantom is designed to be exposed to electromagnetic waves, its shape as well as its dielectric properties play an important role to have relatively accurate reflections and penetrations based on the realistic anatomy of the organ. Therefore, the phantoms must satisfy three main criteria: be realistic, have dielectric constant values consistent with the values of the biological tissues they emulate over the frequency range of interest and be stable over time due to the need of conducting several experiments in the same conditions.

Phantoms are classified into various types. Depending on the electrical properties, they can be low- and high- water-content materials [1, 2].

Here we classified the phantoms based on their physical appearance. The head is one of the most complicated structures of the human body. Due to the sophisticated fabrication techniques, there are limited numbers of solid human head phantoms developed. Examples are reported in [2, 3].

Gel head phantoms are made of polymer- or agar-based substances. For broadband applications one can take semi-solid head phantoms as a good candidate. These phantoms are made of several homogenous materials to fabricate layered or heterogeneous semi-solid phantoms [4, 5]. These phantoms are generally agar-based to have a realistic shape of the head [6].

Researchers are interested in liquid phantoms, since adding or removing components can change the phantom properties. Besides sugar-water-salt-based solutions, the mixture of propylene glycol and deionized water has been used for homogeneous liquid phantoms [7]. However, the techniques to produce biological tissue-mimicking liquids suggested in the literature are generally not easy to set up and often require specific equipment.

In the context of the EMERALD project, the frequency bands range from 0.5 to 6 GHz, depending on the intended applications. The remote reproducibility as well as the precision over all the frequency bands range and the simplicity of realization constitute additional requirements. This leads to the choice of 3D printing technology over gels manufactured by molding. The use of 3D printing technology in conjunction with segmentation/CAD software to design the phantom from MRI and CT scans allows for an accurate shape. The cavities are solid, stable over time, and easily remotely reproducible. 3D printing implies the partition of the phantom into cavities that are filled with different liquids whose composition must be adjusted according to the desired dielectric properties. The question is which liquids are the most suitable to mimic living tissues in a wide band, stable in time and easy to produce. In recent work by our group, it has been shown that mixtures based on TritonX-100 and salt water can dielectrically mimic the different tissues of the breast and head in a wide range of frequencies and are easily fabricated [8]. In practice, tissues with water content, such as for example most parts of the brain, have high dielectric permittivity. Therefore, using water as one of the components of the Tissue Mimicking Materials (TMMs) is attractive. For low permittivity tissues such as bone and fat, liquid TMM can still be used by decreasing the percentage of water and adding liquid components that are miscible in water and have low permittivity such as Glycerin [9], Sugar [10] and TritonX-100 [11].

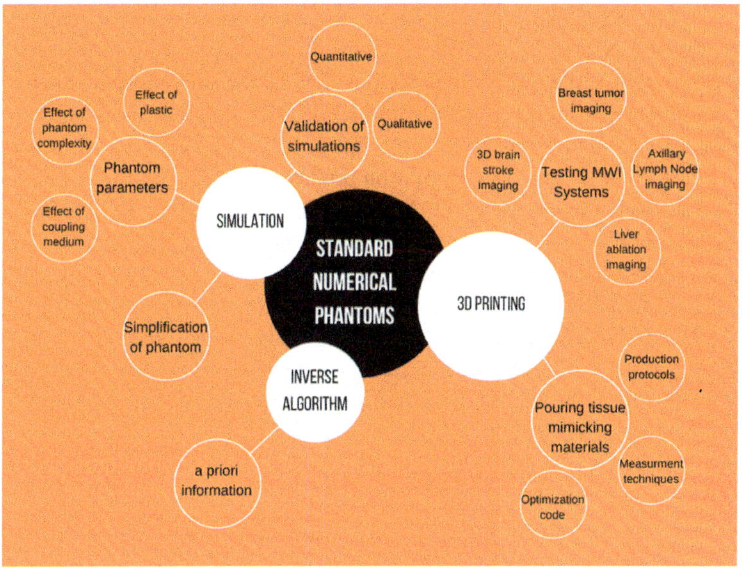

Fig. 1 Complete scheme of what covers the usage of standardized phantoms in this work

Pure TritonX-100 has dielectric properties similar to those of fatty tissues (whose permittivity is the lowest of the biological tissue) on a wide band and is viscous providing stability to the mixture over time. In addition, the use of a binary, i.e., two-component mixture of saltwater and TritonX-100 has several advantages, including the ability to be modeled by a binary mixture law. Therefore, in this study, mixtures based on salt water and TritonX-100 are studied. In summary, 3D printed liquid anthropomorphic human phantoms provide high accurate shape of the organ with dielectric properties in a wide band, reproducible, adjustable, reusable and easy to produce that none of the other types of phantoms can offer.

In addition, 3D printing involves digital files that can be loaded into a direct solver to simulate the data of the microwave imaging problem, for the study of the device (source, frequency, coupling medium…) and the phantom itself (design, cavity material…), as well as into inverse solvers to compare their efficiency or add a priori information from the anthropomorphic models. This is another advantage of this type of phantom: the possibility to perform experimental validations on anthropomorphic objects associated with numerical simulations [12]. A complete scheme of what covers the usage of standard numerical phantoms suggested in this work, is given Fig. 1 and detailed in [13]. This chapter's book is focused on the realization of the phantoms; that corresponds to the 3D printing block. The feasibility of the process is tested and applied to the production of phantoms adapted to the microwave imaging systems developed in EMERALD project at Politecnico di Torino (POLITO), Faculdade de Ciências da Universidade de Lisboa (FCUL) and

Institute of Electromagnetic Sensing of the Environment, National Research Council of Italy (IREA-CNR).

After a review of the motivations for developing anthropomorphic cavities filled with liquid mixtures in the context of the EMERALD project, Sect. 2.2 deals with the fabrication of 3D printed anthropomorphic cavities. From an MRI scan, a STL file is created in two steps: a segmentation process using the 3D-Slicer software and a design process using the Blender CAD software. Both softwares are in open access and thus suitable for use in European projects. The next sections are devoted to the realization of three phantoms adapted to the electromagnetic devices developed by our EMERALD partners that represent the head, Sect. 2.3, the thorax, Sect. 2.4, and the liver with ablation, Sect. 2.5. The last section is devoted to the method used to determine the tissue mimicking materials in the [0.5–6] GHz frequency range. We will show how this issue can be reduced to an optimization problem of two unknowns, corresponding to the concentrations of the mixture components. This section ends with the TMM recipes used for the fabrication of the three phantoms and 10 other tissues to show that the technique can be used to mimic any biological tissue in the frequency range used in the EMERALD project.

2 Development of Anthropomorphic Printable Cavities

2.1 3D Printing for Medical Application

In recent years, some researchers used 3D printing technology to make phantoms. In addition to surgical planning applications, 3D printed phantoms are essential for medical computational models validation, as well as for medical training and patient education. In [14] a comprehensive and recent review of the state of the art as well as new developments and trends in 3D printed functional medical phantoms (i.e., tissue-mimicking medical phantoms, radiologically relevant medical phantoms, and physiological medical phantoms) and 3D bio-printed structures (i.e. hybrid scaffolding materials, convertible scaffolds, and integrated sensors) for regenerated tissues and organs is provided. Further research on 3D printed tumors is helping researchers study metastasis and facilitating complex treatments, surgery and therapies [15]. In this way, 3D printing can significantly improve patient comfort and treatment accuracy [16]. In the field of microwave imaging, anthropomorphic head phantoms have been developed by printing 3D molds to be filled with gel-based head parts [17, 18] or mixtures based on graphite carbon black polyurethane [2, 19]. A review for the development of anthropomorphic breast phantoms is given in [20]. In this work, we took advantage of this recent development in 3D printing technology to engineer anthropomorphic cavities.

2.2 *Modeling and Printing 3D Cavities*

As mentioned above, to develop realistic phantoms for microwave imaging it is critical to get as close to the anatomy as possible. To obtain the outline of a biological organ there are several possibilities. We can take advantage of online libraries such as [21] or obtain the shape of the organ from the segmentation of MRI/CT scan. Here, we will shed light on segmentation techniques by introducing the 3D Slicer software. Then the files are modified with the CAD software Blender to develop printable cavities in such a way that the resulting phantom is fillable and compatible with the experimental system. The numerical version of the phantom obtained at the end of the process will be readable by 3D printers, in STL format.

2.2.1 Segmentation with 3D Slicer

3D Slicer [22] is a software application for the visualization and analysis of medical image computing data sets. All commonly used datasets are supported, such as images, segmentation, surfaces, annotations, transformations, etc., in 2D, 3D, and 4D. Visualization is available on the desktop and in virtual reality. The analysis includes segmentation, registration, and various quantifications.

3D Slicer is a research software platform that allows researchers to quickly develop and evaluate new methods and distribute them to clinical users. The software provides a product development platform, which also allows companies to rapidly prototype and release products. Developers can focus on developing new methods and do not have to spend time redeveloping basic data import/export, visualization, interaction features. The application is designed to be highly customizable (with custom branding, simplified user interface, etc.). 3D Slicer is free and there are no restrictions on its use. It is the responsibility of the software distributor to ensure that the application developed is suitable for the intended use [22].

Uploading the DICOM (Digital Imaging and Communications in Medicine) images in the software, gives access to all layers of these images in three different views. Figure 2 shows MRI of head in three views.

Using 3D slicer, different parts of the head, detected on the basis of the contrast difference, can be selected and the selected area extracted. In Fig. 2, there is a tumor inside the brain. By setting the right threshold to distinguish the contrast of the tumor tissue from the brain tissue, we can obtain a 3D model of the tumor. This is shown in Fig. 3.

The same technique has been applied to extract the brain from the MRI image given in Fig. 4. The result of the extracted brain is shown in Fig. 5.

This technique is applicable to different parts of the human body. After extracting the 3D model of the target organ, an STL format of the model can be exported from 3D Slicer software. This file format is widely used for rapid prototyping, 3D printing, and computer-aided manufacturing. The STL format specifies both ASCII and binary

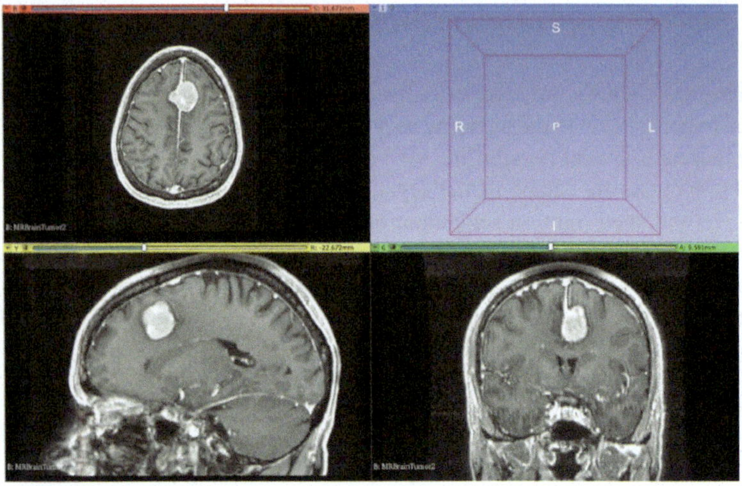

Fig. 2 MRI of head in 3D Slicer for three views

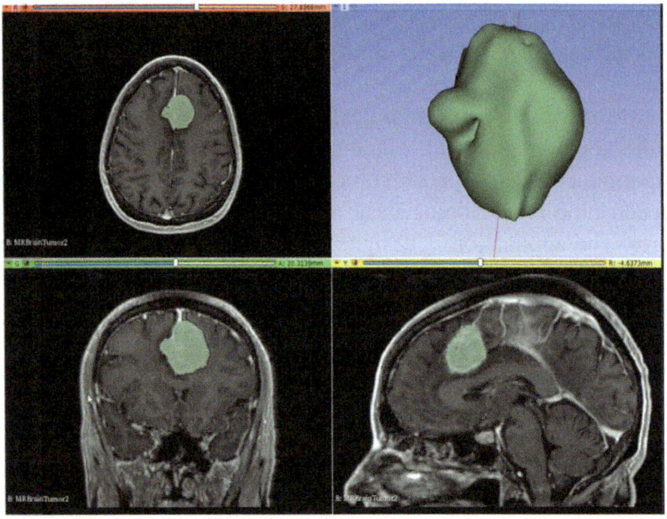

Fig. 3 3D model of segmented tumor

representations. Binary files are more common since they are more compact. This means these numerical versions of the model can be used in microwave imaging simulations.

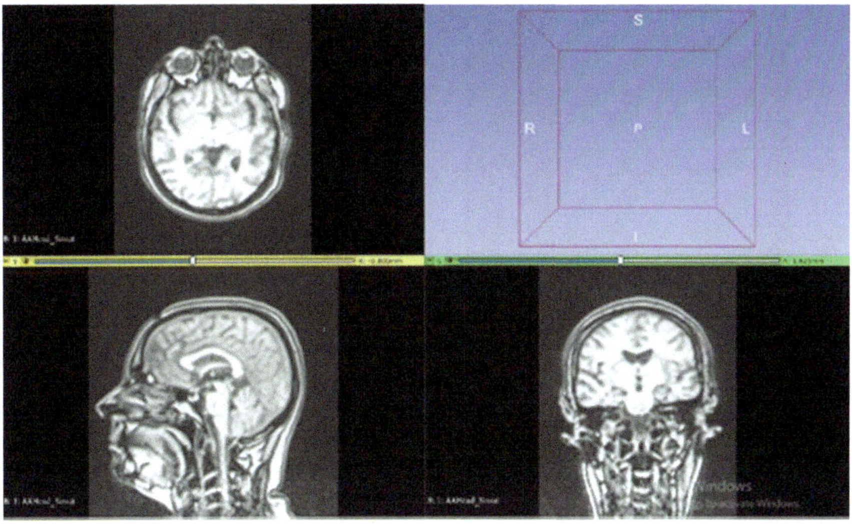

Fig. 4 MRI of the head, files were imported to the 3D slicer for segmentation

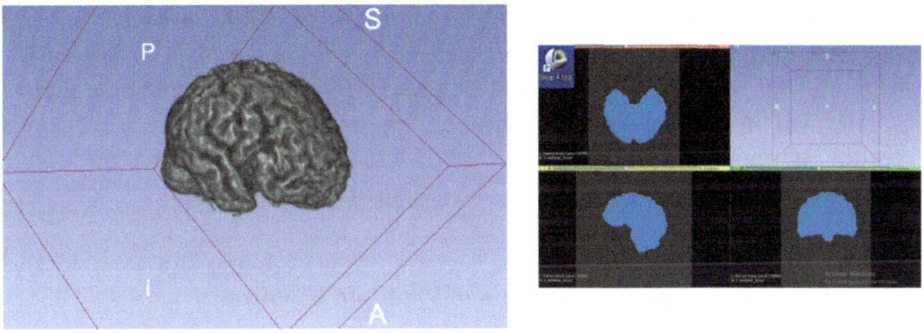

Fig. 5 Segmented brain and the outcome of the software for extracting the brain form the MRI of the head

2.2.2 Design and Realization with Blender

Blender is a free and open-source graphics software that allows to adjust 3D models and prepare them for 3D printing. Blender is not the most used software for printing parts, as it only works on the surfaces of objects and does not take volumes into account. On the other hand, it is perfectly adapted to work and adjust complex shapes such as those of anthropomorphic models. This is why this software was chosen to develop anthropomorphic cavities from STL files provided by 3D Slicer.

Blender offers several interaction modes. *Object mode* represents the shape of the model after 3D printing. By entering the *edit mode* one can observe that the model is meshed by triangles and several tools provided by Blender make it possible to change the shape and all corners of the 3D model precisely (see Fig. 6b). After

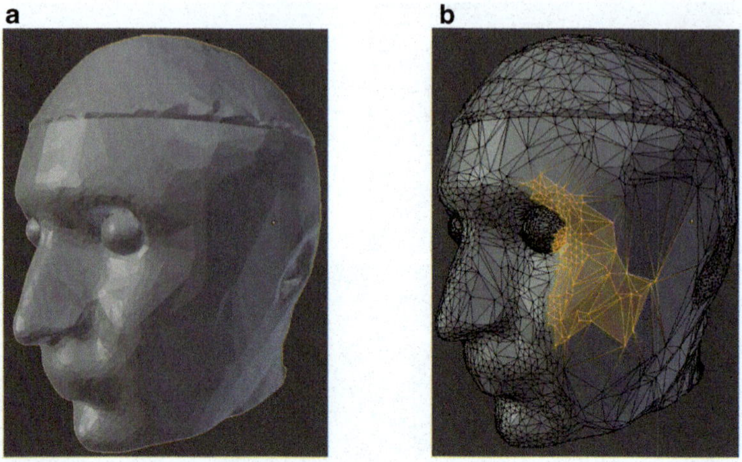

Fig. 6 **a** Imported a 3D model of head (object mode) and **b** the meshed version of the model prepared by Blender (edit mode) ready for further adjustments

applying these changes, by switching back to object mode, you can see the effect of the adjustments. Blender provides a variety of tools for editing meshes. These are tools used to add, duplicate, move and delete elements (each triangle is an element), which allow us to shape the 3D model as desired. Boolean is an operation that can be performed on selected objects in Blender. This tool is necessary to create cavities from the 3D model exported by 3D slicer. Moreover, solidifier is another tool to define a thickness to the outer shell of the designed cavity (Fig. 7).

2.2.3 Application to the Realization of Phantom Cavities

As it is defined in the framework of the EMERALD project, the manufacturing process used to build up accurate ultra-wideband (UWB) phantoms must be easily reproducible by an electrical engineer in a non-specific environment without extreme precautions and, to some extent, with a low cost. In the next sections, we present three phantoms: head, thorax and ablated liver, adapted to some of the experimental setups developed in the EMERALD project. These phantoms have been designed and developed according to the process described previously.

2.3 Head Phantoms Designed for POLITO System

Brain stroke is one of the most common cardiovascular diseases causing permanent injuries or even death. There are two types of strokes: ischemic (blocked vessel) or hemorrhagic (ruptured vessel), and the treatments are completely different. It is therefore essential to understand the type of stroke before starting any treatment.

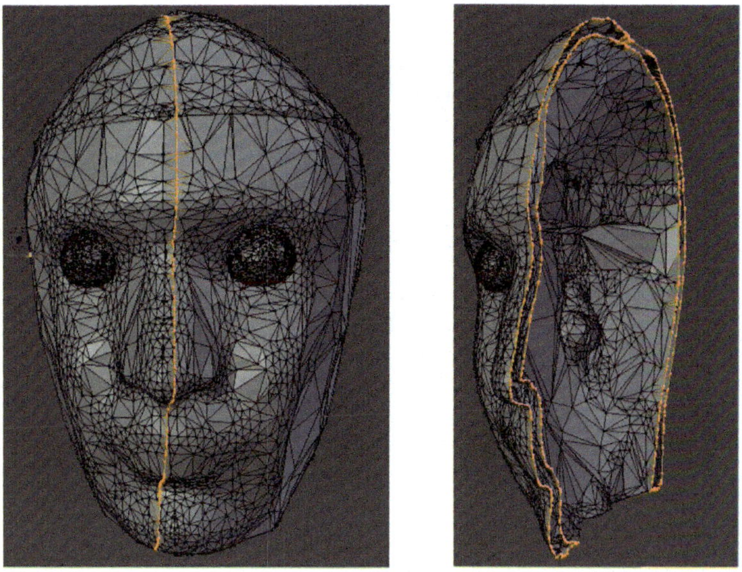

Fig. 7 Selection of one cut of the model to indicate the thickness of the head phantom

In [23, 24] a novel MWI prototype device for brain stroke diagnosis is presented and characterized. This MWI device is developed at POLITO and as a part of our collaboration to the EMERALD project we have designed, and 3D printed head phantoms for testing this MWI system.

2.3.1 POLITO System

For providing 3D images of head, 24 antennas are organized in an anatomically conformal shape forming a helmet. Each antenna is embedded in a box of graphite-silicon material, acting as the coupling medium, and connected to a two-port VNA through a 24 × 24 switching matrix, which allows the whole differential scattering matrix required for imaging to be acquired [23, 24]. The anthropomorphic head phantoms designed in our group are compatible with the system, as shown in Fig. 8. The phantom will be inserted in the blocks of antennas upside down. More details on the system are available in Chap. "Towards a Microwave Imaging Device for Cerebrovascular Diseases Monitoring: from Numerical Modeling to Experimental Testing" of this book.

2.3.2 Head Cavities

Since these cavities must be fitted in the POLITO microwave imaging system, they are designed in upside down configuration (Fig. 8). Figure 9 shows 3D printed phantoms with two and three cavities. The material used to print these phantoms is acrylonitrile

Fig. 8 POLITO MWI system. From left to right: VNA, switching matrix and head phantom wearing a helmet interconnected to the switching matrix by means of coaxial cables [23, 24]

butadiene styrene (ABS), which is resistance to Triton-X 100. The thickness of the different cavities of the head phantom varies between 2 and 8 mm.

The reason we have multiple cavities is that we need to fill each cavity with a mixture with different dielectric properties. These cavities will be filled with pre-prepared mixtures to mimic the dielectric properties of different parts of the head such as the brain, CSF, and muscles.

By using Blender software, a pair of cylinders are added to some cavities. One can fill them and, as it is shown in Figs. 9 and 10, put any object inside the brain for testing different microwave imaging systems and algorithms in a differential mode, according to the time evolution of the stroke. The brain is cut into two slices then the inside of it is accessible for cleaning after each experiment and the position of the stroke can be easily fixed. Each part is designed as Lego® so after putting two parts of the brain together there will be no leakage.

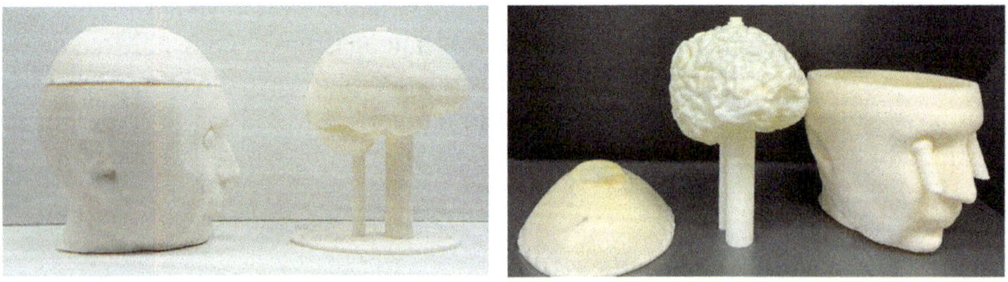

Fig. 9 Head phantoms including on left: two cavities and on right: three cavities with a circumvolution's brain

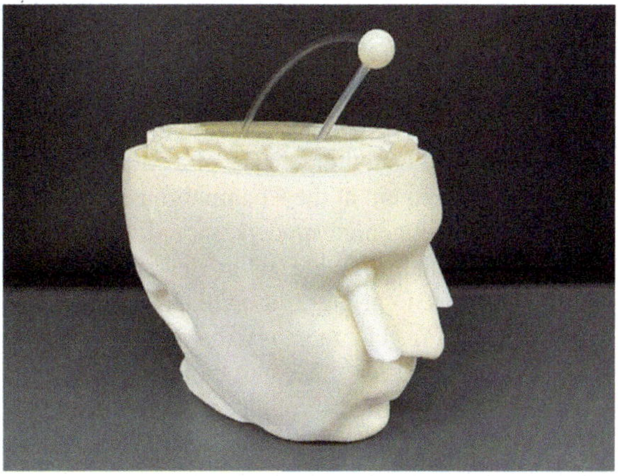

Fig. 10 A sphere located in the head phantom as a stroke

2.4 Axillary Phantom Designed for FCUL Microwave Imaging System

Breast cancer is the most frequently diagnosed and cause of cancer death among women [25]. Metastases (i.e., cancer's spread to secondary locations) are the leading cause of death for patients suffering from breast cancer. Seventy-five percent of the lymph from the breast drains into Axillary Lymph Nodes (ALNs), making those vulnerable areas to breast metastases.

2.4.1 FCUL System

The device developed by FCUL in the framework of the EMERALD project is a new MWI system designed to be a complimentary screening modality helping breast cancer staging by detecting axillary lymph nodes [26] An ultra-wideband low power radar pulses in a monostatic configuration is used. Besides the VNA, a single Vivaldi antenna with a working frequency of [2–5] GHz is implemented in the system. This single antenna covers a cylindrical area since it is attached to an angular positioner.

To make the axillary region more accessible, the patient would be asked to be with the arm raised in order to place an antenna under it. The antennas should cover the area in a way that receivers get access to enough information for image reconstruction. The proposed positioning of the patient has a direct impact on the phantom development and the axillary phantom is designed based on a CT scan with the same gesture from the patient [27]. More details on the system can be found in Chap. "The Dielectric Properties of Axillary Lymph Nodes" of this book.

2.4.2 Thorax Cavities

The microwave imaging system developed by FCUL needs to be tested on an anthropomorphic phantom before any clinical trial. The first version of thorax cavities is developed based on a thoracic CT of a patient diagnosed with breast cancer and was undergoing treatment at the Champalimaud Foundation [28].

Using the segmented models, including skin, muscles, lung, bones and lymph nodes, we created printable 3D models and exported them as STL files. Importing these data in a 3D printer, the axillary phantom can be inserted into the setup. In this phantom (Fig. 11), there is a thin layer of plastic, which defines the shape of the body and encompasses the rest of the cavities. The lung cavity is fixed inside the skin cavity and can be filled through two cylinders attached to it. The bone part, which represents the scapula and humerus (upper arm), is made of ABS. The fat cavity is between the outer layer and muscle cavity and has lymph nodes inside of it. Lymph nodes are elliptical cavities of different sizes. The difference in size represents an anomaly. These elliptical cavities are connected by a plastic pipe in order to be filled.

As mentioned in previous sections, it is also possible to use online phantom library in which you have access to different 3D models related to the anatomy of human body. For instance, to have the second version of axillary phantom we had to choose couple of lymph nodes and cut the part of the shoulder which is the area of interest for us. From the prepared 3D model of the human body available online, the second version of thorax cavities has been designed (Fig. 12). We considered only scapula and humerus and lung, so we had to get rid of the rest of the organs present in the phantom obtained from the library. Blender is used to add the cylinders to fill the cavities with the appropriate TMM and to make the adjustments to fit the phantom to the experimental set-up. Inside the larger cavity, slots are placed to insert the ABS bone and lung cavity and all elements are connected to a plate. In this way, the phantom is stable and resistant to movement. The plastic pipe of lymph nodes of different sizes to simulate healthy and pathological ones is also here fixed by slots in the large cavity, as shown in the Fig. 13. Compared to the previous one, this phantom has a more regular shape, does not contain a layer of fat but is easier to print and

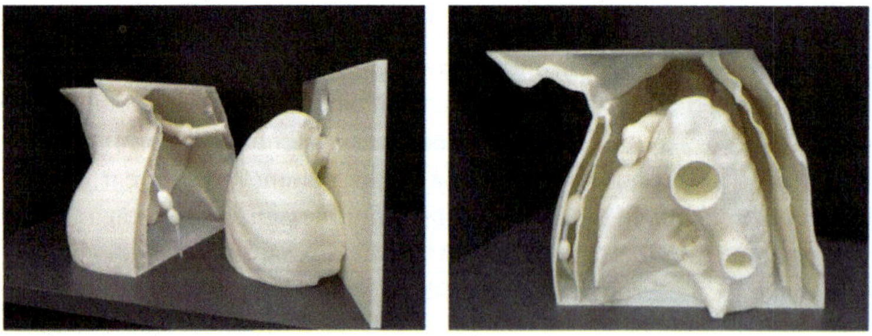

Fig. 11 3D printed axillary phantom. This phantom includes an outer shell with the shape of skin, fat cavity below the skin, lymph nodes embedded inside the fat cavity, and lung cavity and shoulder bone fixed in the muscle cavity

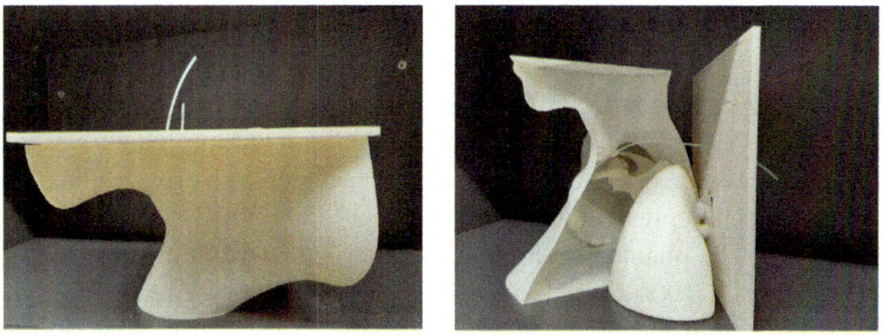

Fig. 12 3D printed axillary phantom based on online phantom library

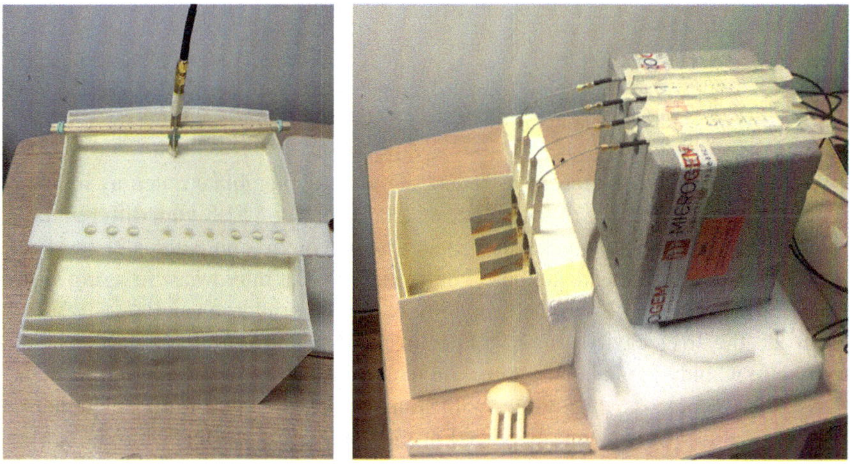

Fig. 13 MWI system for liver ablation with one and four antennas

develop as it does not require the use of 3D slicer for the segmentation and a MRI scan.

2.5 *Liver Phantom Designed for CNR-IREA System*

Liver cancer is a major health problem, ranking fifth in men and eighth in women among all cancer diseases [29]. This fatal health issue is increasing yearly, despite the advanced technologies. In recent decades, thermal ablation techniques are grown as an alternative to surgical approaches. In fact, microwave thermal ablation (MTA) is a cancer treatment targeting focal malignancies, in which an area of tissue is targeted by high temperatures at microwave (MW) frequencies (typically 915 MHz or 2.45 GHz) [30].

Our partners at CNR-IREA in Naples are developing a microwave imaging system for continuous monitoring of liver tumor ablation [31].

2.5.1 CNR-IREA System

MWI system developed at CNR-IREA is shown in Fig. 13. The set-up includes of a tank made of ABS material filled with a coupling medium which was designed in a way to maximize the electromagnetic power delivered to the liver [32]. An ellipsoidal phantom representing the liver tissue after the ablation treatment is located inside the tank. The phantom is a 3D-printed structure made of ABS material and filled with Triton-X-100/water mixture as tissue-mimicking material (Fig. 16). Slot-loaded Vivaldi antennas which were designed specifically for this MWI system are placed 35 mm away from the phantom, moving in a linear direction measuring the scattering parameters [33]. More details on the system can be found in Chap. "An Initial Assessment of a Microwave Imaging System to Monitor Microwave Ablation Treatments" of this book.

2.5.2 Liver Cavities

The phantom should represent different sections of the ablated area as indicated. After agreeing on the size of the phantom, a tank made of ABS with different layers representing skin, fat and liver is designed. The size of this container is $21 \times 24 \times 21$ cm^3. These dimensions are set based on the size of the microwave imaging system designed in CNR-IREA and the limitation of the 3D printer at GeePs.

Inside the liver cavity, there is an ellipsoidal phantom mimicking the ablated region. In a study the ablated part of the liver is divided to 4 zones (Fig. 14a) [34]. After ablation process, those zones with similar tissue damage are considered as one zone. The tip of the antenna is placed at the center of zone zero where the tumor center is located. The main changes of dielectric properties are among carbonized, coagulated and healthy liver. For the sake of simplification of the phantom and avoiding using several interlocking cavities in a small space, the ablation part of the phantom is designed as it is shown in Fig. 14b, c. Thus, ablation phantom has two cavities representing carbonized and ablated tissue.

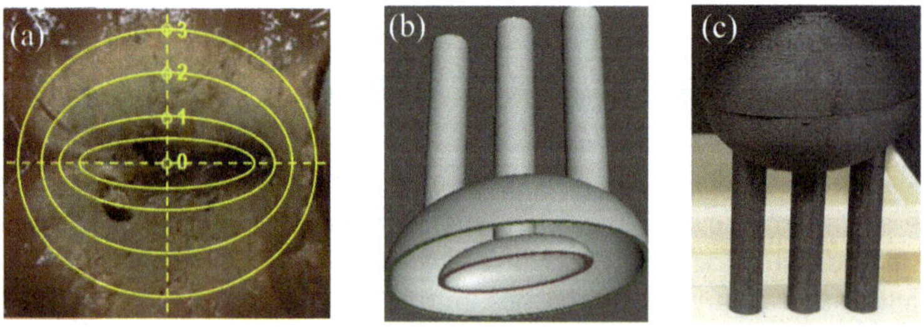

Fig. 14 Elliptical phantom representing ablated tissues. **a** Different zones of ablated tissue, **b** CAD model, **c** PA printed cavities

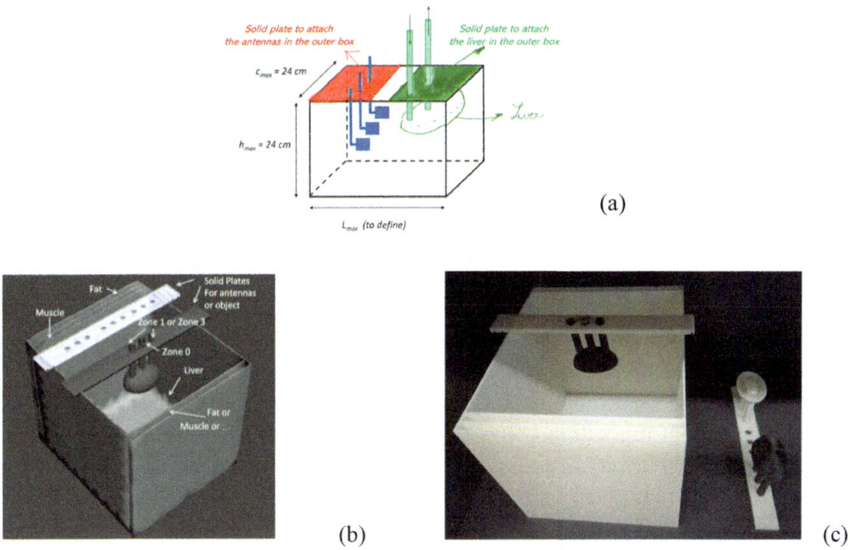

Fig. 15 First prototype of a multilayer abdomen with the ablated liver cavity. **a** The draft from CNR-IREA/CNRS/Sapienza University discussions, **b**, **c** the realization

The ablated phantom is hanging from a plate fixed on the top side of the tank and that can move along the tank. This plate has several holes on it (Fig. 15). This allows the elliptical phantom to be moved and the experiments to be performed with different antenna positions relative to the phantom.

3 Tissue Mimicking Materials' (TMMs) Fabrication

This section is devoted to the prediction of the TritonX-100 and salt concentration in binary mixtures for several types of tissues over wide range of frequency whose dielectric properties are given by the Gabriel data base (Sect. 3.1). The use of a two-component mixture, salt water and TritonX-100, to mimic the tissue has the advantage that it can be modelled by a binary mixture law. In this way, the prediction of the TritonX-100 and salt concentration in the mixtures is obtained for a given tissue by minimization in an iterative way (Gauss–Newton technique [35]), of a cost function that represents the difference between the reference values (Gabriel data) and those given by a binary law.

Several binary laws are proposed in the literature; Sect. 3.2 is devoted to comparing the results of three binary laws in order to identify the one that gives the highest accuracy over wide frequency range. The law is then introduced into the optimization code (Sect. 3.3).

Next an experimental validation that includes the production and the dielectric characterization of the wideband mimicking mixtures for several tissues is presented in Sect. 3.4.

3.1 Gabriel Reference Data

The first step to produce a mixture with dielectric properties close to the living tissues is having the dielectric properties of that organ as the reference values. The need for extensive data on the dielectric properties of human tissues as a reference is strongly felt among scientists and researchers involved in the interactions of electromagnetic fields and biological systems. Dielectric data from biological tissues commonly exhibit relaxation behavior. Relaxation models are used to extrapolate the measured permittivity of biological tissues to higher frequencies. The outcomes of these models are used as a reference in the process of producing tissue mimicking materials.

The human body is made of between 50 and 70% water. This is why biological tissues show lossy characteristics when illuminated by EM waves. The Debye model has been extensively used for modeling the frequency dependence of the complex permittivity of polar materials such as water and salted water [36]. However, dielectric data from biological tissues commonly exhibit relaxation behavior that is sometimes far broader than the simple Debye type. This arises from the superposition of several relaxation processes.

Among works done in the literature on this topic, Gabriel et al. provides detailed measurements of a variety of tissues up to 20 GHz [37]. They proposed a four-pole Cole–Cole model with the parameters obtained from the best fit to measured data.

According to Gabriel's work, a database [38] is provided on dielectric properties of different human tissues over the full frequency span (this span goes from 10 Hz up to 100 GHz in 10 decades). The four-pole Cole–Cole model selected for the dielectric data across the frequency range is:

$$\varepsilon_t(\omega) = \varepsilon_\infty + \sum_{i=1}^{4} \frac{\Delta\varepsilon_i}{1 + (j\omega\tau_i)^{(1-\alpha_i)}} - j\frac{\sigma_i}{\omega\varepsilon_0} \tag{3.1}$$

where $\Delta\varepsilon$ is the magnitude of dispersion (magnitude of relaxation) between the "static" and "infinite frequency" dielectric constants, ω is the angular frequency and τ is a time constant (relaxation time) and α is an empirical parameter used to adjust the dispersion width.

This model has been chosen to compute the reference values over the frequency range [500 MHz–6 GHz] of all the biological tissues considered in this study.

3.2 *Tissue Model with a Binary Mixture Law*

The use of a binary mixture, i.e., made of two components, to model a biological tissue has several advantages, including the possibility of analytically expressing the dispersive behavior of the tissue using a mixture law. There are several binary mixture laws developed in the literature that are capable of modeling the dielectric behavior of tissue over a wide range of frequencies [39]. The question is which law gives the best result so that the dielectric properties of the mixture obtained by applying this binary law are the closest to those given by the relaxation model of the target biological tissue.

Models based on the law of mixtures are intended to express the effective permittivity of a multiphase medium as a function of the permittivity and volume fraction of each component. Predicting the dielectric properties of mixtures composed of known components, has been a great challenge over the years and several formulas has been proposed depending upon shape of inclusions, the conductivity of the inclusions, etc. [39, 40]. In the following, three of the most well-known theories used in the literature were investigated to model biological tissue: Böttcher, Lichtenecker and Looyenga.

Considering the studies based on the static permittivity of emulsions, any passive and homogeneous medium is made of two phases, a dispersed one denoted (D) and a continuous one (C). The volume fraction V_D and V_C are occupied by phase D in the total volume and the continuous phase C in the same volume, relatively. Hence, the aggregation of these two volume fractions is a unit, i.e. $V_D + V_C = 1.$. The dielectric properties of components are defined as the dielectric constant of the dispersed phase (ε_D) and continuous phase (ε_C). The term, ε_E, denotes the dielectric properties of the emulsion. Considering these notations, we obtained the following equations for the three different laws [41, 42, 43]:

$$\frac{(\varepsilon_E - \varepsilon_C)(2\varepsilon_E + \varepsilon_D)}{3\varepsilon_E} = V_D(\varepsilon_D - \varepsilon_C) \tag{3.2}$$

$$\ln \varepsilon_E = V_D \ln \varepsilon_D + (1 - V_D) \ln \varepsilon_C \tag{3.3}$$

$$\varepsilon_E^{\frac{1}{3}} = V_D \varepsilon_D^{\frac{1}{3}} + (1 - V_D) \varepsilon_C^{\frac{1}{3}} \tag{3.4}$$

The question is how to choose the two components of the TMM. Since a high percentage of the human tissues is made of water, using water seems to be relevant and by adding salt, the conductivity of the mixture can be controlled. Therefore, saltwater mixtures will be considered as one component of the mixture. The other component will need to be miscible in water (to simplify the fabrication), with low loss and with a low permittivity, in such a way that the mixtures have dielectric properties in the range of those of biological tissues. TritonX-100 seems to be a good candidate.

To find the most suitable mixture law to model the tissue-mimicking liquid we proceed as follows. First, the mixture composed of TritonX-100 and salt water with

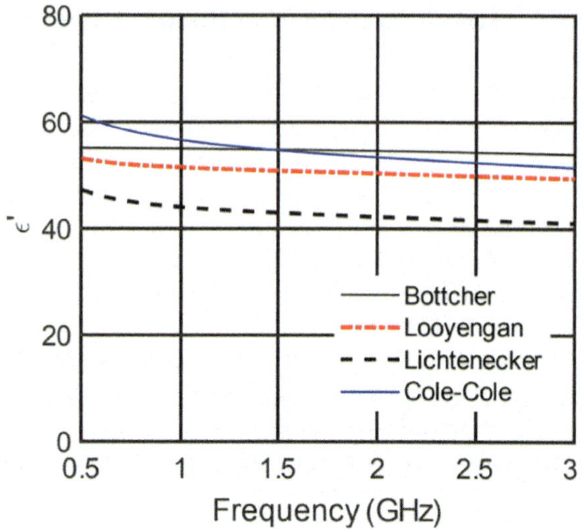

Fig. 16 Dielectric constant of the Triton X-100 salted water mixtures obtained with different binary laws and the Cole–Cole model for the spleen tissue

dielectric properties close to the values given by Cole–Cole's model for a given tissue, was empirically generated. In this way, experimental values of TritonX-100 and salt concentrations were obtained; by introducing these concentrations into the different binary laws, the one that best fits the Cole–Cole model has been selected. From this empirical process, we learn that the best-fitting mixture law is the Bötcher's law [Diagnostics 2018]. An illustration is given in Fig. 16 for the spleen tissue. A binary liquid mixture of 21% TritonX-100 and 7 g/L salt has similar dielectric properties in the [0.5–3] GHz band, considering the Cole–Cole model for spleen tissue as reference. The three binary laws were tested using these experimental values, among them, the Böttcher's law shows the best results over this frequency range. This result was observed for all biological mimic tissues produced in our study. Further investigations showed that the Krasweski's law [39] provides also satisfactory results. The equations with the latter are given in [44].

3.3 *Optimization Code*

Cole–Cole reference models have been developed for almost all the biological tissues from Gabriel and Gabriel's work. The IFAC database [38] provides those models. On the other hand, binary laws provide equations in which by proper tuning of its variables, the dielectric properties of the modeled tissue get closer to the reference dielectric properties of the target tissue. Hence, an optimization code can be developed to adjust the concentration of the components of mixtures for each tissue by minimization of a cost function which represents the quadratic error between the complex permittivity values ε_E given by a selected binary law and the reference values ε_t in a specific frequency range and at room temperature for a given tissue. This issue is then reduced to an optimization problem of two unknowns, corresponding to the concentrations of the mixture components.

We assume that the TMM is a binary mixture composed of salted water and TritonX-100, for which both the Debye's models are valid and known from literature and given respectively for salt water [36] and for the Triton X-100 (group three [8, 45]). The following gives the equations when a binary Böttcher law expressed by the Eq. 3.2 is used.

The complex permittivity of the mixture is calculated using the Böttcher formula rewritten in Eq. 3.5, to point out the salt concentration dependency of the continuous phase and compared to the reference values given by the Cole–Cole model for a given tissue, in the specific frequency range.

$$\frac{(\varepsilon_E - \varepsilon_C(S_C))(2\varepsilon_E + \varepsilon_D)}{3\varepsilon_E} = V_D(\varepsilon_D - \varepsilon_C(S_C)) \tag{3.5}$$

Therefore, the volume fraction of TritonX-100 and the concentration of salt concentration in water (V_D, S_C) required to mimic a specific tissue can be calculated by fitting the complex permittivity of the mixture model ε_E to that of the tissue ε_t over a specific frequency range. This is done by using a Gauss–Newton process [35], where at each iteration, the NaCl concentration and volume fraction V_D are determined at discrete frequencies over [0.5–3 GHz], by minimizing the cost functional:

$$\mathrm{J} = \sum_{\mathrm{f}} \omega_{\mathrm{f}} |\varepsilon_{\mathrm{E}} - \varepsilon_{\mathrm{t}}|_{\mathrm{f}}^2, \text{ where } \omega_{\mathrm{f}} = 1/|\varepsilon_{\mathrm{t}}|_{\mathrm{f}}^2 \tag{3.6}$$

The quadratic cost function is approached by its first (gradient **g**) and second (approximate Hessian **H**) derivatives with respect to the NaCl mass concentration and volume fraction, which are computed analytically.

$$\mathbf{g} = 2\sum_f \omega_f \Re e\Big[(\mathcal{E}_E - \mathcal{E}_t)_f^* \boldsymbol{\varepsilon}_E'\Big]_f \tag{3.7}$$

$$\mathbf{H} = 2\sum_f \omega_f \Re e\Big(\boldsymbol{\varepsilon}_E'^{*} \boldsymbol{\varepsilon}_E'^{\dagger}\Big)_f \tag{3.8}$$

In the above equations $\boldsymbol{\varepsilon}_E' = (\partial\varepsilon_{\mathrm{E}}/\partial \mathrm{V_D}, \partial\varepsilon_{\mathrm{E}}/\partial \mathrm{S_E})^\dagger$ where $\mathrm{S_E}$. indicates the NaCl concentration of the mixture, V_D the volume fraction of TritonX-100, * the conjugate, and † the transposition.

Then by considering Eq. 3.5, $\boldsymbol{\varepsilon}_E'$ becomes:

$$\boldsymbol{\varepsilon}_E' = \begin{pmatrix} 3\gamma(\varepsilon_D - \varepsilon_C)/4\delta \\ \big[(\gamma(3V_C - 1) + 4\varepsilon_D)/4V_C\delta\big]\frac{\partial\varepsilon_C}{\partial S_C} \end{pmatrix}, \tag{3.9}$$

With:

$$\gamma = \delta - \eta, \delta = \left(\eta^2 + 8\varepsilon_D\varepsilon_C\right)^{1/2} \quad and \quad \eta = \varepsilon_D - 2\varepsilon_C - 3V_D(\varepsilon_D - \varepsilon_C)$$

As it is mentioned in [8] the term $\partial\varepsilon_C/\partial S_C$ can be inferred from the salted water parametric model [36], where S_C represents the salt concentration saline water.

At iteration step k + 1, $\mathbf{x} = (V_D, S_E)^\dagger$ the solution reads:

$$\mathbf{x}^{k+1} = \mathbf{x}^k - \mathbf{H}^{-1}\left(\mathbf{x}^k\right)\mathbf{g}\left(\mathbf{x}^k\right) \tag{3.10}$$

This process converges towards a stable solution in a few iterations, most of the time independently of the initial NaCl concentration S_E and volume fraction V_D, by inverting at each iteration the approximate Hessian matrix **H** of rank 2.

3.4 Experimental Validation

3.4.1 Production of the Mixture

A mass addition protocol is considered to produce the mixtures. Therefore, standard laboratory equipment is used, including a digital balance scale, heater and stirrer, magnetic bars and several beakers (Fig. 17).

The masses of TritonX-100 (m_{TX}), NaCl (m_{NaCl}), and H_2O $\left(m_{H_2O}\right)$ are calculated from the output of the optimization code, for a given volume of the mixture V_{tot}, according to the following.

Let denote S_{NaCl} the concentration of *NaCl* in the mixture and V_{TX}/V_{tot} the volume concentration of TritonX-100 of the mixture. From those quantities, we obtain straightforward the masses of each component:

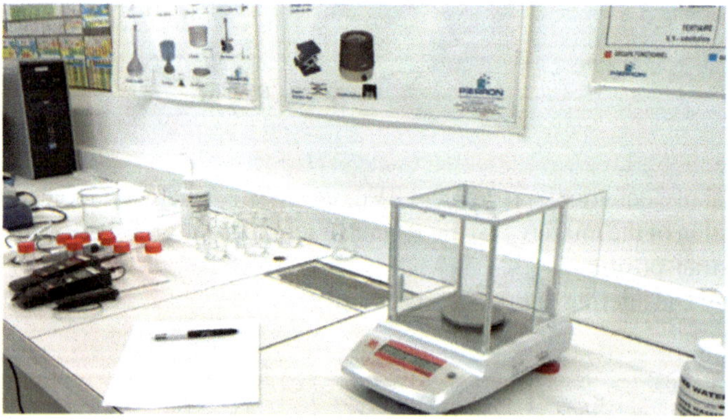

Fig. 17 The simple equipment to produce the mixture

$$m_{\text{NaCl}} = S_{\text{NaCl}} \times V_{tot} \tag{3.11}$$

$$m_{TX} = \rho_{TX} \times V_{TX} \tag{3.12}$$

$$m_{\text{H}_2\text{O}} = (V_{tot} - V_{TX} - m_{\text{NaCl}}/\rho_{\text{NaCl}}) \times \rho_{\text{H}_2\text{O}} \tag{3.13}$$

where, the values of the volume density of water, TritonX-100 and salt are:

$$\rho_{\text{H}_2\text{O}} = 1\,\text{g/ml},\ \rho_{TX} = 1.061\,\text{g/ml} \quad \textit{and} \quad \rho_{\text{NaCl}} = 2.16\,\text{g/ml}$$

The process can be summarized as follows:

- Tare the balance with an empty beaker.
- Add successively the masses of NaCl deionized water and tare the balance between each step.
- Stir the solution with a magnetic bar—when the salt is dissolved, tare the balance.
- Add the mass of TritonX-100. Note that, the TritonX-100 is very viscous at room temperature, and then it must be heated first separately in a 45 °C water bath.
- Place the beaker in a water bath and stir the solution with a magnetic bar. The bath water must cover all the mixture. Note that the 40–60% solutions of TritonX-100, depending on their salt concentration, can be gelled at room temperature. It is therefore necessary to increase the temperature to obtain homogeneous mixtures in any case.
- Once the solution is homogeneous, pour it into an airtight container and keep the sample at room temperature and safe from light.

Next, an open-ended coaxial is used for non-invasive measurement to check if the produced TMMs can mimic the biological tissues over a wide frequency range. This will be done by comparison of the reference values given by IFAC [38] with the measured dielectric properties of the manufactured mixtures.

3.4.2 Dielectric Characterization with a «Homemade» Open-End Coaxial System

Measured results are all obtained using a homemade open-ended coaxial probe of our laboratory GeePs (Fig. 18a) coupled to a portable vector network analyzer (VNA) Rhode and Schwartz ZVB8 VNA.

The open-ended coaxial cable reflection method is based on the measurement of the complex reflection coefficient at a single network port [46]. The coaxial cable is built up from a 3.6-mm-diameter, 15-cm-long, Teflon-filled copper rigid coaxial cable (Fig. 18).

To calculate the dielectric properties of the tissue from the measured reflection coefficient, it is useful to use an equivalent circuit of an open-ended coaxial line. The end of the probe is then considered as two capacitors in parallel (Fig. 19). C_f

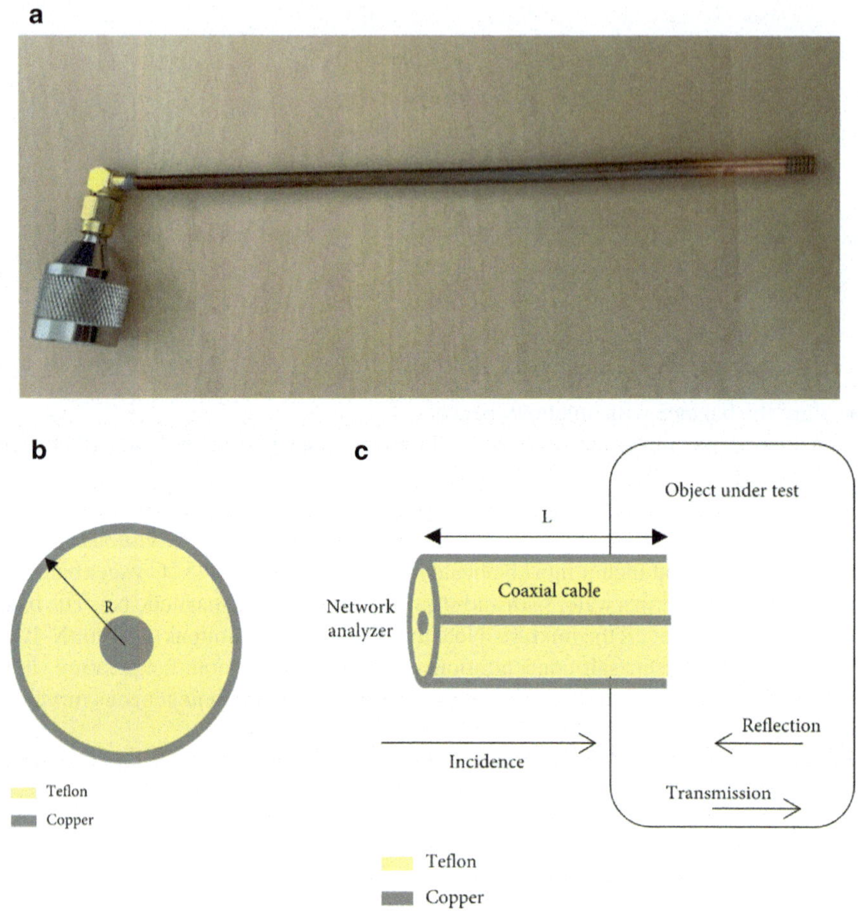

Fig. 18 **a** The probe used for measurements. **b** The configuration of a coaxial cable. **c** Configuration of a permittivity measurement system using reflection methods

is the capacitance representing the electric field concentrated inside the Teflon and $C(\omega)$ is the capacitance related to the fringing field in the unknown medium which is the product of C_0 (capacitance of the air-filled parallel plate) and ε (the relative permittivity of the object under test). The conductance $G(\varepsilon)$ models the radiation into the dielectric surrounding the cable and is frequency dependent.

The equivalence admittance of open-ended coaxial cable can be calculated as:

$$Y = j\omega C_f + j\omega C(\varepsilon) + G(\varepsilon) = j\omega C_f + j\omega C_0 \varepsilon_r + G(\varepsilon) \tag{3.14}$$

From transmission theory, terminal admittance can be defined from the complex reflection coefficient (Γ) as it is shown below:

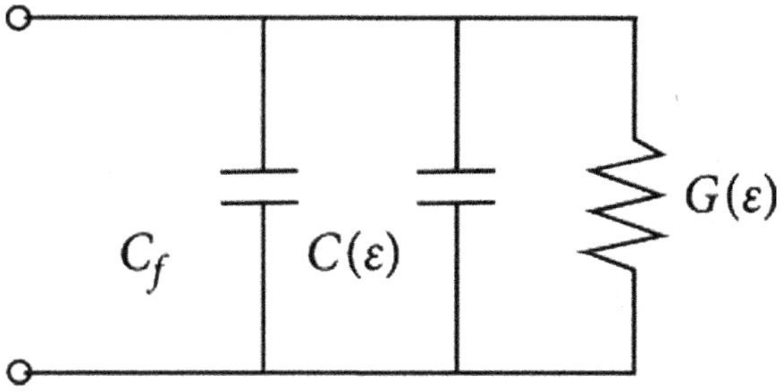

Fig. 19 Equivalent circuit of the probe

$$Y = \left(\frac{1-\Gamma}{1+\Gamma}\right)\frac{1}{Z_0} \tag{3.15}$$

where, Z_0 represents the characteristic impedance of the coaxial line (50 Ω). Considering the negligible external radiation of the coaxial cable, $G(\varepsilon)$ is assumed zero, which leads to an equation with complex permittivity $(\varepsilon' + j\varepsilon'')$ on one side and the reflection coefficient on the other side. Thus, the relative dielectric constant and the loss factor can be calculated using Eqs. (3.14) and (3.15).

$$\varepsilon' = \frac{1}{2\pi f Z_0 C_0} \times \frac{-2|\Gamma|\sin(\varphi)}{1+2|\Gamma|\cos(\varphi)+|\Gamma|^2} - \frac{C_f}{C_0} \tag{3.16}$$

$$\varepsilon'' = \frac{1}{2\pi f Z_0 C_0} \times \frac{1-|\Gamma|^2}{1+2|\Gamma|\cos(\varphi)+|\Gamma|^2} \tag{3.17}$$

where, $|\Gamma|$ and φ are the modulus and phase of the reflection coefficient measured by the probe, respectively.

The reflection coefficients measured by VNA (Γ_m) at its port is not the one which should be given in input in the equations. Due to the variation of the potential along the probe, it is needed to calculate the reflection coefficient Γ at the probe-mixture interface. This can be done by considering the probe equivalent to a Quadrupole (Fig. 20).

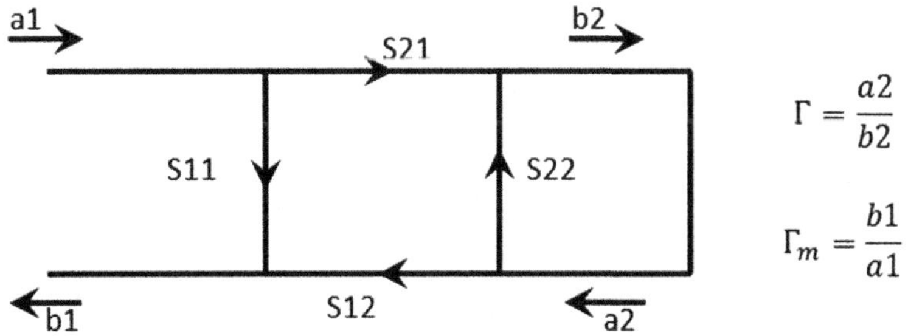

Fig. 20 Quadrupole equivalent of the probe

Thus, the reflection coefficient measured at the tip of the probe from the one measured by VNA is given by the following equation:

$$\Gamma = \frac{\Gamma_m - S_{11}}{S_{22}\Gamma_m + S_{12}S_{21} - S_{11}S_{22}} \tag{3.18}$$

Γ_m is known from the VNA measurement while the values of (S_{11}, $S_{21}S_{12}$, and S_{22}) on the right side of the formula are unknown. Therefore, to find the three unknown, three calibrations with three standards are necessary. Open circuit ($\Gamma = 1$), short circuit ($\Gamma = -1$), and calibration using distilled water are calibrations needed for the calculating Γ at the probe-mixture interface. It is noteworthy that the use of distilled water makes the system more suitable for measuring water-based liquids. In this case, Γ is obtained from the dielectric properties computed by a Stogryn model for a given temperature and frequency range [36].

3.4.3 TMM Recipes for the EMERALD Phantoms

Table 1 displays the results obtained for the head TMMs at 1 GHz. The comparison between the measured dielectric properties of each mixture and the values given by Cole–Cole model ensures that the mixtures mimic the dielectric properties of the target tissue. In this study, the brain is considered as a mixture of white and grey matters (75% of white matter and 25% of grey matter).

The measured mean value shown in Table 1 uses 9 measurements: the mixture was characterized with different systems at several randomly selected locations [8]. At this point, we would like to comment on the number of significant digits relevant for writing the measured and reference values. In addition to the sources of uncertainty related to the system, the dielectric properties of biological tissues vary from their average. Indeed, the biological structure of tissues is generally not homogeneous [47]. Furthermore, measurements can be performed under various conditions: in vivo, where living tissue is measured in situ, ex vivo, where samples are measured after being removed from the model body, and in vitro. This involves changes in blood flow and temperature for which the dielectric properties are sensitive. Finally, the dielectric properties are usually measured using animal models, which vary among age and

Table 1 Concentration values and properties of head TMMs at 1 GHz and 37 °C versus the values inferred from Cole–Cole models [8]

	Mixture composition	Measurement	Cole–Cole
Tissue	TX-100 (vol%), NaCl (g/L)	ε_r, σ (S/m)	ε_r, σ (S/m)
Brain	(38, 5.2)	44 ± 1, 0.84 ± 0.02	42, 1.0
CSF	(6, 13.7)	70 ± 4, 2.7 ± 0.2	68, 2.5
Muscle	(24, 5.0)	54 ± 1, 0.97 ± 0.02	55, 1.0
Blood	(14, 9.6)	61 ± 2, 1.72 ± 0.05	61, 1.6

Table 2 Mixture composition and properties of axillary phantom TMMs for frequency band of 1–8 GHz [46], dielectric properties @ 2.5 GHz

Tissue	TMM composition TX-100 vol%, NaCl g/L	Measurement ε′, σ (S/m)	Reference ε', σ (S/m)
Muscle	24, 5	52, 2	53, 2
Lung inflated	56, 4	25, 1.4	20, 0.8
Fat	100, 0	5, 0.2	5, 0.1
ALN [ref]	25, 8	50, 2	50, 2

species [48]. The tissue reference value is therefore known with great variability. Considering that this large variability is common to all tissues, the importance is to produce mixtures that reproduce dielectric contrasts as closely as possible around an approximate mean value of the dielectric properties of the tissue. Therefore, in the following the writing of reference and measured values when compared will be given to a maximum of 2 significant digits.

To fill cavities of the axillary phantom introduced in Sect. 2.4.2 with TMMs, the concentrations of TritonX-100 and salt are calculated using the optimization code. Muscle, fat, lung and lymph nodes are the four mixtures provided for the axillary phantom. Based on the protocols explained in Sect. 3.4.1 the mixtures are produced and measured. The measurement values displayed in Table 2 and Fig. 21 bis refer to the mean value of the three measurements made at different locations of the mixture at 2.5 GHz and over the frequency range [0.5–6 GHz].

Values given in [34] at 2.45 GHz, for ex vivo liver tissues, are used as the input of the optimization algorithm and the extrapolation is done on the frequency range of [0.5–6] GHz. Table 3 shows the result at 2.5 GHz, room temperature. It gives the recipes of the different TMMs for the liver phantom that contained ablated, coagulated and carbonized parts, according to the layers present in the phantom

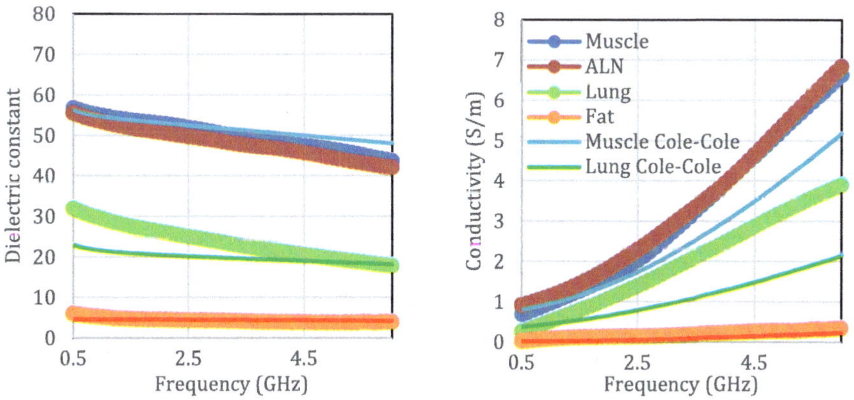

Fig. 21 Measured dielectric properties of the produced mixtures over the [500 MHz–6 GHz] frequency range, for the axillary phantom

reports Fig. 14. The mixtures were produced and characterized. Figure 22 shows their dielectric properties over the frequency range [0.5, 6] GHz. Note that the ablated liver mixture is a gel at room temperature. The values Table 3, consider measurements made at 22 °C and 34 °C, the temperatures when the mixture is gelled and liquefied respectively.

Extension to 10 Other Tissues and Reproducibility

An initial run of the optimization algorithm allowed adjusting numerically the concentration of TritonX-100 and salt, required to produce TMMs with dielectric properties similar to 10 biological tissues, at room temperature, over a wide frequency range [49]. The results obtained at 2.5 GHz are given in Table 4.

The mixtures including different concentration of TritonX-100 and saline water for 10 different human tissues, based on the outcome of the optimization code, were produced. After each measurement, results were compared with the Cole–Cole

Table 3 Mixture composition and properties of liver ablated phantom TMMs at 2.5 GHz

Tissue	TX-100 vol%	NaCl g/L	Measurement ε', σ (S/m)	Reference ε', σ (S/m)
Healthy liver	31	5	46, 2	43, 1.7
Ablated liver	44	8.5	33, 1.9	32, 1.4
Coagulated liver	60	9	21, 1.4	22, 1.1
Carbonized liver	88	4.5	7, 0.4	8, 0.4

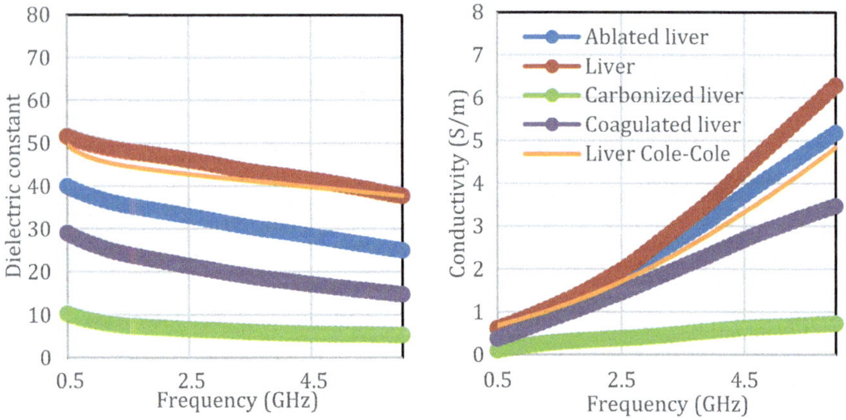

Fig. 22 Measured dielectric properties of the produced mixtures over the [500 MHz–6 GHz] frequency range, for the liver ablated phantom

Table 4 Outcome of the optimization code for 10 new biological tissues at 2.5 GHz, by using Böttcher's law as the binary law

@ 25 °C, 2.5 GHz	Mixture		Böttcher		Cole–Cole	
Tissue	TX-100 vol%	NaCl g/L	εr	σ S/m	εr	σ S/m
Stomach	12	7	62.2	2.1	62.1	2.3
Spleen	20	7	52.8	2.1	52.3	2.3
Liver	30	5	43.4	1.5	43.0	1.7
Small intestine	16	14	54.6	3.0	54.3	3.2
Gall-bladder	16	7	57.7	2.1	57.6	2.1
Kidney	19	8	53.2	2.2	52.6	2.5
Colon	19	6	54.2	1.9	53.8	2.1
Heart	17	7	55.2	2.1	54.7	2.3
Nerf	44	4	30.4	1.1	30.1	1.1
Lung inflated	56	4	20.7	0.8	20.4	0.8

Last two columns belong to Cole–Cole model, which is the reference model of dielectric properties of biological tissues in the literature [49]

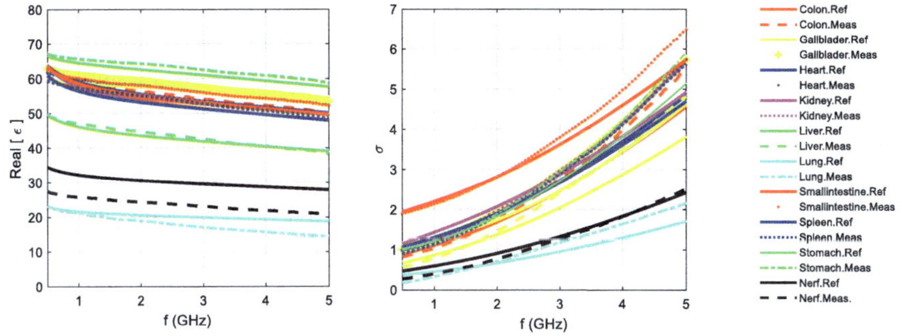

Fig. 23 Comparison of the dielectric constant and conductivity of 10 different biological tissues of the values obtained from the Cole–Cole model as the reference values with the measured dielectric properties of produced mixtures at room temperature, over the [500 MHz–5 GHz] frequency range

model. The error between the measured results and Cole–Cole model, was reduced by changing slightly (maximum 10% with respect to the theoretical value) the amount of TritonX-100 or salt. Results presented in Fig. 23 validate the generalization of the process to any biological part of the body and show that it is possible to make mixtures mimicking the dielectric contrasts between the different tissues on a wideband.

One of the reasons that liquid phantoms seem to be more practical is that the mixtures are easily reproducible remotely in different labs. As a demonstration of that, those results were validated by the reproduction of the mixtures and their dielectric characterization using a different set up used at National University of Ireland Galway (NUIG).

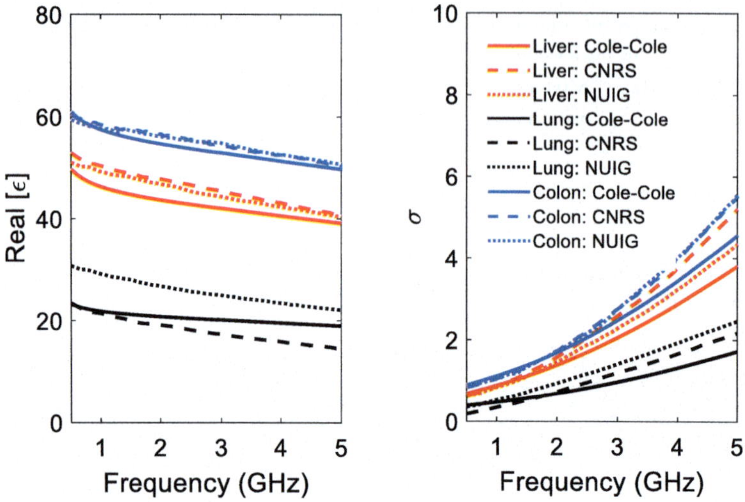

Fig. 24 Dielectric properties of liver, lung, and colon measured in France and Ireland compared with Cole–Cole model

Figure 24 shows that the results of our measurements on mixtures produced in different labs (Supelec-FR/NUIG-IR) are quite similar and that the dielectric contrast between the different tissues is reproduced quite well. Note that the mixture for the lung tissue can be gelled depending on the room temperature. This may account for the difference in the dielectric constant of the mixture for this tissue. The set-ups used different probes (homemade/Keysight) and different network analyzers (Rhode and Schwartz ZVB8 VNA/Keysight E5063A). This proves the fact that the process and the results are reliable in different lab and with different set-ups.

4 Conclusions

To set up a MW imaging system, several simulations need to be done before building up the system in the lab. Having a numerical version of the human body is essential for these simulations. Moreover, after setting up the imaging system many tests are needed to analyze the performance of the system and its feasibility. These tests cannot be done on a human, which is why a 3D printed human phantom with dielectric properties of target tissues is desirable. This chapter addressed the tools to produce anthropomorphic phantoms for numerical and experimental validation of microwave imaging systems. Liquid phantoms are proposed as anthropomorphic phantoms including 3D printed cavities filled with tissue mimicking materials (a mixture of TritonX-100 and salted water in this case). To produce those TMMs, knowledge of relaxation models, dielectric characterization techniques, fluid mixture laws, optimization process and some practical protocols is essential. Within the framework of the EMERALD network, we designed a protocol for other groups in the project to be able to test their MWI systems and algorithms. This process is simple, and an electrical engineer should be able to remotely reproduce the mixtures and obtain same results as we did. From this work, a website named SUPELEC RECIPES based on the Gabriel database is under development [50]. An extension of the database is in progress, including the work on dielectric properties of living tissues from literature [45, 47, 48, 51–56], etc.

Another important part of this chapter is modeling and printing 3D cavities. General open-source tools for creating STL files of the human body from MRI and CT scans of the organ are given. These tools allow the designer to segment the target organ and then design it to fit the device. In the end, a 3D model of the organ can be exported as a STL file.

To elaborate on it, several cases are studied in the frame of the EMERALD project. The numerical files of different head, axillary and liver phantoms are developed and used for 3D printing of these phantoms. These 3D printed phantoms have been sent to our partners in other countries since they are compatible to MW imaging systems developed in POLITO (Italy), FCUL (Portugal) and IREA (Italy). Numerical versions will give the ability to perform numerical simulations as well as experiments for studying the behavior of the phantoms in the microwave imaging systems.

Acknowledgements This work was funded by the EMERALD project funded from the European Union's Horizon 2020 research and innovation program under the Marie Skłodowska-Curie grant agreement No. 764479.

References

1. S. Abedi, N. Joachimowicz, B. Duchêne, H. Roussel, J. Tobon, D. Rodriguez-Duarte, R. Scapaticci, F. Vipiana, L. Crocco, Benchmark head phantoms for microwave imaging of brain strokes, in *Conference PIERS 2019* (2019)
2. B. McDermott, E. Porter, A. Santorelli, B. Divilly, L. Morris, M. Jones, B. McGinley, M. O'Halloran, Anatomically and dielectrically realistic microwave head phantom with circulation and reconfigurable lesions. Progr. Electromagn. Res. B **78**, 47–60 (2017)
3. T. Kobayashi, T. Nojima, K. Yamada, S. Uebayashi, Dry phantom composed of ceramics and its application to SAR estimation. IEEE Trans. Microw. Theory Tech. **41**(1), 136–140 (1993)
4. A.T. Mobashsher, A.M. Abbosh, Y. Wang, Microwave system to detect traumatic brain injuries using compact unidirectional antenna and wideband transceiver with verification on realistic head phantom. IEEE Trans. Microw. Theory Tech. **62**(9), 1826–1836 (2014)
5. B.J. Mohammed, A.M. Abbosh, S. Mustafa, D. Ireland, Microwave system for head imaging. IEEE Trans. Instrum. Meas. **63**(1), 117–123 (2013)
6. Z. Gong, Y. Chen, X. Lin, K. Wu, M.J. Cree, Generic wide-band phantom design methodology for microwave medical applications. IEEE Antennas Wirel. Propag. Lett. (2021)
7. D.B. Rodrigues, P.F. Maccarini, S. Salahi, T.R. Oliveira, P.J. Pereira, P. Limão-Vieira, B.W. Snow, D. Reudink, P.R. Stauffer, Design and optimization of an ultra-wideband and compact microwave antenna for radiometric monitoring of brain temperature. IEEE Trans. Biomed. Eng. **61**(7), 2154–2160 (2014)
8. N. Joachimowicz, B. Duchêne, C. Conessa, O. Meyer, Anthropomorphic breast and head phantoms for microwave imaging. Diagnostics **8**(4), 85 (2018)

9. P.M. Meaney, C.J. Fox, S.D. Geimer, K.D. Paulsen, Electrical characterization of glycerin: water mixtures: implications for use as a coupling medium in microwave tomography. IEEE Trans. Microw. Theory Tech. **65**(5), 1471–1478 (2017)
10. G. Hartsgrove, A. Kraszewski, A. Surowiec, Simulated biological materials for electromagnetic radiation absorption studies. Bioelectromagn. J. Bioelectromagn. Soc. Phys. Regul. Biol. Med. **8**(1), 29–36 (1987)
11. N. Joachimowicz, C. Conessa, T. Henriksson, B. Duchêne, Breast phantoms for microwave imaging. IEEE Antennas Wireless Propag. Lett. **13**, 1333–1336 (2014)
12. N. Joachimowicz, S. Abedi, H. Roussel, Numerical wideband anthropomorphic phantom for biomedical microwave imaging, in *CAMA* (2021)
13. S. Abedi, *Standard Phantoms for EM Device Testing*, Ph.D. Thesis (Sorbonne Université, 2022)
14. K. Wang, C.C. Ho, C. Zhang, B. Wang, A review on the 3D printing of functional structures for medical phantoms and regenerated tissue and organ applications. Engineering **3**(5), 653–662 (2017)
15. A. Haleem, M. Javaid, R. Vaishya, 3D printing applications for the treatment of cancer. Clin. Epidemiol. Glob. Health **8**(4), 1072–1076 (2020)
16. A. Asfia, J.I. Novak, M.I. Mohammed, B. Rolfe, T. Kron, A review of 3D printed patient specific immobilization devices in radiotherapy. Phys. Imaging Radiation Oncol. **13**, 30–35 (2020)
17. A.T. Mobashsher, A.M. Abbosh, Artificial human phantoms: human proxy in testing microwave apparatuses that have electromagnetic interaction with the human body. IEEE Microw. Mag. **16**(6), 42–62 (2015)
18. T. Pokorny, J. Tesarik, 3D printed multi-layer molds of human head tissue phantoms for microwave stroke detection, in *2019 PhotonIcs & Electromagnetics Research Symposium-Spring (PIERS-Spring)* (IEEE, 2019), pp. 1424–1427
19. J. Garrett, E. Fear, A new breast phantom with a durable skin layer for microwave breast imaging. IEEE Trans. Antennas Propag. **63**(4), 1693–1700 (2015)
20. I.E. Khuda, A comprehensive review on design and development of human breast phantoms for ultra-wide band breast cancer imaging systems. Eng. J. **21**(3), 183–206 (2017)
21. Plasticboy, Blender Rigged V07 Anatomy 3D models store [online]. Available https://www.plasticboy.co.uk/store/index.html . Accessed on 22 Nov 2022
22. 3D slicer, image-computing platform [online]. Available https://www.slicer.org. Accessed on 22 Nov 2022
23. J.A. Tobon Vasquez, R. Scapaticci, G. Turvani, G. Bellizzi, D.O. Rodriguez-Duarte, N. Joachimowicz, B. Duchêne, E. Tedeschi, M.R. Casu, L. Crocco, F. Vipiana, A prototype microwave system for 3D brain stroke imaging. Sensors **20**(9), 2607 (2020)
24. D.O. Rodriguez-Duarte, J.A. Tobon Vasquez, R. Scapaticci, G. Turvani, M. Cavagnaro, M.R. Casu, L. Crocco, F. Vipiana, Experimental validation of a microwave system for brain stroke 3-d imaging. Diagnostics **11**(7) (2021). ISSN: 2075-4418. https://doi.org/10.3390/diagnostics11071232 [Online]. Available https://www.mdpi.com/2075-4418/11/7/1232
25. F. Bray, J. Ferlay, I. Soerjomataram, R.L. Siegel, L.A. Torre, A. Jemal, Global cancer statistics 2018: GLOBOCAN estimates of incidence and mortality worldwide for 36 cancers in 185 countries. CA Cancer J. Clin. **68**(6), 394–424 (2018)
26. D.M. Godinho, J. Felicio, C.A. Fernandes, R.C. Conceicao, Experimental evaluation of an axillary microwave imaging system to aid breast cancer staging. IEEE J. Electromagn. RF Microw. Med. Biol. (2021)
27. M. Savazzi, S. Abedi, N. Ištuk, N. Joachimowicz, H. Roussel, E. Porter, M. O'Halloran, J.R. Costa, C.A. Fernandes, J.M. Felício, R.C. Conceição, Development of an anthropomorphic phantom of the axillary region for microwave imaging assessment. Sensors **20**(17), 4968 (2020)
28. Fundação Champalimaud (Lisbon, Portugal) [online]. Available https://fchampalimaud.org/. Accessed on 22 Nov 2022
29. F.X. Bosch, J. Ribes, J. Borràs, Epidemiology of primary liver cancer, in *Seminars in Liver Disease*, vol. 19, no. 03 (Theme Medical Publishers, Inc., 1999), pp. 271–285

30. V. Lopresto, L. Strigari, L. Farina, S. Minosse, R. Pinto, D. D'Alessio, B. Cassano, M. Cavagnaro, CT-based investigation of the contraction of ex vivo tissue undergoing microwave thermal ablation. Phys. Med. Biol. **63**(5), 055019 (2018)
31. M. Wang, R. Scapattici, M. Cavagnaro, L. Crocco, Towards a microwave imaging system for continuous monitoring of liver tumor ablation: design and in silico validation of an experimental setup. Diagnostics **11**(5), 866 (2021)
32. M. Wang, M. Cavagnaro, S. Abedi, N. Joachimowicz, H. Roussel, S. Costanzo, L. Crocco, *Initial Experimental Validation of a Microwave Imaging System to Monitor Liver Microwave Thermal Ablation* (EUCAP, 2022)
33. M. Wang, R. Scapaticci, S. Abedi, H. Roussel, N. Joachimowicz, S. Costanzo, M. Cavagnaro, L. Crocco, *A Microwave Imaging System Prototype for Liver Ablation Monitoring: Design and Initial Experimental Validation* (iWAT, 2022)
34. V. Lopresto, R. Pinto, G.A. Lovisolo, M. Cavagnaro, Changes in the dielectric properties of ex vivo bovine liver during microwave thermal ablation at 2.45 GHz. Phys. Med. Biol. **57**(8), 2309 (2012)
35. E. Walter, *Numerical Methods and Optimization* (Springer, Cham, 2014)
36. A. Stogryn, Equations for calculating the dielectric constant of saline water (correspondence). IEEE Trans. Microw. Theory Tech. **19**(8), 733–736 (1971)
37. S. Gabriel, R.W. Lau, C. Gabriel, The dielectric properties of biological tissues: III. Parametric models for the dielectric spectrum of tissues. Phys. Med. Biol. **41**(11), 2271 (1996)
38. IFAC, Calculation of the Dielectric Properties of Body Tissues in the frequency range 10 Hz–100 GHz [online]. Available http://niremf.ifac.cnr.it/tissprop/htmlclie/htmlclie.php. Accessed on 22 Nov 2022
39. A. Kraszewski, Prediction of the dielectric properties of two-phase mixtures. J. Microw. Power **12**(3), 216–222 (1977)
40. J.B. Hasted, *Aqueous Dielectrics*, vol. 122 (Chapman and Hall, London, 1973)
41. C.J.F. Böttcher, P. Bordewijk, *Theory of Electric Polarization*, vol. 2 (Elsevier Science Limited, 1978)
42. K.V. Lichtenecker, Die Herleitung des logarithmischen Mischungsgesetzes aus allgemeinen Prinzipien der stationaren Stromung. Phys. Z **32**, 255–260 (1931)
43. H. Looyenga, Dielectric constants of heterogeneous mixtures. Physica **31**(3), 401–406 (1965)
44. S. Abedi, N. Joachimowicz, N. Phillips, H. Roussel, A simulation-based methodology of developing 3D printed anthropomorphic phantoms for microwave imaging systems. Diagnostics **11**(2), 376 (2021)
45. M. Lazebnik, M. Okoniewski, J.H. Booske, S.C. Hagness, Highly accurate Debye models for normal and malignant breast tissue dielectric properties at microwave frequencies. IEEE Microw. Wirel. Compon. Lett. **17**(12), 822–824 (2007)
46. J. Wang, E.G. Lim, M.P. Leach, Z. Wang, K.L. Man, Open-Ended Coaxial Cable. Selection for Measurement of Liquid Dielectric Properties via the Reflection Method. Mathematical Problems in Engineering (2020)
47. N. Ištuk, E. Porter, D. O'loughlin, B. McDermott, A. Santorelli, S. Abedi, N. Joachimowicz, H. Roussel, M. O'halloran, Dielectric properties of ovine heart at microwave frequencies. Diagnostics **11**(3), 531 (2021)
48. A. Fornés, *Dielectric Characterization of Biological Tissues for Medical Applications*, Ph.D. Thesis (Universitat Politècnica de Valencia, 2019)
49. S. Abedi, N. Joachimowicz, O. Meyer, D. Picard, H. Roussel, Phantoms for a novel generation of medical microwave imaging devices, in *2019 13th European Conference on Antennas and Propagation (EuCAP)* (IEEE, 2019), pp. 1–4
50. SUPELEC RECIPES, website under development
51. J. Boparai, M. Popović, Heterogeneous skin phantoms for experimental validation of microwave-based diagnostic tools. Sensors **2022**, 22 (1955). https://doi.org/10.3390/s22051955
52. A. Bottiglieri et al., Dielectric characterization of ex vivo ovine and human adrenal glands for microwave thermal ablation applications. IEEE J. Electromagn. RF Microw. Med. Biol. (2021)

53. S. Gavazzi et al., Comparison of microwave dielectric properties of human normal, benign and malignant thyroid tissues obtained from surgeries: a preliminary study. Biomed. Phys. Eng. Express**4**, 047003 (2018)
54. E. Porter, S. Salahuddin, A. La Gioia, E.M. Adnan, A. Shahzad, A. Kumar, D. Kilroy, M. O'Halloran, Characterization of the dielectric properties of the bladder over the microwave range. IEEE J. Electromagn. RF Microw. Med. Biol. **2**(3), 208–215 (2018)
55. S. Salahuddin et al., An anatomically accurate dielectric profile of the porcine kidney. Biomed. Phys. Eng. Express **4**(2), 025042 (2018)
56. A. Shahzad, D. Clausing, P. Prakash, M.C. Dennedy, M. O'Halloran, Broadband dielectric properties of adrenal gland for accurate anatomical modelling in medical applications, in *2017 International Conference on Electromagnetics in Advanced Applications (ICEAA)*, Sept 2017

Hardware Acceleration of Microwave Imaging Algorithms

Mohammad Amir Mansoori and Mario R. Casu

Abstract Microwave Imaging (MWI) is a technique that allows to reconstruct an image of the internal structure of an object by irradiating the object with a known incident field and by acquiring and processing the scattered field. As such, MWI can be used as a diagnostic technique for various medical issues, such as brain stroke or breast cancer. However, the image reconstruction process entails a sequence of compute-intensive algorithms, which we call "kernels." The kernels execution time might become an issue when the time to diagnosis is a key factor. To speed up the kernels execution, we can use hardware acceleration techniques. In this chapter, we will identify the complex computational kernels that are recurrent in MWI. Then, we present the methodologies to design specific hardware accelerators in Field-Programmable Gate Arrays (FPGAs) for these kernels by means of an efficient design method called *High Level Synthesis (HLS)* and by several HLS-based hardware optimization techniques. In addition, we will present new methodologies for design automation in HLS-based hardware designs. The results show that the presented HLS optimizations and design automation techniques can significantly improve the efficiency of the hardware accelerator designs for MWI.

1 Introduction

Microwave Imaging (MWI) in medical applications can be used to obtain some information about the internal structure of the body by using electromagnetic fields at microwave frequencies. Although other imaging modalities such as CT-scan or MRI are more common in medical diagnosis, MWI has attracted attention in recent years because of its non-invasive, non-ionizing, and low-cost characteristics. There

M. A. Mansoori (✉) · M. R. Casu
Department of Electronics and Telecommunications, Politecnico di Torino, Turin 10129, Italy
e-mail: mohammadamir.mansoori@polito.it

M. R. Casu
e-mail: mario.casu@polito.it

F. Vipiana and L. Crocco (eds.), *Electromagnetic Imaging for a Novel Generation of Medical Devices*, Lecture Notes in Bioengineering,
https://doi.org/10.1007/978-3-031-28666-7_2

are mainly two methods to obtain the properties of an object by MWI: *Qualitative Imaging* and *Quantitative Imaging* methods.

In ***Qualitative Imaging***, only the information such as shape or location of any anomalies in the object are determined. For example, in brain stroke imaging, the anomaly is a brain stroke that must be ultimately classified into ischemic or hemorrhagic stroke. By means of Qualitative Imaging it is possible to identify the position, and shape of the stroke, but not the type (ischemic or hemorrhagic), as this information is related to the value of the dielectric properties of the stroke.

On the contrary, in ***Quantitative Imaging*** a complete characterization of the targets in terms of morphology and values of the electromagnetic properties is achieved by processing the microwave measurements and by solving an *Inverse Scattering* problem. In spite of the non-linear nature of inverse scattering, one of the common approaches to tackle this problem is to linearize it by using the Born approximation. However, in highly heterogeneous objects, the linearized solutions are not effective and non-linear algorithms can provide better results. A common non-linear method is to iteratively use a Forward Solver to estimate the microwave measurements and a linear Inverse Solver to update the dielectric profile.

In addition, ***Machine Learning (ML)*** as a new category is recently introduced, which is used to learn both qualitative and quantitative information from the observations. ML algorithms are useful to obtain Qualitative or Quantitative information in MWI. The general processing steps of ML methods is as follows. After pre-processing the data (i.e., the microwave measurements), feature extraction techniques, such as for example Principal Component Analysis (PCA), can be used to obtain the most important information from data. The final stage in ML processing steps can be an image reconstruction, or the detection followed by classification of an anomaly. Different ML classifiers exist including Support Vector Machine (SVM), Random Forest (RF), and Neural Networks (NN), while Generative Adversarial Networks (GANs) can be used to reconstruct an image from microwave data.

1.1 Problem Statement

High execution time of MWI algorithms (especially for Quantitative, or ML methods) makes it difficult to use them in real-time biomedical systems or, in general, when the time-to-diagnosis is a critical factor. The complexity of each MWI algorithm is due to some compute-intensive parts that we call "kernels." These kernels can be implemented in hardware accelerators to increase the speed of execution and so to improve the overall performance of an MWI-based biomedical system.

Field Programmable Gate Arrays (FPGAs) are powerful hardware accelerators used in different applications and that can be also used in MWI [1–3]. Compared to Graphical Processing Units (GPUs), FPGAs have the advantage of lower power consumption for a given performance level [4]. However, they are notoriously more difficult to program, as in general programming an FPGA requires an expertise in digital hardware design that GPU programmers do not need to possess. For the implementation of an algorithm in FPGA, however, the recent trend is to use High Level Synthesis (HLS) tools that are able to convert a software code written in C or C++ into its corresponding hardware implementation [5]. The advantage is the faster design and development process compared to the traditional approach of manual hardware design.

In this chapter, we first identify a comprehensive set of computational kernels that are recurrent in MWI algorithms. After that, we present the methodologies to design specific hardware accelerators in FPGAs for these kernels with the HLS approach. To design efficient hardware accelerators, we introduce several HLS-based hardware optimization techniques. In addition, we present new methodologies for design automation in HLS-based hardware designs that can increase the design efficiency by allowing designers to find the optimum hardware configurations in a relatively short time. For the evaluation of hardware accelerators, we compare the performance of FPGAs with CPU and GPU-based designs in terms of processing time, power consumption, cost, form factor, and resource usage. In addition, the impact of different HLS optimization methods on the hardware performance is thoroughly explored. The results show that the presented HLS optimizations and design automation techniques can significantly improve the efficiency of the hardware accelerator designs for microwave imaging.

2 Microwave Imaging System

An MWI system uses a set of antennas emitting microwave radiations towards a given body part and collecting the reflected electromagnetic field. The discontinuity caused by the difference between the dielectric properties of different body tissues is the root cause of these reflections, a phenomenon termed *scattering*. By acquiring and processing by means of various methods the reflected signals, it is possible to either reconstruct an image that, in turn, can highlight the presence of an anomaly within the irradiated body part, or to directly detect and classify such anomaly.

The components of an MWI system are illustrated in Fig. 1. The first component is a set of antennas arranged around the body part under consideration. A switching matrix is connected to these antennas, specifying which pairs of antennas must be active during the measurements. The switching matrix is connected to a Vector Network Analyzer (VNA) that measures the microwave radiations in the form of a scattering matrix. The last component of the device is a processing system that is used to convert the captured microwave radiations into the required information for medical diagnosis. The algorithms used in the processing system are usually compute-intensive and can benefit from hardware acceleration.

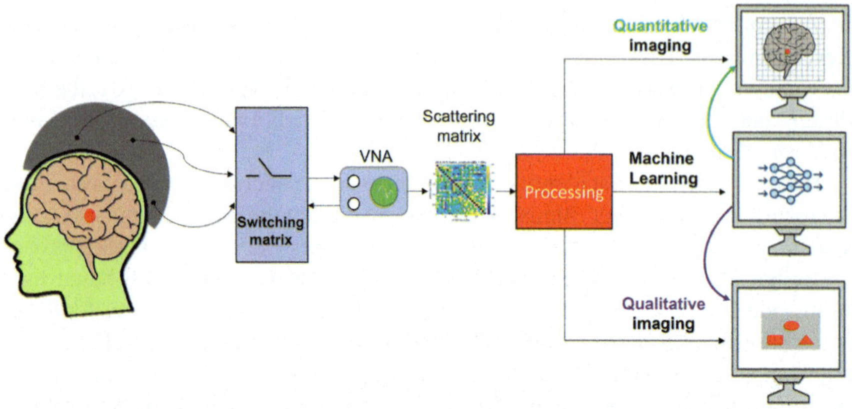

Fig. 1 General diagram of a microwave imaging system

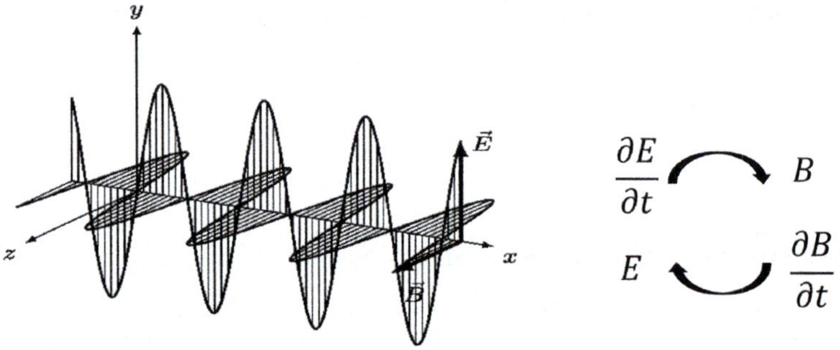

Fig. 2 Propagation of electromagnetic fields and the mutual dependency between electric and magnetic fields

3 Compute-Intensive Kernels in Microwave Imaging

In this section, the compute-intensive kernels that are commonly used in biomedical MWI algorithms are introduced. In addition, a brief overview of the relevant previous works and their limitations will be presented. In the subsequent sections, the details of the proposed hardware accelerators for these kernels will be explained.

3.1 3D FDTD

Finite Difference Time Domain (FDTD) is a method to solve the Maxwell equations, which model the propagation of electromagnetic fields, by means of finite differences. As well known, the Maxwell equations state that any variation in the magnetic field creates an electric field and vice versa, as illustrated in Fig. 2.

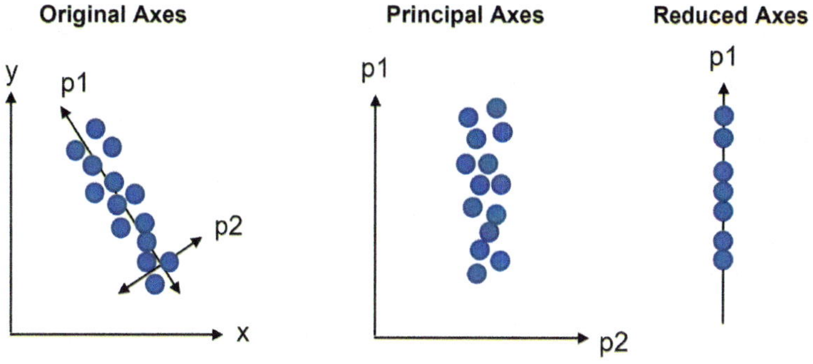

Fig. 3 PCA for dimensionality reduction to remove redundant information

Several recent works propose GPU as FDTD hardware accelerator [6–9]. High power consumption of GPUs pushes hardware designers to use FPGA accelerators instead. More details on FPGA implementations of FDTD can be found in [10–15].

The above-mentioned FDTD accelerators have some limitations. Most of them considered simple boundary conditions for the Maxwell equations, such as Periodic Boundary Conditions (PBC), which cannot be used in medical MWI because they lead to a lower accuracy: more complex boundary conditions like Convolutional Perfectly Matched Layer (CPML) are required instead. In previous FPGA accelerators of FDTD designed in high level description language, CPML conditions were considered only in two directions (top and bottom) of the domain in which the equations are solved, while for the other directions PBC conditions were used. In addition, none of the previous works considered the impact of polarization currents in *dispersive* materials, which is important for a more complex and accurate modeling of propagation but complicates the design of hardware accelerators for FDTD.

3.2 *PCA Using SVD/EVD*

PCA is used in ML for feature extraction and to eliminate redundant information from data. Its computation consists of different steps, which include Singular value or Eigenvalue Decomposition (SVD/EVD). When the dimensions of data is large, PCA is highly beneficial to reduce the data dimension. As shown in Fig. 3, by transforming the original data into its corresponding principal axes, it is possible to ignore the horizontal axis and reduce the dimensions of data. In MWI, PCA can remove redundancy in scattering matrices collected at different frequencies.

Different FPGA accelerators have been recently proposed for PCA, but most of them have been implemented starting from a manually obtained Register Transfer Level (RTL) design [16–19]. A few works concentrated on the acceleration of only some components of PCA such as EVD or SVD [20–26]. Other references used High Level Synthesis for the design of PCA accelerator [27–30].

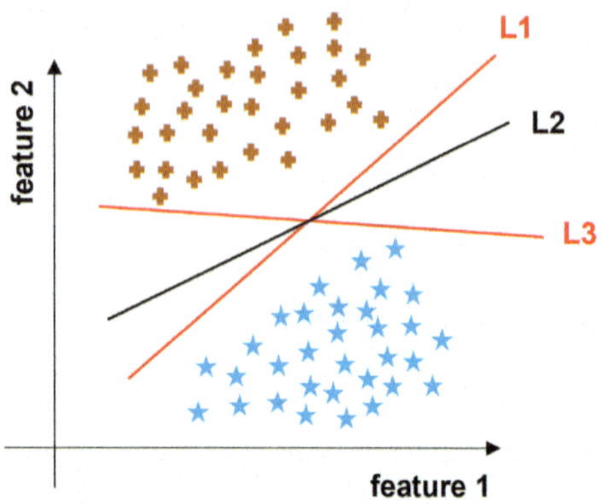

Fig. 4 SVM for binary classification finds the decision boundary with maximum margin between two classes (L2)

Differently from our flexible PCA accelerator, most of the previous works did not propose to implement PCA in its entirety in a way to support different data dimensions. Instead, our FPGA accelerator can be used in different applications including MWI supporting large data dimensions.

3.3 SVM

One of the widely-used classification techniques in ML is Support Vector Machine (SVM). As illustrated in Fig. 4, SVM finds a decision boundary that can maximize the margin between two classes. Among all the possible decision boundaries (L1–L3 in Fig. 4), only one line is optimum (L2) and can be obtained after the *training phase* of SVM. In the *inference phase,* new data points will be classified based on the previously obtained decision boundary. In addition to binary classification, SVM can be used in multi-class classification by using a *Kernel* function which transforms the data points into a higher dimension space in which the samples are linearly separable. In MWI, SVM can be used for the detection and classification of different anomalies such as breast cancer or brain stroke.

Several FPGA accelerators have been proposed for SVM [31–42]. Recent SVM accelerators in FPGA are reviewed in [43]. The main limitation of almost all the previous works is the lack of scalability, which would allow the same design to be used for larger data dimensions. These conventional accelerators assumed that all the SVM coefficients and input data are stored in an on-chip memory, which prevents these designs to be used for large data dimensions. Instead, we proposed a flexible and scalable SVM accelerator in FPGA that can be used for different data dimensions when we have a limited amount of on-chip memory.

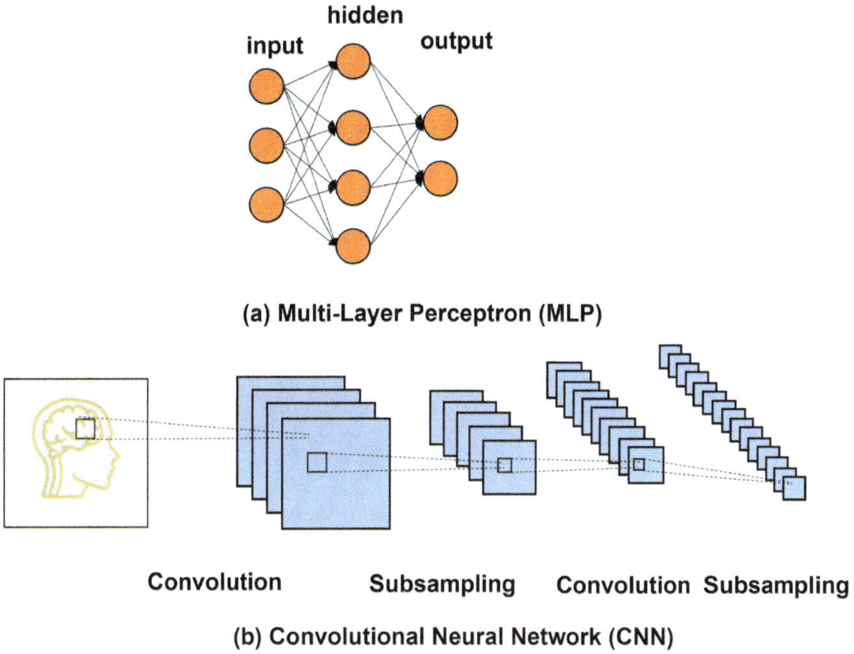

Fig. 5 Network structure for **a** MLP and **b** CNN

3.4 Neural Networks

Artificial Neural Networks (ANNs) are progressively used in different application areas, including MWI. ANNs are inspired from the human brain structure and are organized in layers of artificial neurons: the input layer, several hidden layers, and the output layer. Each layer receives the data from the previous layer and processes them before feeding the next layer. The operations performed by each neuron are relatively simple (multiplication, accumulation, and the application of a non-linear function), but the number of neurons in a layer, their connections, and the number of layers complicate in a substantial way the hardware implementation of ANNs.

Multi-Layer Perceptrons (MLPs) and Convolutional Neural Networks (CNNs) are two common structures of ANNs. In MLP, the layers are fully connected to each other. In CNNs, there are several filters in each layer that are convolved with the input data from the previous layers. Figure 5 shows these two network architectures. One of the main challenges in designing an efficient FPGA accelerator for ANNs is how to optimize at the same time the training hyper-parameters and the hardware configuration to achieve the best performance in terms of training accuracy, speed, hardware resource usage and power consumption. Although there have been several recent works tackling this problem [44–49], there is still a great potential to further enhance and optimize the previous works. One of the areas that has not been

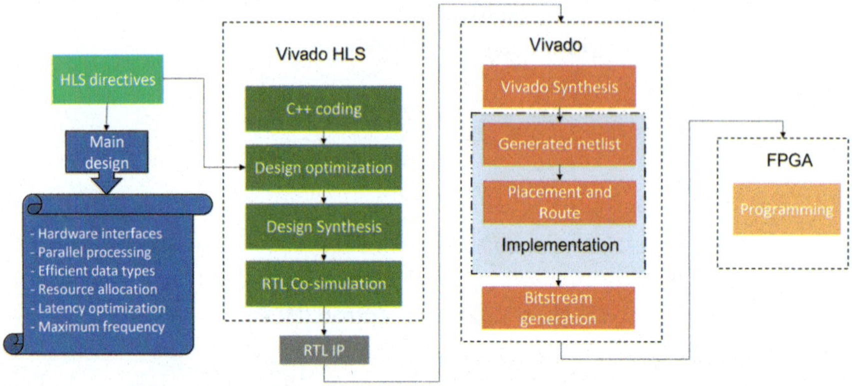

Fig. 6 Hardware design flow in Vivado [51]

fully explored in the context of co-optimization of network training parameters and hardware configurations is Bayesian Optimization which we consider in this work.

4 HLS-Based Hardware Acceleration in FPGA

The traditional method to design a hardware accelerator in FPGA is manual RTL design using Hardware Description Languages (HDLs) like VHDL or Verilog. Although being still the dominant approach, RTL manual design using HDLs requires a significant amount design and development time. This precludes the exploration of multiple alternatives in search of the optimum design, which is especially challenging for very large designs due to a large hardware design space. A recent approach that simplifies the design space exploration and can achieve a similar performance compared to the HDL languages with the advantage of faster development phase is High Level Synthesis (HLS) approach. HLS receives a software code written in C or C++ and transforms it into the corresponding RTL design by properly scheduling the tasks, binding the operations, and allocating the resources [50]. Since the RTL is automatically generated and the generation is fast, exploring multiple design options becomes viable even for very large designs.

Figure 6 shows the general HLS design flow in FPGAs for Xilinx devices. The input to the *Vivado HLS* tool is a software code in C or C++. In the next step, based on the design constraints such as timing and available resources, different hardware optimizations can be applied to the code to achieve the desired performance. This can be obtained by using so-called HLS directives [52]. The next step is the synthesis in which the netlist corresponding to the optimized code is generated. The RTL co-simulation step checks if the generated RTL is functionally identical to the software code. Finally, after generating an RTL Intellectual Property (IP) block, the

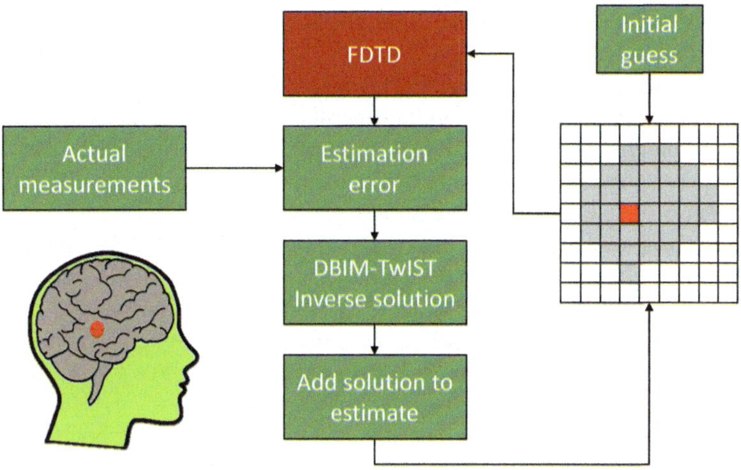

Fig. 7 General diagram of a non-linear image reconstruction iterative algorithm in MWI, with the compute-intensive FDTD step

implementation steps follow in the Vivado tool in the same way as if the RTL was manually generated [51].

5 FPGA Acceleration of 3D FDTD for Multi-antenna MWI

Two main computational problems often found in MWI are the so-called *Forward Scattering* and *Inverse Scattering*. In the former, the dielectric profile is already known and the goal is to compute the scattered fields using this information. Instead, in the latter, the electromagnetic properties of the body tissues are unknown and the goal is to retrieve them from the known scattered fields. In iterative non-linear MWI algorithms, usually the Forward and Inverse scattering are combined in such a way that the Inverse solver calls the Forward solver to iteratively update the solution.

One of these cases is shown in Fig. 7. It is a non-linear iterative algorithm in which a forward electromagnetic solver, in this work an FDTD solver, is executed iteratively to update the initial guess of the dielectric profile based on the difference between the FDTD output and the actual microwave measurements received from a set of antennas. The error between the actual and simulated electromagnetic fields is minimized in multiple iterations by using a piece-wise linear inverse solution (in this work we used DBIM-TwIST algorithm developed in [53] for the inverse solution). As highlighted in Fig. 7 in red, FDTD is the bottleneck for the execution time of this algorithm, thus it is highly beneficial to have an efficient hardware accelerator.

Initially, a GPU implementation for the critical FDTD part was used. Although a speed-up can be achieved with GPU compared to the software design in CPU, the overall processing time is still high. This is why we focused on FPGA acceleration to further improve the performance.

The discretized domain in which FDTD solves the Maxwell equations, especially for 3D cases, contains so many unknowns that it is impossible to store them in the limited on-chip memory of the hardware accelerator, be it a GPU or an FPGA. Therefore, moving data between an external memory (i.e., outside the FPGA) and the local memory (inside) is necessary and can become a performance limiter. The FPGA accelerator proposed in this work uses a *spatial* blocking approach to reduce the data transfer time between the external memory (i.e., outside the FPGA) and the local memory (inside the FPGA). In this approach, a new plane from the 3D simulation space is read from the memory while the previous plane is being processed. The computations of FDTD for each plane consists of the famous Yee's update equations [54], in which we include also the effect of the boundaries.

In particular, the boundary regions are modeled with absorbing boundary conditions (CPML). Therefore, in addition to the variables representing the electric and magnetic fields in the three directions (E_x, E_y, E_z, H_x, H_y, H_z), additional variables in these equations are used in the boundary regions. In addition, we model dispersive materials, which requires an extra variable termed J_P ((J_{Px}, J_{Py}, J_{Pz}) in FDTD equations that complicates the hardware design and has not been considered in previous works. The main features of our FDTD accelerator are:

- the impact of dispersive materials and complex CPML boundary conditions on FDTD equations are considered;
- to increase the flexibility we designed two hardware architectures, Large and Small, from which the designer can choose based on the specific characteristics of their design;
- we used High Level Synthesis (HLS) and several hardware optimization methods to implement the accelerator in an efficient way;
- for multiple antennas, we could design single- and multi-FPGA platforms that can process FDTD equations in parallel for different antennas;
- the high level design of our FDTD algorithm can be implemented in GPU as well with a performance comparable with a commercial GPU implementation.

5.1 FPGA Design of an FDTD Compute Unit

Figure 8 shows a block diagram of the HLS code for a single Compute Unit (CU) related to the FDTD design. Multiple CUs can be instantiated to work concurrently on different antennas. Each block in a CU is a function corresponding to an update equation (electric or magnetic update equations). Note that the J_P equations related to the polarization currents in dispersive materials are merged with the Update E and E boundary equations in Fig. 8: this avoids multiple accesses to fetch the field values from the external memory. Table 1 shows the HLS-based hardware optimization methods used in our design. They will be explained in more detail in Sect. 5.3.

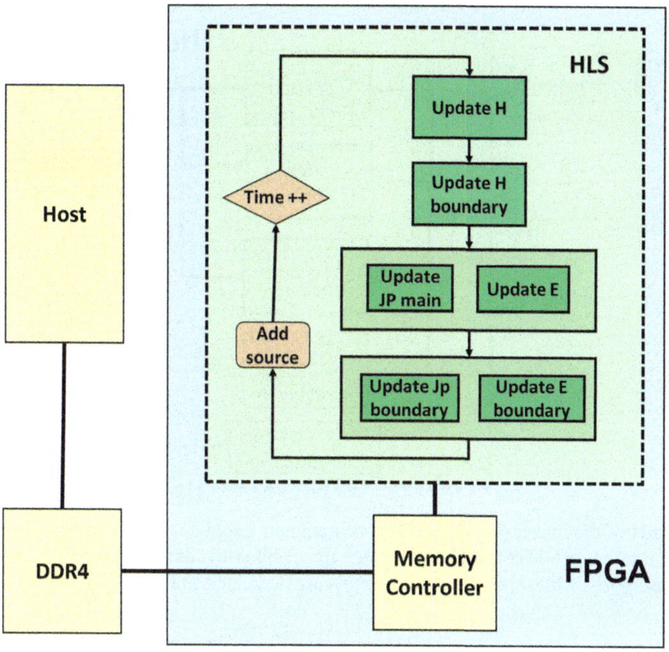

Fig. 8 FDTD accelerator in FPGA, JP is the polarization current

Table 1 HLS hardware optimization strategies for a FDTD CU

Method	Functions
Blocking	Update E and Update H
Merge JP	Update E and E boundary
Loop merge	Update E boundary and H boundary
Storage for boundaries	Update H boundary and/or Update E boundary
Storage for constant coefficients	top-level function
Loop pipeline	all
function inline	all

5.2 *Two Architectures: Large and Small*

The difference between *Large* and *Small* designs is the number of AXI I/O ports toward the external memory and the amount of local resources in on-chip memory.

The Large design uses more AXI ports and compute resources resulting in higher speed for a single CU whereas the Small design consumes less resources and AXI ports. Nonetheless, since more Small CUs can fit in one FPGA, the Small design can lead to a higher throughput. Therefore, depending on the number of available FPGAs and the number of CUs to parallelize, the designer can select either Small

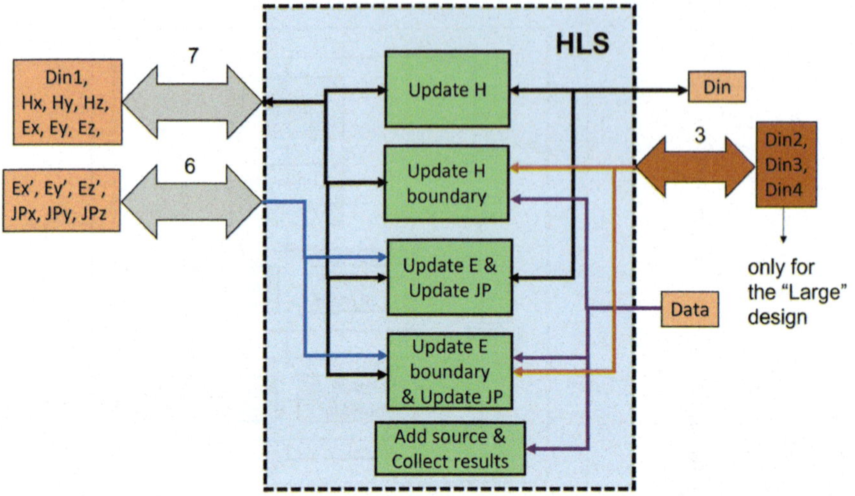

Fig. 9 Details of the interfaces of the CU for Small and Large designs. Numbers near the arrows indicate the number of variables read or written from and to the external memory ports. Din-n are multi-purpose ports, while specific memory ports are dedicated to E and H variables

or Large design. Note that the larger number of AXI ports in the Large design (18 ports) is still compatible with the maximum memory bandwidth in the target FPGA, whereas in the Small design there are 15 AXI ports, as shown in Fig. 9.

5.3 HLS Optimizations

We briefly explain the HLS optimizations in Table 1 used in the proposed FDTD accelerator.

The main optimization is the spatial blocking approach used to reduce the memory access time in the main update equations. As shown in Fig. 10, processing of 2D planes in the simulation space can be pipelined by using a shift register in the local memory that can store one row (in 2D) or one plane (in 3D) of the simulation space plus an additional cell. After the shift register is full, the computation of the previous plane and the storage of the new plane in the shift register can occur in parallel. In this way, the shift register helps reduce the number of accesses to the external memory and the throughput can increase.

Loop merging was used for the loops in the boundary regions as well as the loops related to the polarization currents (J_P loops). In addition, we used local RAM memories available in Xilinx FPGAs, namely Block-RAMs (BRAMs) and Ultra-RAMs (URAMs), for the storage of Boundary fields and constant coefficients, respectively. Finally we used "unrolling" and "pipelining" for the innermost loop in FDTD algorithm.

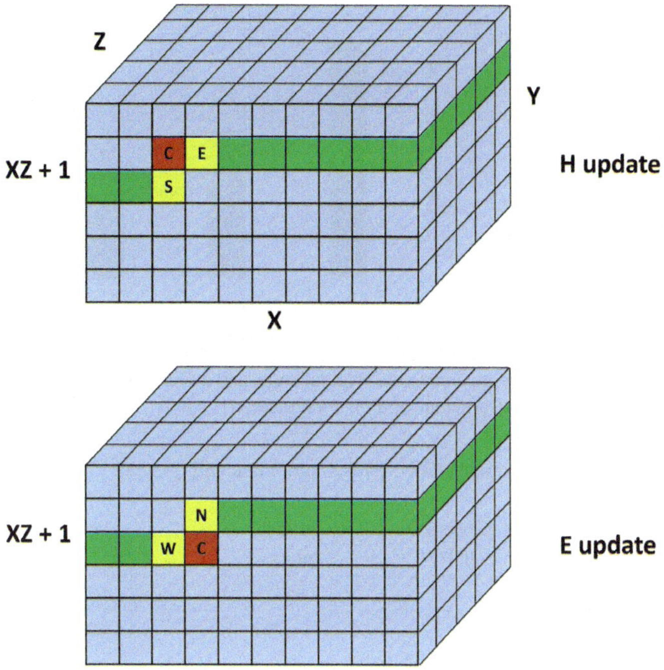

Fig. 10 Blocking method for 3D FDTD

5.4 FDTD Result

The hardware platform used in our experiments is the Virtex UltraScale+ used in the Amazon EC2 F1 instance (vu9p-flgb2104-2-i). This FPGA consists of 3 Super Logic Regions (SLRs). There are 4 DDR4 memory interfaces. Each DDR interface can access a 16 GiB external memory.

The impact of the HLS optimization techniques described in Sect. 5.1 on the total latency is shown in Fig. 11. These optimizations are added incrementally starting from the original code without any HLS directives. The performance of the FDTD Compute Unit (CU) after each optimization is measured in terms of latency. The results show that each directive reduces the latency until a minimum latency of 8.6 s is obtained after applying all the directive in the Large design.

Table 2 presents a comparison between the performance of the FDTD accelerator on a CPU, three GPU designs, and FPGA. To have a fair comparison, the *time per antenna* is selected as the performance metric (total processing time divided by the number of antennas processed in parallel). The power consumption for FPGA in the table is the total consumed on-chip power that is calculated by the post-route report in the *Vivado Power Analysis* tool, while for the CPU and GPU, the maximum Thermal Design Power (TDP) is reported. The maximum TDP in our FPGA target is 128*W* which is still less than the TDP of CPU or GPU designs. By multiplying

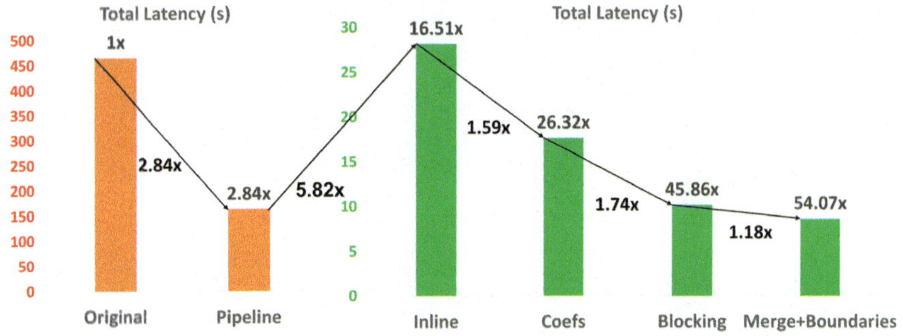

Fig. 11 Impact of different HLS optimization methods on the total latency. Numbers on top of the bars show the improvement compared to the original code, and numbers below the arrows show the improvement compared to the previous optimization method

Table 2 Performance comparison: CPU (Intel Xeon), three GPU designs (Tesla K20C and P40), and FPGA (UltraScale+)

Hardware	Ant. processed in parallel	Time per ant. (s)	Power (W)
Intel Xeon Gold 5120	1	24.97	105
GPU1 Tesla K20c (Matlab)	1	12.06	225
GPU2 Tesla P40 (Acceleware)	1	5.64	250
GPU3 Tesla K20c (This work)	1	4.88	225
UltraScale+ (This work, 15-ports)	3	3.38	16.24
UltraScale+ (This work, 18-ports)	2	4.3	14.5

either the on-chip power or the maximum TDP to the Processing time per antenna, it is observed that the total energy consumption of our proposed FPGA design is the lowest.

A comparison between similar related works are presented in Table 3. It is observed that the previous works did not consider the impact of polarization currents which creates extra computations and complicates the data dependencies. It can be inferred that the performance in Mcells/s is not high in our work. However, due to the extra operations created by the polarization currents, the performance in GFLOP/s is superior to other implementations. In addition, our proposed accelerator operates in higher clock frequency than traditional works. Please refer to [55] for more details about the results of the proposed hardware accelerator for FDTD.

Table 3 Performance comparison between our single Small FPGA design and other FPGA implementations

	Work [15]	Work [11]	Work [14]	This work Small design
Boundary Conditions	Dirichlet (zero)	CPML (limited)	CPML (Full)	CPML (Full)
Polarization	No	No	No	Yes
Language	MaxCompiler (HLS)	MaxCompiler (HLS)	Verilog	Vivado HLS (C++)
Mcells/s	325	100	336	101
GFLOP/s	11.7	5.7	29.2	34.2
Freq (MHz)	100	100	100	167

6 High Level Design of PCA Accelerator in FPGA

PCA can be broken down into a sequence of computations performed over the incoming data. These are *Mean*, *Covariance*, *EVD* (or *SVD*), and *Projection*. Each of these can be designed as an HLS function, and by using the Dataflow optimization approach all the functions can be executed concurrently in a pipelined manner. Array partitioning, loop unrolling, loop pipelining, and function inlining are among other HLS optimizations that are used in the hardware design to achieve a high performance in our FPGA accelerator.

The hardware design of our PCA accelerator is illustrated in Fig. 12, which can be interpreted as the hardware implementation of the software code in C++. The software code consists of a top function including two sub-functions: *Dispatcher* and *PCA core*. In the Dispatcher, the required data are read and sent to the PCA core through First-in First-out (FIFO) channels. In the PCA core, the main processing steps of the PCA algorithm are executed. The final results are written back to the external DDR memory.

The input to the PCA algorithm is a matrix with N samples and F features, and usually N is much greater than F. The first step in PCA algorithm is Mean computation which computes the mean values of all the samples. These mean values are stored in an internal on-chip memory to be used in Covariance (Cov) and Projection units. The first one computes the covariance matrix of the normalized input data and the SVD unit computes the Singular Values (SVs) of such covariance matrix: after sorting the SVs, the most important information from the data is preserved by keeping the highest values of the SVs as the Principal Components (PCs). In the Projection Unit, the normalized input data is multiplied with the PCs to obtain the output data with lower dimensions than the input data.

Since the covariance matrix has size $F \times F$, which is much smaller than the input matrix, computing the singular values is not as time-consuming as computing the covariance matrix itself. Therefore, hardware acceleration of Cov unit can signif-

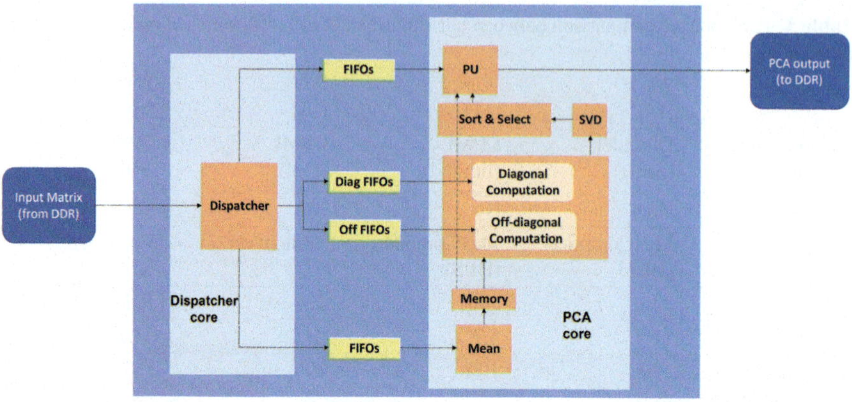

Fig. 12 Architecture of the proposed hardware accelerator for Principal Component Analysis (PCA) in Field-Programmable Gate Arrays (FPGA). The dispatcher distributed the input features to the PCA core units: mean, diagonal and off-diagonal computation, and projection unit. This last unit generates the PCA output and sends it to the external DDR memory

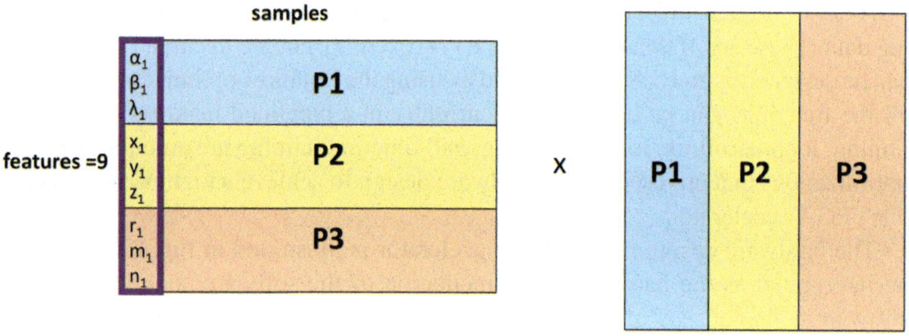

Fig. 13 Example of partitioning of input data into blocks. The total number of features is 9 and the block size is $B_{\max} = 3$

icantly enhance the overall performance. For this reason, a new Block-streaming method is introduced for the Cov unit that can access the large input data in an efficient way to obtain the diagonal and off-diagonal elements of the covariance matrix. Optimization of the other units is done by applying the proper HLS *directives* such as loop unrolling, loop pipelining, and array partitioning.

6.1 Block-Streaming for Covariance Computation

The block-streaming method is useful for large data dimensions when there is limited amount of hardware resources available. The main idea is to partition the input data into multiple blocks, and then stream these blocks through the FIFO channels in a specific order to compute the diagonal and off-diagonal elements of the covariance

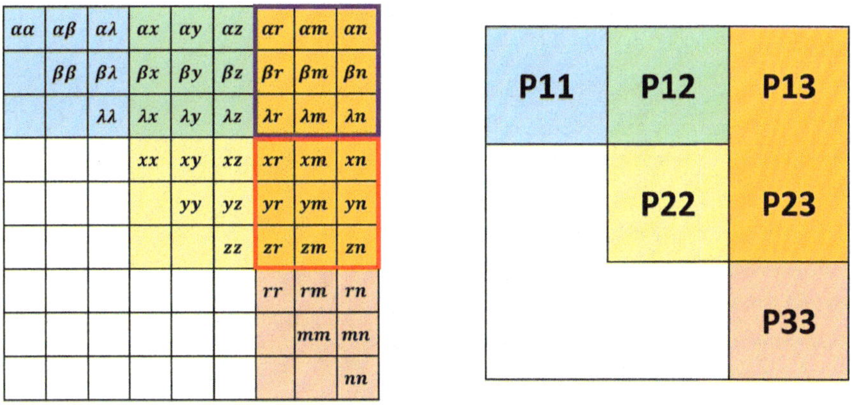

Fig. 14 Illustration of an example of covariance computation using the block-streaming method with 3 blocks ($B = 9$, $B_{\max} = 3$)

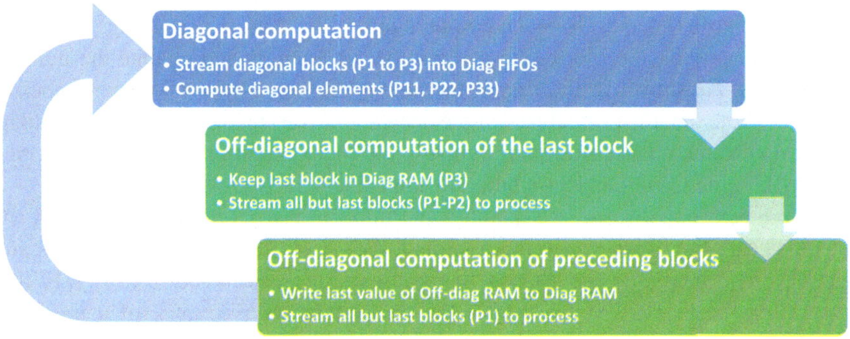

Fig. 15 Block-streaming algorithm in covariance computation

matrix. This specific order is shown with an example in Figs. 13 and 14. The number of features in this example is set to 9 and the Block size is set to 3. Therefore, there are 3 data blocks shown in the figure as P1 to P3. In addition, there are two internal memories to store the streaming blocks of data; these memories are used in the Diagonal and Off-diagonal computations of the covariance matrix, so they are termed "Diag" and "Off-Diag" RAM respectively. The flowchart of the proposed blocking method is shown in Fig. 15.

6.2 *PCA Results*

In our experiments, the data dimensions are considered to be large enough to correspond to the Microwave Imaging scenarios. To evaluate our performance, we used

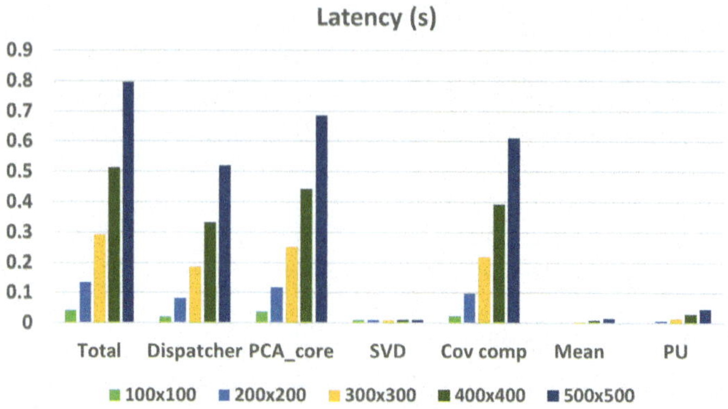

Fig. 16 Processing time for PCA compute units in FPGA

the Virtex FPGA in the VC709U evaluation board as the hardware platform. The processing time (latency) for each compute unit of PCA is shown in Fig. 16. In this experiment, we fix the block size and number of features and change the number of samples. As shown in the figure, the latency is increased for larger number of samples. Note that the critical part of the design is Covariance computation due to its large input dimensions. However, by using the block-streaming strategy described in Sect. 6.1, the Covariance computation can achieve its optimal performance.

In the next experiment, we quantify the impact on performance of varying the block size in the block-streaming approach. We fixed the number of features and samples, and increased the block size from 4 to 16. We can see from Fig. 17 that increasing the block size can reduce the total latency in exchange of using more hardware resources. The critical compute units in terms of latency are the Covariance computation and Projection Unit. In addition, the Dispatcher's latency, which corresponds to the time to write the input data into the corresponding FIFO channels, is the limiting factor if we use the largest block size. This shows that we can speed up the PCA computation by using larger block size until the algorithm becomes memory-bound (the time to read data from external memory and write to the FIFOs exceeds the computation time). More details about the proposed hardware design for PCA algorithm can be found in [56] .

7 Dataflow Hardware Architecture for SVM Using HLS

Support Vector Machine (SVM) is a classification (or regression) technique that is useful in different application areas including microwave medical diagnosis. SVM inference can be briefly explained with the following equation:

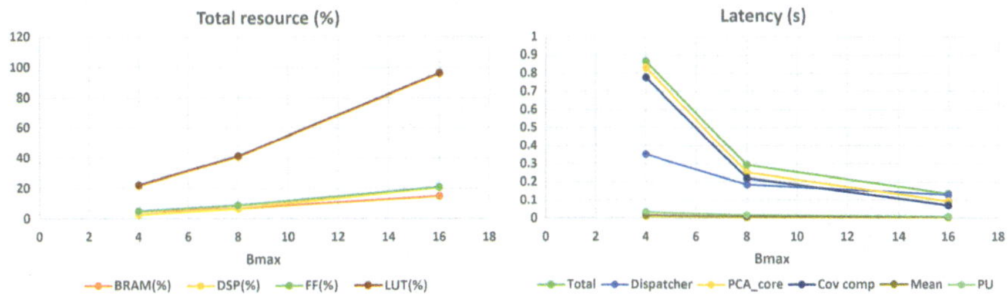

Fig. 17 Impact of block size (B_{max}) on the resource usage and latency for the Virtex7, features = 48, samples = 300 × 300, floating-point design

$$\text{Decision} = \sum_{i=1}^{N_{SV}} \big(\alpha_i K(x, SV_i) + b_i\big), \tag{1}$$

in which N_{SV} is the number of Support Vectors, α_i and b_i are the SVM coefficients and bias values obtained after training, and $K(x, SV_i)$ is the Kernel function applied to the input data (x) and each Support Vector (SV_i).

Due to the heavy computations in SVM algorithm, hardware acceleration of SVM is of great importance in real-time embedded systems. For this purpose, several hardware accelerators have been recently proposed based on FPGAs. In [43], recent FPGA accelerators for SVM are reviewed. As opposed to previous SVM accelerators, we propose a scalable hardware architecture that uses the Dataflow HLS method to obtain concurrency between different computational parts of SVM algorithm, hence improving the throughput. In addition, thanks to the dataflow design, there is no need to store all the SVM coefficients in the internal memory, hence reducing the on-chip memory requirements.

Figure 18 shows the proposed hardware design consisting of three main building blocks: *Read*, *Kernel*, and *Decision*. The main idea is to partition our input features into multiple chunks of data, and while the current data chunk is being processed, we can read the next chunk by the Read function. In this way we can pipeline the execution of SVM computations and reading the inputs from the external memory. By using the Dataflow and Stream directives, this concurrent execution can be achieved in such a way that the *Read* function reads the data chunks and sends them through FIFO channels to the *Kernel* function. After Kernel computation, the *Decision* function receives the *Kernel* output and computes the final classification output (prediction or vote).

7.1 Read SVM Inputs

In the *Read* function, we read N data features and send them through BF FIFO channels to the *Kernel* computation. These parameters (N and BF) can be adjusted

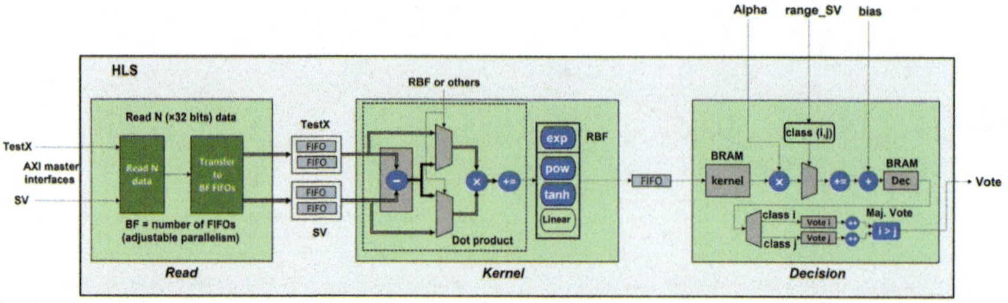

Fig. 18 Proposed SVM accelerator in HLS. The Support Vectors (SV) and the input values (TestX) are read and distributed to the middle block in order to compute the kernel function. The result is sent to the last block for the prediction (Vote), which needs to read from the memory three data: Alpha and bias are α_i and b_i in (1), while Range_SV is the range of the SV. The three blocks operate in pipeline fashion thanks to the FIFO queues

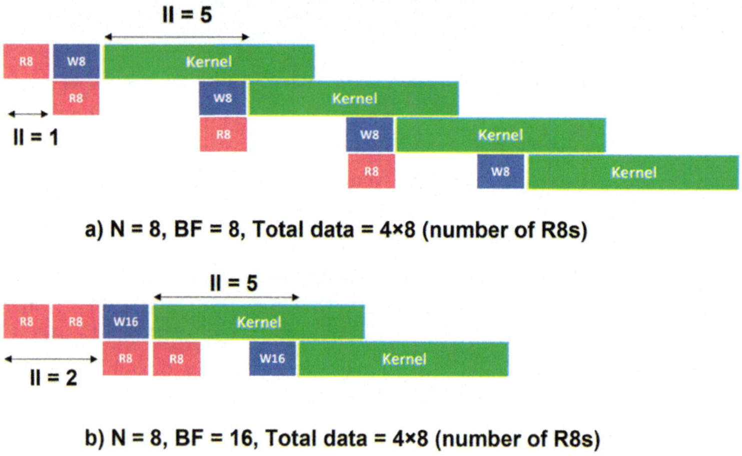

Fig. 19 Timing diagram showing the PCA execution. R8 stands for a read operation of 8 data, while W8 and W16 stand for a write operation (to the FIFOs) of a block of 8 or 16 data, respectively. The diagram shows the impact of the number of FIFO channels (BF) with a total of $4 \times 8 = 32$ data: **a** $BF = N$, II=5 **b** $BF = 2N$, II=2, hence the overall latency is reduced

based on the available resources and the desired performance. Ideally, the number of data to read (N) must be equal to the FIFO channels (BF) to have a balanced pipeline. However, as shown in Fig. 19, when the Initiation Interval (II[1]) of the *Kernel* function is higher than the *Read* function, increasing number of FIFOs (BF) to more than N can help reducing the overall latency. This is due to the fact that although we are slowing down the *Read* function, this degradation of the performance in the *Read* function is compensated by the higher parallelism (higher number of FIFOs) that is achieved by the *Kernel* function.

[1] The number of clock cycles between the start of subsequent iterations of a loop.

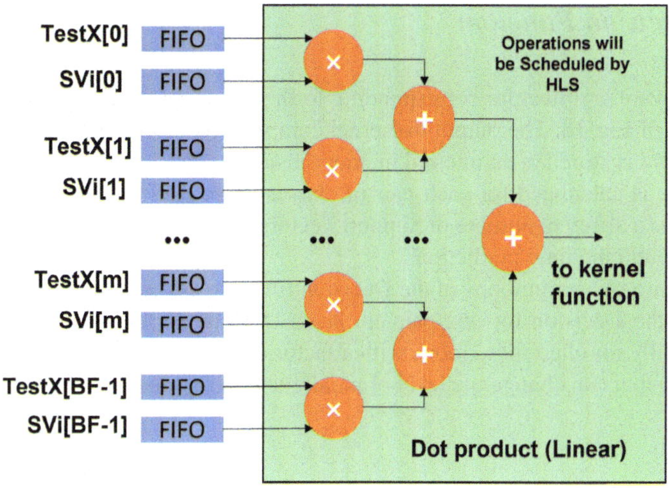

Fig. 20 Manual unrolling for kernel computation. Inputs (TestX) and support vectors (SV) are sent in parallel (depending on the level of parallelism BF) to the unit that computes their dot product

7.2 Kernel Computation

The middle part of Fig. 18 shows the architecture of the hardware design for the *Kernel* computation. It consists of a dot product or square function between the input features and Support Vectors or their differences, respectively. Note that in HLS we can define the desired functionality by using proper C++ *macros*. Therefore, only the required kernel is implemented in hardware which is an advantage of HLS over manual RTL design.

To achieve a high performance for the Kernel computation, we increase the parallelism by using an adjustable parameter to specify the number of FIFOs (BF). As shown in Fig. 20, the computation of the dot product in the Kernel function can be partially unrolled with a factor of BF to match the parallelism of kernel computation to the number of FIFOs. To obtain this hardware architecture, we used manual unrolling with a factor of BF instead of the HLS *array partitioning* directive. This was necessary because accumulating the results of the kernel computation on a scalar variable (i.e., after the $+=$ symbol in Fig. 18) resulted in data dependencies that the HLS tool was not able to solve in an efficient way, leading to keeping the accumulation not parallelized. Therefore, instead of a scalar variable, an array of size BF must be defined to hold the output of each multiplication, and this array must be fully partitioned to have concurrent access to all its elements. These elements are efficiently added together outside the loop to obtain the final scalar result.

7.3 Decision Function

The hardware architecture corresponding to the Decision function is shown in the right part of Fig. 18. The output of Kernel computation is streamed through a FIFO channel and is stored in an internal memory inside the Decision function. The Decision value is calculated for each pair of classes based on (1). Note that each class might have a different number of Support Vectors. The last step is to use the Majority Vote to obtain the final prediction.

There are two main loops in the Decision function corresponding to the computation of the Decision for one pair of classes. Similar to the Kernel computation, we manually unrolled these loops with a factor of DF, which is another adjustable parameter that can change the level of parallelism in our design.

7.4 SVM Results

To show the scalability of our hardware accelerator, we evaluated our design on a low-cost Zynq FPGA which has a low amount of hardware resources. We show that using two different data sets we can obtain optimal performances by adjusting the hardware configurations (N, BF, DF). The first data set is the MNIST and the second is a medical microwave data set of S-parameters provided by the biomedical microwave research team in Politecnico di Torino [57].

Table 4 shows the impact of HLS-based hardware parameters on the resource usage and latency for the MNIST data set. After training SVM with this data set, 16036 SVs with size 784 are obtained. The first three experiments use floating-point data precision while the last two are in fixed-point precision. As we can see, the resource usage increases by increasing the number of FIFOs (BF) while the latency is reduced. It should be noted that in the floating-point design, the main limitation is Kernel computation and the design is compute-bound. However, in the fixed-point design the time-consuming part is the Read function, meaning that the design is memory-bound. That is why we could increase the number of data to read (N) in the last two experiments. Note the low accuracy loss in the fixed-point design compared to the floating-piint design. For the MNIST data set, the DF has negligible impact on the performance due to the low value of latency of the Decision function.

The medical microwave data set contains 462 features originally, but they are reduced to 110 feature after applying PCA algorithm for feature extraction. The data set contains 9 classes that represent the type and positions of the brain stroke: stroke types (*ischemic, hemorrhagic, no stroke*) and locations (*top-right, top-left, bottom-right, bottom-left*). After the training step, 2009 Support Vectors are obtained.

Tables 5 and 6 represent the results of the proposed hardware accelerator for the floating-point and fixed-point data types, respectively. The optimal latency of the design can be obtained by setting the HLS parameters to $N = 2$, $BF = 8$, $DF = 2$ which results in the optimal latency to be 3 ms. Note that with these configurations,

Table 4 MNIST dataset: performance and resource usage. BRAM is a block of Xilinx RAM; DSP48 is the Xilinx blocks for 48-bit multiplications and additions; FF is a flip-flop; LUT is a Look-Up Table for storing the truth table of logic functions

Experiment	**1**	**2**	**3**	**fix1**	**fix2**
N	4	4	4	8	8
BF	4	8	16	8	16
BRAM (%)	15	16	19	12	15
DSP48 (%)	20	23	28	7	11
FF (%)	8	10	13	11	16
LUT (%)	24	28	35	40	62
Latency (ms)	166.93	91.57	58.69	18.12	15.72
Accuracy (%)	98.56 (float)			98.55 (fixed)	

Table 5 Performance analysis of SVM accelerator for medical microwave data set using floating-point data precision ($N_{SV} = 2009$, $N_{features} = 110$, $N_{samples} = 900$)

Experiment	float1	float2	float3	float4
N	2	2	8	8
BF	2	8	8	16
DF	2	2	8	8
BRAM (%)	4	3	3	3
DSP48 (%)	19	22	22	28
FF (%)	8	10	20	23
LUT (%)	22	28	48	55
Read Latency (ms)	1.1	1.1	0.3	0.3
Kernel Latency (ms)	6.5	3	3	3.1
Decision Latency (ms)	1.3	1.3	0.6	0.6
Total Latency (ms)	6.5	3	3	3.1

For the meaning of the acronyms see Table 4

the Kernel computation is still the dominant part of the design. Note that although increasing N and DF can reduce the latency of *Read* and *Decision* functions, the total latency will not change because of the higher latency of the *Kernel* function which is the dominant part.

By using fixed-point data precision, the optimal latency can be reduced to $0.55ms$ (a reduction of 5 times). Since in the fixed-point design the Decision latency is only limited by the number of memory ports that are used to store α_i parameter, Increasing DF will not reduce the total latency. Because of the random access pattern to the α_i memory, partial partitioning of this array is also not helpful. Therefore, changing

Table 6 Performance analysis of SVM accelerator for medical microwave data set using fixed-point data precision ($N_{SV} = 2009$, $N_{features} = 110$, $N_{samples} = 900$)

Experiment	fix1	fix2	fix3	fix4
N	2	4	8	8
BF	2	4	8	16
BRAM (%)	3	3	3	3
DSP48 (%)	5	5	7	10
FF (%)	7	9	11	17
LUT (%)	23	29	40	62
Read latency (ms)	1.1	0.56	0.28	0.28
Kernel latency (ms)	1.36	0.84	0.58	0.44
Decision latency (ms)	0.55	0.55	0.55	0.55
Total latency (ms)	1.36	0.84	0.58	0.55

For the meaning of the acronyms see Table 4

DF does not have any impact on the total latency. In contrast, with floating-point data precision, due to a higher II_D (= 5), increasing DF can also reduce the total latency as well.

8 Hardware Design and Optimization of Neural Networks

Machine Learning (ML) models, including Deep Neural Networks (DNNs), are being progressively used in a wide range of applications, one of which is medical Microwave Imaging. Any ML model contains a set of training hyper-parameters that must be tuned to achieve the maximum accuracy. For example, the number of hidden layers and neurons, or the filter size in DNNs are among these parameters. When a hardware designer aims at the hardware implementation of a neural network, the optimal hardware performance must be obtained while keeping the highest training accuracy. However, the objectives in the hardware design such as low latency and low resource usage are in contrast with the objectives in the training step including the accuracy. Therefore, the designer must find a trade off between these objectives, and this is usually done by solving a multi-objective optimization problem.

In this work, we present a multi-objective framework, termed *joint optimization* to co-optimize the ML training parameters and hardware configurations in order to achieve the optimal performance during both training and hardware design. The proposed framework is based on Multi-Objective Bayesian Optimization with Constraints (MOBOC) which is designed on top of HLS, and is evaluated on two network architectures: Multi-Layer Perceptron (MLP) and a Convolutional Neural Network (CNN) inspired from Lenet5. As opposed to conventional approaches that either use

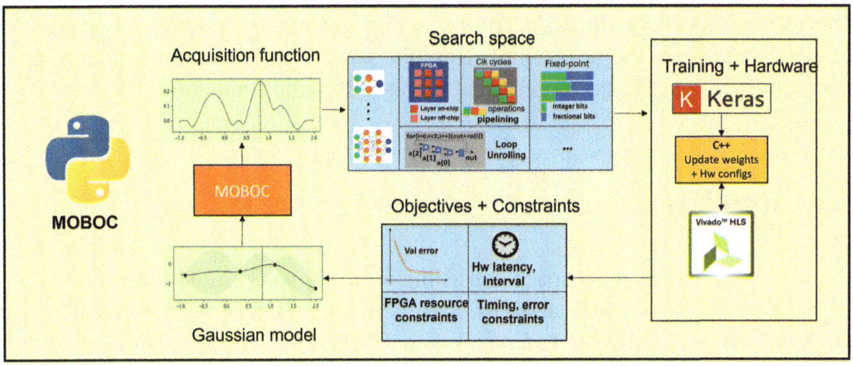

Fig. 21 Proposed methodology for training and hardware co-optimization in FPGA devices

Separate optimization, or use Bayesian Optimization (BO) for the multi-objective optimization, our design supports multi-hardware configurations in which the search space consists of both network architectures and hardware configurations. In addition, our design supports truly multi-objective optimization without combining multiple objectives in a single objective function.

Figure 21 shows the proposed joint optimization flow which can be divided into various parts as thoroughly described in the following.

8.1 Multi-objective BO with Constraints (MOBOC)

Bayesian Optimization is a technique used to optimize functions which are difficult to evaluate. Each function in BO is assumed to be a black-box function that can be modeled with a Gaussian distribution. In MOBOC, after an initialization phase necessary to obtain a few samples from the function evaluations, a new function is created based on the Gaussian models that is termed *acquisition function*. The maximum value of the acquisition function determines the next sample from the search space that must be evaluated in the next iteration of the algorithm. By repeating this process, the uncertainties of the Gaussian models will be reduced until the convergence of the algorithm [58].

In our joint optimization, the MOBOC functions to evaluate and optimize and the constraints are hardware-related and ML-related: we used the prediction error on the validation set, throughput, and the hardware latency as the objectives, and we used the maximum available FPGA resources, clock period, and maximum error threshold as the constraints. We used Keras for the training function evaluation and HLS C-simulation and Synthesis tool for the hardware function evaluation.

After extraction of the objective functions, the Gaussian models can be updated. In this work, we used Predictive Entropy Search for Multi-objective Optimization with Constraints (PESMOC) [59] for the acquisition function which updates the expected improvement of the estimated Pareto set. The maximum value of the acquisition function determines the next sample from the search space. A maximum number of iterations or a convergence criteria can be set to stop the simulation.

8.2 Search Space

We consider a large search space consisting of the network training parameters and HLS-based hardware configurations. Two network architectures are considered in this work which are MLP and a Convolutional Neural Network (CNN) inspired from Lenet-5. The search space for the MLP contains training hyper-parameters such as regularization parameter and learning rate as well as configurations for network architecture including number of hidden layers and neurons. For the CNN based on Lenet-5, the training parameters include the number of neurons, filters, and kernel size. In addition, the hardware configurations in FPGA are considered in the search space.

To design the neural network accelerator in FPGA, we used a high level description of the network in HLS. The HLS-based configurations include the choice between using off-chip or on-chip memory (the memory can be used to store the bias and weight values), loop pipelining and unrolling with adjustable unroll factor, Dataflow pragma[2], precision of the fixed-point data type, and the clock frequency. The optimal values of hardware configurations together with the training parameters can be obtained by the proposed framework based on MOBOC.

8.3 Evaluation on Neural Networks

We evaluated our framework on the two network architectures by using the MNIST data set as an example.[3] The target hardware is a Zynq7000 FPGA (XC7Z020-CSG484). For the implementation of MOBOC, the latest *Spearmint* package is used [59].

8.3.1 Multi-layer Perceptron

The training hyper-parameters and the ranges of the HLS configurations are shown in Table 7.

---

[2] A pragma is a HLS directive; a Dataflow pragma instructs the synthesizer to have multiple units work concurrently—i.e., in pipeline fashion—by means of FIFO buffers inserted between those units. Without the Dataflow pragma, the units operate sequentially [52].

[3] The MNIST database (Modified National Institute of Standards and Technology database) is a large database of handwritten digits that is commonly used for training various image processing systems [60]. Microwave imaging data sets can also be used for the evaluation.

Table 7 Ranges of parameters for the joint training/hardware optimization method

Inputs	Clk (ns)	Hidden Layers	Neurons	Precision #total	Precision #Integer	Reuse factor	Array Partition	Learning rate	Regularization rate
Ranges	4–7	1–3	32–256 step = 32	12–16	4–6	1–4	2^x $x = [1-8]$	$1 \times 10^{(-x)}$ $x = [2-7]$	$1 \times 10^{(-x)}$ $x = [2-7]$

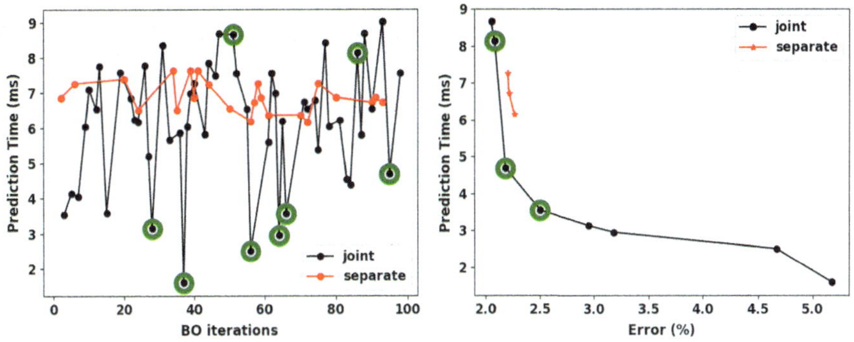

Fig. 22 Comparison of **a** prediction time and **b** pareto fronts

Figure 22 shows a comparison between the traditional separate method and the proposed Joint optimization. The prediction time in the Separate method is relatively higher than the proposed joint method. In addition, the Pareto-optimal points in the Joint optimization approach outperforms the other method. Please refer to [61] for more details about the proposed co-optimization framework for MLP network.

8.3.2 Convolutional Neural Network

The search space for the CNN accelerator inspired from Lenet-5 [60] is described in Table 8, in which *On Chip* refers to the option of selecting internal on-chip memory to store the weights and biases, so there are 5 options available for 5 layers. Total and Integer bits in Table 8 specify the fixed-point data precision. It should be noted that the search space is quite large and it would take years for an exhaustive search to achieve the optimum configurations (with total configurations around 10^{16}).

The Pareto-optimal points in terms of prediction error (after hardware implementation) and execution latency are reported for three different methods (Separate, Random Search, and proposed Joint methods) in Fig. 23. For all the methods, the search space is the same and the total number of iterations is set to 100. Note that in Random search, as opposed to other two methods, it is not possible to set a constraint for hardware error, so there is one point with error larger than the threshold ($> 10\%$). With the exception of one single random point that is randomly detected

Table 8 Search space featuring network architecture and hardware configurations (UF: Unroll Factor)

Configs	Range	Configs	Range
#Filters Conv1	[2, 20]	Dataflow	active/inactive
		On Chip	active/inactive
#Filters Conv2	[2, 20]	Clk (ns)	[8, 20]
		Total bits	[10, 20]
Kernel size	2*([1, 4]) + 1	Integer bits	[1, 9]
#Neurons Dense1	[50, 150]	Conv1 UF	2^([0, 4])
#Neurons Dense2	[50, 150]	Conv2 UF	2^([0, 4])

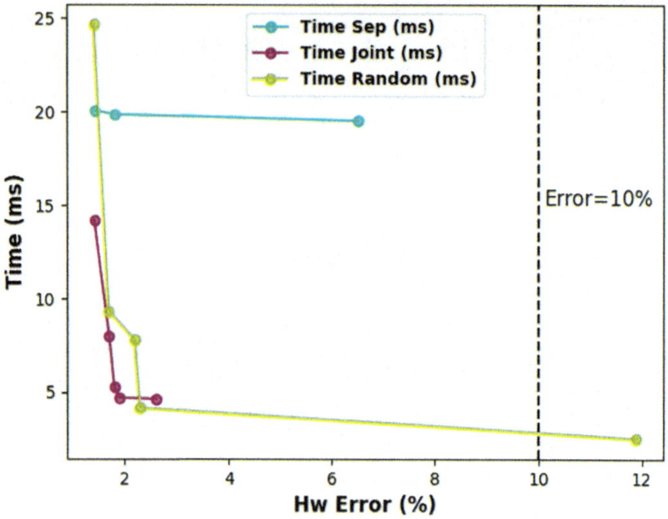

Fig. 23 Pareto-points found by the joint approach, random search, and conventional separate method in the space of prediction error (Hw error) and execution latency (Time). Total number of iterations is 100 for all methods

by the Random search, all the other points in both the Random and Separate method are Pareto-dominated by the points found by the Joint method.

In Fig. 24 all the suggested points in the three optimization methods are illustrated with three different colors. For the hardware error, a log scale is considered for better clarity. Due to the error threshold in MOBOC, the points are more concentrated below the threshold for the joint method which helps for better convergence of the algorithm.

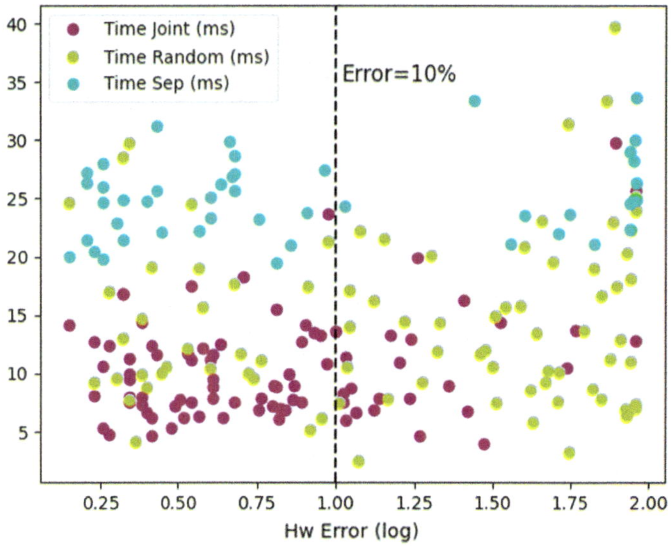

Fig. 24 Total points suggested by the joint, separate, and random search methods; note the concentration of the joint method on low errors ($< 10\%$)

Table 9 Evaluation of MLP performance in microwave anomaly detection

Training accuracy (%)	Test accuracy (%)	Hardware accuracy (%)	Latency (ms)
99.6	98	97.5	1.12

Fixed-point precision is selected in hardware with total and integer widths of 16 and 10, respectively

8.3.3 Evaluation on Microwave Data Set

For the anomalies detection and classification in MWI, MLP networks are useful [62]. The dataset that is used in this part is the same microwave data that we used in Sect. 7. It contains 4500 samples of scattering matrix with 462 elements. There are 9 classes representing the presence, type, and location of the brain stroke. For feature extraction, PCA is used that results in the reduction of features to 110. In this part we focus on the training and hardware performance of an MLP network using the MWI dataset assuming that the optimal parameters of the network are obtained. We selected an MLP with 3 hidden layers and number of neurons per layer 220, 64, 64, respectively. The target device is a Zynq SoC (ZedBoard), and we used fixed-point precision for the hardware implementation. The accuracy before and after hardware implementation, resource usage, and processing time are depicted in Table 9 and Figs. 25 and 26. Note the negligible accuracy loss in the hardware accelerator due to the reduced precision.

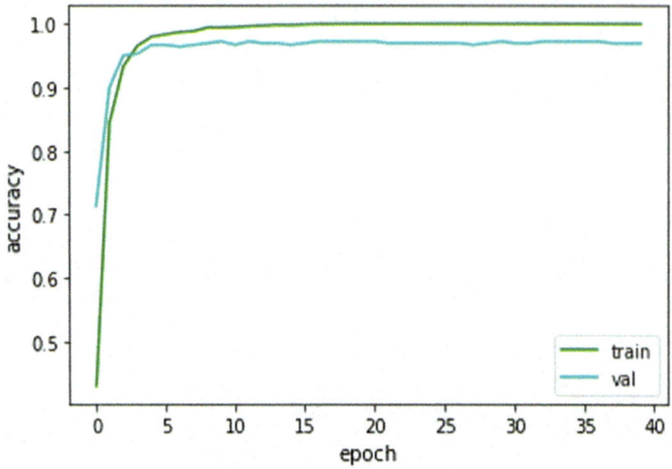

Fig. 25 Accuracy of MLP during training by MWI dataset

Fig. 26 Resource usage for MLP in Zynq FPGA

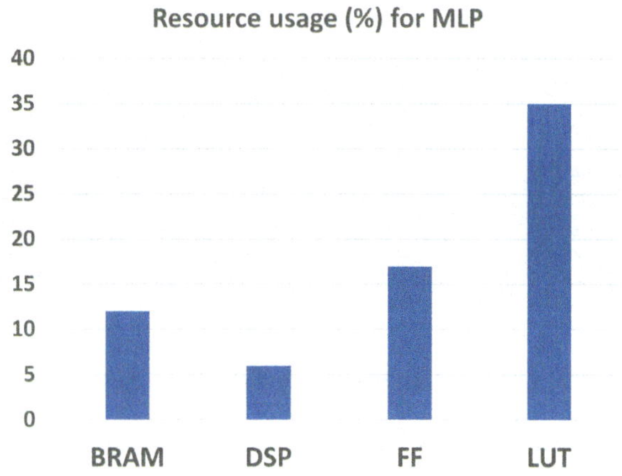

9 Conclusions

In this work we presented efficient hardware design methodologies to accelerate the execution of compute-intensive algorithms that are recurrent in biomedical MWI techniques. Specifically, we proposed different hardware accelerators for 3D FDTD, PCA, SVM, as well as design optimization techniques for ML models, including Multi-Layer Perceptrons (MLPs) and Convolutional Neural Networks (CNNs). We used High Level Synthesis (HLS) as the most efficient approach to design these hardware accelerators and optimize their performance. The flexibility, completeness, efficiency, and high-level design techniques and hardware optimization methodologies are the distinguishing features of the proposed accelerators compared to previous related works. Specifically, these are the main contributions and achievements in our work:

- The accelerator for 3D FDTD considers the CPML boundary conditions for all directions, models the dispersive materials, and is designed as an internal step for a non-linear iterative MWI reconstruction algorithm. Two architectures, called Small and Large, were proposed based on the number of interfaces and the amount of consumed hardware resources. For the reference data set with the dimensions of $70 \times 70 \times 70$, each FPGA in the Small and Large design can be used for 3 and 2 antennas, respectively. The execution time per antenna for our FPGA accelerator is $3.38s$ for the Small design which is slightly better than the conventional GPU design with Acceleware library ($4.88s$). In addition, we presented a multi-FPGA design in which multiple antennas can be assigned to multiple FPGAs. The system level evaluation of our multi-FPGA design with 24 antennas and 8 UltraScale+ FPGAs shows a $13\times$ speed-up compared to the single GPU design on Tesla P40 (Acceleware).
- The proposed PCA accelerator in FPGA consists of several computing elements, including the computations of *Mean*, *Covariance*, *SVD or EVD*, and *Projection of inputs*. Although separate acceleration techniques for each element have been already proposed, their integration in an efficient hardware accelerator has not been considered before. We proposed a flexible FPGA accelerator for PCA that can be used for different data dimensions, thanks to the HLS design methodology. In addition to the floating-point design, a fixed-point implementation was proposed to efficiently use the hardware resources in FPGA. For the computation of Covariance matrix, an efficient block streaming methodology was introduced to read blocks of input data from the external memory, store them in on-chip memory and process them inside the hardware accelerator, while reading the next block of data from external memory. Different data dimensions were analyzed in the evaluation, and compared to the GPU or multi-core CPU designs we could achieve either power efficiency or performance speed-up. In addition, we compared our HLS design an RTL implementation of PCA using VHDL and achieved a $2.3\times$ speed-up as well as significant reduction of resource usage.
- FPGA accelerator for SVM uses a novel dataflow architecture in which the required input data are transferred to the accelerator while the SVM computations are performed. The proposed accelerator is scalable, meaning that it is not limited by the number of dimensions of Support Vectors and can be used for large data dimensions when there is limited on chip memory. In addition, a fixed-point implementation was presented for the SVM accelerator which could improve the speed. The entire accelerator was designed using the HLS tool and we could apply different hardware optimization techniques. One of the main characteristics of our design is having *adjustable parallelism* which enables the designer to specify the amount of parallelism in hardware depending on the available resources. As opposed to most of the conventional SVM accelerators in FPGA, multi-class classification is also supported in our design. In addition, we explored the impact of different SVM kernels on the hardware performance. For the evaluation, we compared our

HLS-based hardware accelerator with recent works which were designed using either HLS or the manual RTL approach. A 4.4× latency improvement could be achieved compared to the RTL design, and a minimum of 10× improvement in the latency was obtained compared to a similar HLS-based design.

- Finally, for the optimum design of Machine Learning models and specifically Neural Networks in FPGAs, a new methodology was proposed that could jointly optimize the training hyper-parameters and HLS-based hardware configurations based on multi-Objective Bayesian optimization. Although the proposed approach can be used for any ML model, we did a primary evaluation on two network architectures, which are Multi-Layer Perceptron (MLP) and Convolutional Neural Networks (CNNs). Compared to other solutions based on Bayesian Optimization, the search space in our method supports *multi-Hw* configurations and consists of a combination of network architectures and hardware configurations. In addition, the available hardware resources in the FPGA can be set as *Constraints* in the optimization problem together with different contrasting *Objectives* such as hardware latency and accuracy loss. We considered a large search space and compared our joint optimization methodology with the separate optimization and the Random search approach. From the comparison, we could notice the improvement in the Pareto sets obtained by the proposed joint approach, with 1.7× and 1.4× improvement in the execution time for the minimum error compared to the random and separate methods, respectively, in a Lenet5-inspired CNN architecture, and 1.43× improvement in the prediction time without an increase in the prediction error compared to the separate design, in the MLP network architecture.

The above-mentioned hardware accelerators are highly beneficial for high-performance embedded computing systems. As stated previously, 3D FDTD is used in non-linear iterative medical microwave image reconstruction, and its acceleration helps in reducing the amount of time needed to reconstruct the final image. FPGA acceleration of SVD/EVD (included in PCA) is useful in linear microwave image reconstruction algorithms. Hardware acceleration of PCA, SVM, and Neural Networks are advantageous not only in MWI, but also in other ML applications which use these algorithms.

Acknowledgements This work was supported by the EMERALD project funded by the European Union's Horizon 2020 research and innovation programme under the Marie Skłodowska-Curie grant agreement No. 764479.

References

1. M.R. Casu, F. Colonna, M. Crepaldi, D. Demarchi, M. Graziano, M. Zamboni, Uwb microwave imaging for breast cancer detection: many-core, gpu, or fpga? ACM Trans. Embedded Comput. Syst. (TECS) **13**(3s), 1–22 (2014)
2. I. Sarwar, G. Turvani, M.R. Casu, J.A. Tobon, F. Vipiana, R. Scapaticci, L. Crocco, Low-cost low-power acceleration of a microwave imaging algorithm for brain stroke monitoring. J. Low Power Electron. Appl. **8**(4), 43 (2018)

3. I. Saied, T. Arslan, R. Ullah, C. Liu, and F. Wang, "Hardware accelerator for wearable and portable radar-based microwave breast imaging systems," in *2021 IEEE International Symposium on Circuits and Systems (ISCAS)*, pp. 1–5, 2021
4. J. Cong, Z. Fang, M. Lo, H. Wang, J. Xu, S. Zhang, Understanding performance differences of fpgas and gpus, in *2018 IEEE 26th Annual International Symposium on Field-Programmable Custom Computing Machines (FCCM)*, pp. 93–96, IEEE (2018)
5. J. Cong, J. Lau, G. Liu, S. Neuendorffer, P. Pan, K. Vissers, Z. Zhang, Fpga hls today: successes, challenges, and opportunities. ACM Trans. Reconfigurable Technol. Syst., **15** (2022)
6. X. Wang, S. Liu, X. Li, S. Zhong, Gpu-accelerated finite-difference time-domain method for dielectric media based on cuda. Int. J. RF Microwave Comput.-Aided Eng. **26**(6), 512–518 (2016)
7. Z. Bo, X. Zheng-Hui, R. Wu, L. Wei-Ming, and S. Xin-Qing, Accelerating fdtd algorithm using gpu computing," in *2011 IEEE International Conference on Microwave Technology and Computational Electromagnetics*, pp. 410–413, IEEE (2011)
8. S. Liu, B. Zou, L. Zhang, S. Ren, Heterogeneous cpu+ gpu-accelerated fdtd for scattering problems with dynamic load balancing. IEEE Trans. Antennas Propag. **68**(9), 6734–6742 (2020)
9. H. Zhang, Y. Lei, H. Ye, Y. Gong, Bistatic radar cross section prediction of 3-d target based on gpu-fdtd method, in *2018 12th International Symposium on Antennas, Propagation and EM Theory (ISAPE)*, pp. 1–4, IEEE (2018)
10. T. Kenter, J. Förstner, C. Plessl, Flexible fpga design for fdtd using opencl, in *2017 27th International Conference on Field Programmable Logic and Applications (FPL)*, pp. 1–7, IEEE (2017)
11. H. Giefers, C. Plessl, J. Förstner, Accelerating finite difference time domain simulations with reconfigurable dataflow computers. ACM SIGARCH Comput. Architect. News **41**(5), 65–70 (2014)
12. Y. Takei, H. M. Waidyasooriya, M. Hariyama, M. Kameyama, Fpga-oriented design of an fdtd accelerator based on overlapped tiling," in *Proceedings of the International Conference on Parallel and Distributed Processing Techniques and Applications (PDPTA)*, The Steering Committee of The World Congress in Computer Science, Computer Engineering and Applied Computing (WorldComp), p. 72 (2015)
13. H. M. Waidyasooriya, T. Endo, M. Hariyama, Y. Ohtera, Opencl-based fpga accelerator for 3d fdtd with periodic and absorbing boundary conditions. Int. J. Reconfig. Comput. **2017** (2017)
14. C. Kong, T. Su, Parallel hardware architecture of the 3d fdtd algorithm with convolutional perfectly matched layer boundary condition. Progress Electromagn. Res. C **105**, 161–174 (2020)
15. K. Okina, R. Soejima, K. Fukumoto, Y. Shibata, K. Oguri, Power performance profiling of 3-d stencil computation on an fpga accelerator for efficient pipeline optimization. ACM SIGARCH Comput. Architect. News **43**(4), 9–14 (2016)
16. D.G. Perera, K.F. Li, Embedded hardware solution for principal component analysis, in *Proceedings of 2011 IEEE Pacific Rim Conference on Communications, Computers and Signal Processing*, pp. 730–735, IEEE (2011)
17. D. Fernandez, C. Gonzalez, D. Mozos, S. Lopez, Fpga implementation of the principal component analysis algorithm for dimensionality reduction of hyperspectral images. J. Real-Time Image Process. **16**(5), 1395–1406 (2019)
18. A. Das, D. Nguyen, J. Zambreno, G. Memik, A. Choudhary, An fpga-based network intrusion detection architecture. IEEE Trans. Inf. Forensics Security **3**(1), 118–132 (2008)
19. U.A. Korat, A. Alimohammad, A reconfigurable hardware architecture for principal component analysis. Circuits Syst. Signal Process. **38**(5), 2097–2113 (2019)
20. M.U. Torun, O. Yilmaz, A.N. Akansu, Fpga, gpu, and cpu implementations of jacobi algorithm for eigenanalysis. J. Parallel Distribut. Comput. **96**, 172–180 (2016)
21. S. Kasap, S. Redif, Novel field-programmable gate array architecture for computing the eigenvalue decomposition of para-hermitian polynomial matrices. IEEE Trans. Very Large Scale Integration (VLSI) Syst., **22**(3), 522–536 (2013)

22. S. Zhang, X. Tian, C. Xiong, J. Tian, D. Ming, Fast implementation for the singular value and eigenvalue decomposition based on fpga. Chinese J. Electron. **26**(1), 132–136 (2017)
23. X. Wang, J. Zambreno, An fpga implementation of the hestenes-jacobi algorithm for singular value decomposition," in *2014 IEEE International Parallel and Distributed Processing Symposium Workshops*, pp. 220–227, IEEE (2014)
24. Y. Ma, D. Wang, Accelerating svd computation on fpgas for dsp systems," in *2016 IEEE 13th International Conference on Signal Processing (ICSP)*, pp. 487–490, IEEE (2016)
25. Y.-L. Chen, C.-Z. Zhan, T.-J. Jheng, A.-Y. Wu, Reconfigurable adaptive singular value decomposition engine design for high-throughput mimo-ofdm systems. IEEE Trans. Very Large Scale Integration (VLSI) Syst., **21**(4), 747–760 (2012)
26. M.V. Athi, S.R. Zekavat, A.A. Struthers, Real-time signal processing of massive sensor arrays via a parallel fast converging svd algorithm: latency, throughput, and resource analysis. IEEE Sens. J. **16**(8), 2519–2526 (2016)
27. A.A.S. Ali, A. Amira, F. Bensaali, M. Benammar, Hardware pca for gas identification systems using high level synthesis on the zynq soc, in *2013 IEEE 20th International Conference on Electronics, Circuits, and Systems (ICECS)*, pp. 707–710, IEEE (2013)
28. M. Schellhorn, G. Notni, Optimization of a principal component analysis implementation on field-programmable gate arrays (fpga) for analysis of spectral images, in *2018 Digital Image Computing: Techniques and Applications (DICTA)*, pp. 1–6, IEEE (2018)
29. M.A. Mansoori, M.R. Casu, Efficient fpga implementation of pca algorithm for large data using high level synthesis, in *2019 15th Conference on Ph.D. Research in Microelectronics and Electronics (PRIME)*, pp. 65–68, IEEE (2019)
30. M.A. Mansoori, M.R. Casu, Hls-based flexible hardware accelerator for pca algorithm on a low-cost zynq soc, in *2019 IEEE Nordic Circuits and Systems Conference (NORCAS): NORCHIP and International Symposium of System-on-Chip (SoC)*, pp. 1–7, IEEE (2019)
31. C. Kyrkou, T. Theocharides, A parallel hardware architecture for real-time object detection with support vector machines. IEEE Trans. Comput. **61**(6), 831–842 (2011)
32. C. Kyrkou, C.-S. Bouganis, T. Theocharides, M.M. Polycarpou, Embedded hardware-efficient real-time classification with cascade support vector machines. IEEE Trans. Neural Networks Learn. Syst. **27**(1), 99–112 (2015)
33. M. Qasaimeh, A. Sagahyroon, T. Shanableh, Fpga-based parallel hardware architecture for real-time image classification. IEEE Trans. Comput. Imaging **1**(1), 56–70 (2015)
34. S. Saurav, R. Saini, S. Singh, Fpga based implementation of linear svm for facial expression classification, in *2018 International Conference on Advances in Computing, Communications and Informatics (ICACCI)*, pp. 766–773, IEEE (2018)
35. Y. Jiang, K. Virupakshappa, E. Oruklu, Fpga implementation of a support vector machine classifier for ultrasonic flaw detection," in *2017 IEEE 60th international midwest symposium on circuits and systems (MWSCAS)*, pp. 180–183, IEEE (2017)
36. L. Han, Z. Yue, X. Guo, Image segmentation implementation based on fpga and svm, in *2017 3rd International Conference on Control, Automation and Robotics (ICCAR)*, pp. 405–409, IEEE (2017)
37. M. Ruiz-Llata, G. Guarnizo, M. Yébenes-Calvino, Fpga implementation of a support vector machine for classification and regression, in *The 2010 International Joint Conference on Neural Networks (IJCNN)*, pp. 1–5 (2010)
38. S. Afifi, H. GholamHosseini, and R. Sinha, "A low-cost fpga-based svm classifier for melanoma detection," in *2016 IEEE EMBS Conference on Biomedical Engineering and Sciences (IECBES)*, pp. 631–636, IEEE, 2016
39. S. Afifi, H. GholamHosseini, R. Sinha, Svm classifier on chip for melanoma detection, in *2017 39th Annual International Conference of the IEEE Engineering in Medicine and Biology Society (EMBC)*, pp. 270–274, IEEE (2017)
40. S. Afifi, H. GholamHosseini, R. Sinha, A system on chip for melanoma detection using fpga-based svm classifier. Microprocessors Microsyst. **65**, 57–68 (2019)
41. A. Baez, H. Fabelo, S. Ortega, G. Florimbi, E. Torti, A. Hernandez, F. Leporati, G. Danese, G.M. Callico, R. Sarmiento, High-level synthesis of multiclass svm using code refactoring to classify brain cancer from hyperspectral images. Electronics **8**(12), 1494 (2019)

42. R. Campos, J.M. Cardoso, On data parallelism code restructuring for hls targeting fpgas, in *2021 IEEE International Parallel and Distributed Processing Symposium Workshops (IPDPSW)*, pp. 144–151, IEEE (2021)
43. S. Afifi, H. GholamHosseini, R. Sinha, Fpga implementations of svm classifiers: a review. SN Comput. Sci. **1**(3), 1–17 (2020)
44. A. Mehrabi, A. Manocha, B.C. Lee, D.J. Sorin, Bayesian optimization for efficient accelerator synthesis. ACM Trans. Architect. Code Optim. (TACO) **18**(1), 1–25 (2020)
45. Y. Xiong, R. Mehta, V. Singh, Resource constrained neural network architecture search: Will a submodularity assumption help?, in *2019 IEEE/CVF International Conference on Computer Vision (ICCV)*, pp. 1901–1910 (2019)
46. S.-C. Kao, G. Jeong, T. Krishna, Confuciux: autonomous hardware resource assignment for dnn accelerators using reinforcement learning, in *2020 53rd Annual IEEE/ACM International Symposium on Microarchitecture (MICRO)*, pp. 622–636 (2020)
47. J. Ney, D. Loroch, V. Rybalkin, N. Weber, J. Kruger, N. Wehn, Half: holistic auto machine learning for fpgas, in *2021 31st International Conference on Field-Programmable Logic and Applications (FPL)*, IEEE Computer Society (Los Alamitos, CA, USA, 2021), pp. 363–368
48. H. Lee, S. Lee, S. Chong, S.J. Hwang, Hardware-adaptive efficient latency prediction for NAS via meta-learning," in *Advances in Neural Information Processing Systems*, eds. by A. Beygelzimer, Y. Dauphin, P. Liang, J.W. Vaughan (2021)
49. M.S. Iqbal, J. Su, L. Kotthoff, P. Jamshidi, Flexibo: cost-aware multi-objective optimization of deep neural networks. CoRR **abs/2001.06588** (2020)
50. Xilinx, *Introduction to FPGA Design with Vivado High-Level Synthesis*. Xilinx (2019)
51. Xilinx, *Vivado Design Suite User Guide: Design Flows Overview*. Xilinx (2022)
52. Xilinx, *Vivado Design Suite User Guide: High-Level Synthesis*. Xilinx (2021)
53. Z. Miao, P. Kosmas, Multiple-frequency dbim-twist algorithm for microwave breast imaging. IEEE Trans. Antennas Propag. **65**(5), 2507–2516 (2017)
54. K. Yee, Numerical solution of initial boundary value problems involving maxwell's equations in isotropic media. IEEE Trans. Antennas Propag. **14**(3), 302–307 (1966)
55. M.A. Mansoori, P. Lu, M.R. Casu, Fpga acceleration of 3d fdtd for multi- antennas microwave imaging using hls. IEEE Access **9**, 122696–122711 (2021)
56. M.A. Mansoori, M.R. Casu, High level design of a flexible pca hardware accelerator using a new block-streaming method. Electronics., **9**(3) (2020)
57. V. Mariano, J.A.T. Vasquez, M.R. Casu, F. Vipiana, Model-based data generation for support vector machine stroke classification, in *2021 IEEE International Symposium on Antennas and Propagation and USNC-URSI Radio Science Meeting (APS/URSI)*, pp. 1685–1686 (2021)
58. S. Greenhill, S. Rana, S. Gupta, P. Vellanki, S. Venkatesh, Bayesian optimization for adaptive experimental design: a review. IEEE Access **8**, 13937–13948 (2020)
59. E.C. Garrido-Merchán, D. Hernández-Lobato, Predictive entropy search for multi-objective bayesian optimization with constraints. Neurocomputing **361**, 50–68 (2019)
60. Y. Lecun, L. Bottou, Y. Bengio, P. Haffner, Gradient-based learning applied to document recognition. Proc. IEEE **86**(11), 2278–2324 (1998)
61. M.A. Mansoori, M.R. Casu, Efficient training and hardware co-design of machine learning models, in *Applications in Electronics Pervading Industry, Environment and Society* ed. by S. Saponara, A. De Gloria. (Cham, Springer International Publishing, 2022), pp. 243–248
62. T. Reimer, J. Sacristan, S. Pistorius, Improving the diagnostic capability of microwave radar imaging systems using machine learning, in *2019 13th European Conference on Antennas and Propagation (EuCAP)*, pp. 1–5, IEEE (2019)

Metasurface Technology for Medical Imaging

Eleonora Razzicchia, Navid Ghavami, Olympia Karadima, and Panagiotis Kosmas

Abstract Nowadays, medical imaging and healthcare in general are facing an ever growing number of challenges. Increase in life expectancy, for example, results in a growing economic cost of healthcare services. In particular, the need for medical imaging equipment worldwide is expected to grow driven by the global rise of various pathologies such as cancer, cardiovascular diseases, brain disorders and lung diseases. In this framework, medical imaging technologies play a key role, being the essential clinical tool to deliver accurate initial diagnosis and treatments' monitoring. Similar to other imaging modalities, medical imaging based on microwave technology requires designing systems with increasing number of sensors (i.e. antennas). Using metasurfaces can not only improve the characteristics of these antennas, but also enhance detection by tackling the impedance mismatch problem that electromagnetic waves encounter when probing human tissue. This chapter focuses on two potential clinical applications where the use of metasurfaces can improve the performance of microwave-based systems: stroke detection and liver ablation monitoring.

Keywords Metasurfaces · Split-ring resonators · Medical imaging · Microwave imaging

E. Razzicchia (✉) · N. Ghavami · O. Karadima · P. Kosmas
Faculty of Natural and Mathematical Sciences, King's College London, Strand, London WC2R 2LS, UK
e-mail: eleonora.razzicchia@kcl.ac.uk

N. Ghavami
e-mail: navid.ghavami@kcl.ac.uk

O. Karadima
e-mail: olympia.karadima@kcl.ac.uk

P. Kosmas
e-mail: panagiotis.kosmas@kcl.ac.uk

F. Vipiana and L. Crocco (eds.), *Electromagnetic Imaging for a Novel Generation of Medical Devices*, Lecture Notes in Bioengineering,
https://doi.org/10.1007/978-3-031-28666-7_3

1 Introduction: Metamaterials and Metasurfaces in Medical Sensing and Imaging

Metamaterials (MMs) are engineered materials exhibiting some unique electromagnetic (EM) properties which are not shown by naturally occurring materials. Nowadays, different split-ring resonator (SRR) structures have been developed for several applications including sensing and imaging, because they present many advantages in the design process relative to conventional structures, such as high flexibility [1].

Recently, MM-based technologies have attracted an increased attention in biosensing, from the microwave to the optical frequencies. In the microwave spectrum, layers of split-ring resonators (SRRs) have been used for magnetic resonance imaging (MRI) [2–5], cancer detection [6, 7], and imaging in the near-field [8, 9] and far-field [10]. Computational imaging systems for security-screening applications have also used frequency-diverse metasurface apertures in the K-band frequencies [11–14].

With the advent of innovative devices involving MMs, one of the challenges in microwave imaging (MWI) for biomedical applications is integrating MM technology into current systems [15]. A relevant aspect is the possibility of realizing focusing lenses. Highly efficient gradient metasurface (MTS) lenses based on MM structures can efficiently manipulate the spatial field distribution and focus the incident beam, working efficiently even at wide angles of incidence [16]. In terms of MWI applications, it has been demonstrated that the field generated by a localized source can be focused on a specific point by a flat slab. Thus, a target could be inspected by scanning a focal point, moving the source (in front of the slab) in different directions [17].

Flat MM slabs can be also used as a matching medium to couple EM energy into the region of interest. Impedance mismatch occurs when EM waves propagate between media with different permittivity and permeability values. This phenomenon results in reduction of the transmitted energy. This issue becomes significant in diagnostic imaging and other non-invasive biomedical applications, such as non-invasive glucose sensing [18]. An innovative and efficient way to achieve impedance matching and improve the transmitted power is by placing a MM film that acts as an impedance-matching coating in front of skin.

In this chapter, we present a feasibility study with the goal of enhancing microwave imaging using properly designed metasurfaces. In particular, we present two designs. The first one is a monopole antenna covered by an MTS-superstrate loading which is integrated in a brain scanner for haemorragic stroke detection. The second on is an MTS film proposed as a hardware enahncement to an MWI system for thermal ablation monitoring.

2 Benefits of Employing Metasurfaces in the Design of a Scanner for Haemorrhagic Brain Stroke Detection

Stroke occurs when blood supply to a part of the brain is interrupted or reduced (ischemic stroke) or when a blood vessel breaks and bleeds into the brain (hemorrhagic stroke). Worldwide stroke incidences are predicted to rise considerably due to the increase of population over 65 years old especially in low-to-middle-income countries [19]. Stroke is a medical emergency and prompt treatment is crucial. An early detection can reduce brain damage and other complications. Thus, patient's survival depends on a quick diagnosis through the use of an efficient imaging method.

Whereas MWI was previously focused on breast cancer detection, recent research efforts have been now extended to other applications such as brain imaging. This is because, despite its resolution is lower than MRI, ultrasound, and X-Ray imaging, MWI presents many advantages [20]. First of all, it is a non-invasive technique which is completely harmless, since the involved waves are non-ionizing and used in very low doses (low-power EM waves). Moreover, the short data acquisition time (ranging from milliseconds to a few seconds) can allow a quick assessment of the stroke, which is crucial for proper drug administration [21]. Another appealing property of MWI is its economical sustainability for the health-care system. Due to the progress in mobile industry and microwave devices in recent years, EM imaging has the potential to provide mobile, low-cost imaging platforms well suited to the healthcare needs.

Besides all the advantages, developing a valid MWI scanner for brain imaging is a challenging task. For example, to achieve an operational MWI apparatus for detecting a bleeding in the brain, it is necessary to find a compromise between resolution and penetration depth into the brain tissue. For this reason, it is important to maximize the incident power coupled into the tissue of interest and ensure acceptable spatial resolution images [22]. To maximize the amount of incident power penetrating into the brain tissue, compact antennas operating below 2.0 GHz immersed in a lossy dielectric medium are recommended, due to the strong attenuation of the EM waves propagating inside the head [22]. According to this requirement, there are several possible MWI head scanner designs that could be considered. Headband, helmet structures or specific chambers [23], for example, allow the use of a coupling medium for the operating antennas.

The coupling medium reduces unwanted multipath signals [24] and broadens the frequency range of operation [25], but at the same time affects the detection of useful "weak signals due to the target. One way to increase the transmitted power and improve the received signal is to place a MM film in front of the skin, which can act as a further impedance-matching layer [26]. This approach is motivated by previous numerical studies which have demonstrated that MTSs can suppress unwanted reflections and enhance the transmission near a specific frequency [27, 28].

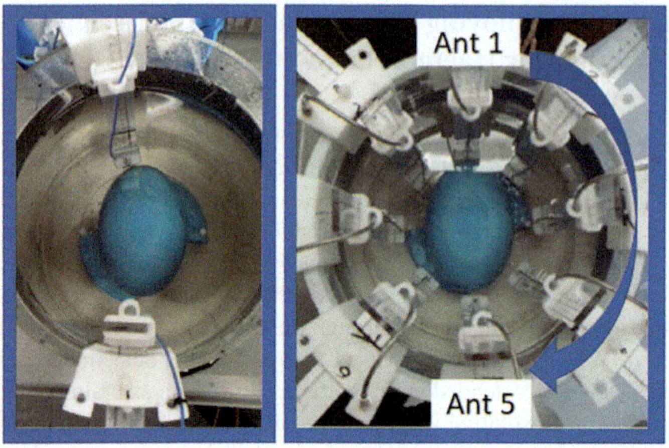

Fig. 1 Measurement hardware and setup for radar (left), and tomography (right)

2.1 Microwave Brain Imaging Prototype for Brain Stroke Detection

Fig. 1 shows the custom-made microwave imaging setup developed by King's College London (KCL) [29], which comprises a 300 mm diameter cylindrical tank, shielded by an RF absorbing sheet from Laird Technologies EMI. This setup is used for both radar and tomographic configurations [30]. The transmitting and receiving antennas are connected to a multi-port vector network analyser (Keysight M9019A). These are positioned in a circular ring inside the acrylic tank and surround a head phantom model. Vertical and horizontal mounts are used to adjust the position of the antennas with precision and accuracy.

For tomography measurements, eight antennas are positioned as close as possible to the head phantom's external surface. These antennas, which perform as receivers and transmitters, create an 8×8 scattering matrix which is fed into the DBIM-TwIST algorithm. For the radar measurements, only two antennas are used, one performing as the transmitter, and the other as receiver. For each transmitter position, the receiving antenna is rotated radially (anti-clockwise) with 15° steps, measuring the external field of the phantom at 24 receiving positions. The transmitting antenna is initially positioned 30 mm away from the larger axis of the elliptical brain phantom, while the receiving antenna is placed at a shorter distance (10 mm) from the phantom. The antennas are immersed in a 90% glycerol-water mixture with permittivity of $\approx$17 at 1 GHz, which is used as matching medium. This dielectric medium is required for the system's antennas to operate efficiently in the 0.5–2 GHz frequency range [29].

The head model is also immersed inside the coupling liquid contained in the imaging tank and consists of a three-dimensional (3D) printed ABS elliptic mould containing a gelatin-oil mixture based on the phantom preparation method described in [31]. The 3D printed mould with a length of 170 mm and width of 120 mm is used as a holder for the liquid phantom, which, once solidified, can mimic the brain. Furthermore, it acts as an additional phantom layer during image reconstruction.

Using this setup, an haemorrhagic stroke can be mimicked by inserting a cylindrical target of 30 mm diameter in the gel-based brain phantom. This is done through extracting a part of the brain mixture using a 30 mm diameter cylindrical mould, resulting in the creation of a cylindrical cavity. This cavity is then filled with gel phantoms resembling the bleeding in haemorrhagic stroke.

2.2 *Imaging Algorithms*

2.2.1 Distorted Born Iterative Method with Two-Step Iterative Shrinkage Thresholding Algorithm

The distorted DBIM is an iterative algorithm for solving the EM inverse scattering problem, which is used to estimate the spatial distribution of the tissue's dielectric properties inside a region V of the human body [32]. The two-dimensional (2D) DBIM considers that the non-linear inverse scattering problems for each transceiver pair is linearized and approximated by the following equation:

$$E_s(r_n, r_m) = E_t(r_n, r_m) - E_b(r_n, r_m) = \omega^2 \mu_0 \epsilon_0 \int_V G_b(r_n, r) E_b(r, r_m) \delta_\epsilon(r) dr \tag{1}$$

where E_t, E_s, E_b are the total, scattered and background electric fields, respectively, r_n and r_m are the transmitting and receiving antenna locations, ω is the angular frequency, μ_0 is the permeability of free space, ϵ_0 is the permittivity of free space, and G_b is the Green's function for the background medium. As we only consider point sources, the Green's functions can be calculated from the electric field as:

$$G_b(r_n, r) = \frac{j}{\omega \mu_0 J} E_b(r, r_n), \tag{2}$$

The difference δ_ϵ between the relative complex permittivity of the reconstructed region, $\epsilon_r(r)$, and the background medium, $\epsilon_b(r)$, is defined as: $\delta_\epsilon(r) = \epsilon_r(r) - \epsilon_b(r)$ whereas amplitude. The scalar integral equation above assumes 2D transverse-mode propagation, and is only an approximation of the 3D inverse problem at hand. Despite this loss in information, the 2D approximation can produce images of acceptable quality in many MWI problems arising in medical applications.

At each DBIM iteration i, the integral equation can be discretized for each transceiver pair as,

$$E_s(r_n, r_m) = j\omega\epsilon_0 \sum \delta_\epsilon(r) E_b(r_n, r) E_b(r, r_m), \tag{3}$$

leading to an ill-posed linear system:

$$\mathbf{A}\delta_\epsilon = b \tag{4}$$

where $\mathbf{A}$ is an $M \times N$ matrix ($M \ll N$) and b is the $M \times 1$ vector of the scattered fields. M is the number of transmitting-receiving pairs and N are the pixels of the reconstruction region V. The matrix $\mathbf{A}$ is calculated at each DBIM iteration by the forward solver, which provides E_b for a known background ϵ_b. The background field is used to build the linear system above, which is then solved by an inverse solver. Finally, the background profile is updated by $\epsilon_b^{i+1}(r) = \epsilon_b^i(r) + \delta_\epsilon(r)$ and the DBIM continues to next iteration $i + 1$. The forward solver uses the finite difference time domain (FDTD) method. This method simulates the EM wave propagation of the direct, "forward" problem based on Maxwell's equations. Furthermore, this implementation uses a single-pole Debye model to model frequency-dependent materials, such as brain tissues, as:

$$\frac{\varepsilon}{\varepsilon_0} = \varepsilon_\infty + \frac{\Delta\varepsilon}{1 + j\omega\tau} + \frac{\sigma_s}{j\omega\varepsilon_0} \tag{5}$$

The TwIST algorithm is employed as the solver of the ill-posed linear inverse problem at each DBIM iteration. Thresholding algorithms solve the ill-conditioned linear system $Ax = b$, by finding a solution x which minimizes the least squares error function as $F(x) = \frac{1}{2}\|Ax - b\|^2 + \lambda\|x\|_1$. The regularization term $\lambda\|x\|_1$ stabilizes the solution by limiting its l_1-norm. The general structure of the TwIST algorithm for solving this minimization problem is given by [33]:

$$\begin{aligned} x_{t+1} &= (1-\alpha)x_{t-1} + (\alpha - \beta)x_t + \beta C_\lambda(x_t) \\ C_\lambda(x) &= \Psi_\lambda(x + A^T(y - Ax)) \end{aligned} \tag{6}$$

The parameters for the TwIST algorithm are calculated as:

$$\kappa = \frac{\xi_1}{\xi_m} \tag{7}$$

$$\rho = \frac{1 - \sqrt{\kappa}}{1 + \sqrt{\kappa}} \tag{8}$$

$$\alpha = \rho^2 + 1 \tag{9}$$

$$\beta = \frac{2\alpha}{\xi_1 + \xi_m} \tag{10}$$

where ξ_1 and ξ_m are the smallest and largest eigenvalues of $A^T A$, respectively. The shrinkage/thresholding operation is a soft-thresholding function, calculated as $\Psi_\lambda(x) = sign(x)\max\{0, |x| - \lambda\}$. At each TwIST step, the new solution is updated based on two previous solutions and the soft-thresholding operation. As the linear system can be extremely ill-conditioned for MWI problems, the TwIST parameters above must be optimised specifically for the problem at hand [34].

The stopping criterion of the TwIST algorithm can be set based on a tolerance value, which is the normalised difference between the previous and current values of $F(x)$ and is defined as $tol = \frac{F(x_{k+1})-F(x_k)}{F(x_k)}$. The TwIST algorithm stops when tol is smaller than a preset value, usually in range between 10^{-4} and 10^{-1} [32].

2.2.2 Huygens Principle Based Algorithm

The radar-based Huygens algorithm, initially presented in [35], has already shown its potential for use in medical applications in [36, 37]. Considering an imaging region of interest such as the human head which is illuminated by a transmitter TX_m (operating at a frequency range f_i) and R receivers (with receiver r at location ρ_r measuring the field $E_{rm}(f_i)$ from transmitter m at frequency f_i), the Huygens algorithm measures the field on the external surface of the medium $E_{rm}(f_i)$ and back-propagates it virtually inside the imaging domain. In this way, the 2D internal field within the medium is reconstructed as:

$$E_{\mathrm{HP}}(\rho, m, f_i) = \sum_{r=1}^{R} E_{rm}(f_i) G(k|\rho_r - \rho|) \tag{11}$$

where $G(k_1|\rho_r - \rho|) = \frac{1}{4\pi} e^{-jk_1|\rho_r - \rho|}$ indicates the Green's function as defined in [35] at locations ρ, while k represents the complex wave number of the background medium.

To calculate the total intensity, all the transmitting positions and all the recorded frequencies are summed incoherently as shown below:

$$I_{\mathrm{HP}}(\rho) = \sum_{m=1}^{M} \left[\sum_{i-1}^{I} E_{\mathrm{HP}}(\rho, m, f_i) \right]^2 \tag{12}$$

2.3 Metasurface Design

2.3.1 Unit Cell Metasurface

The MTS design proposed in this chapter is based on simulation studies to achieve a transmission improvement through lossy biological tissues. To this end, a first MTS unit cell was designed to act as an impedance matching layer in the presence of a simplified planar model of the head [38]. The design of the unit cell was initially based on a plane wave analysis inspired by [26, 39]. This analysis is tailored to the desired operation bandwidth (selected as 0.5–2.0 GHz for our head imaging application)

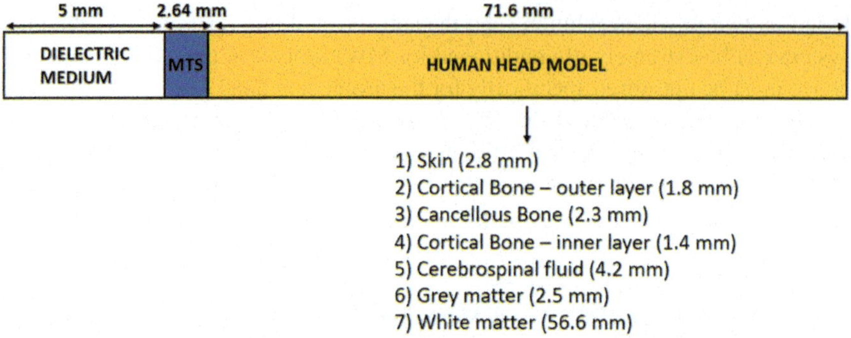

Fig. 2 Simulation setup with planar head model comprising seven tissue layers. The model comprises seven flat tissue layers placed in the following order: skin, cortical bone (outer layer), cancellous bone, cortical bone (inner layer), cerebrospinal fluid (CSF), gray matter and white matter. The dielectric properties of the tissues at 1 GHz were taken from [41]. The thickness of each tissue's slab is indicated in brackets

and is based on a slab model of tissue layers interacting with the MTS. The planar head model is shown in Fig. 2. The interaction of this setup with a linearly polarized plane wave was modelled in CST Microwave Studio [40]. The S-parameters were calculated considering "Port 1" before the lossy dielectric medium (permittivity $\epsilon' =$ 18 and conductivity $\sigma =$ 1.3 S/m at 1 GHz) and "Port 2" after the white matter layer [38].

The first MTS unit cell design is shown in Fig. 3 and comprises a Jerusalem Cross-based metallic lattice (thickness = 0.10 mm) embedded between two Rogers 3010 TM substrates (thickness = 1.27 mm, $\epsilon' = 10.2$ and tan $\delta = 0.0022$). The two Rogers high-dielectric substrates are bonded with Rogers 3001 bonding film ($\epsilon' = 2.28$, tan δ = 0.003), as shown in Fig. 3.

This design was optimized manually until the performance achieved was close to the expected. In particular, as reported in [42], variations in the classic Jerusalem Cross design were introduced in order to reduce the reflection coefficient. After that, CST optimizer tool with a Trust Region Framework algorithm was used to tune the wires' dimensions. In particular, the wires' thickness was varied up to 10% of its initial value, setting maximum transmission across all the frequency range as a goal.

The plots in Fig. 4 show the reflection and transmission parameters, with and without MTS, for the setup of Fig. 3. The transmission coefficient is enhanced by 1.7 dB at 1 GHz, where the reflection coefficient is reduced.

2.3.2 MTS-Enhanced Antenna

After optimization of the single MTS element, an MTS superstrate loading based on the unit cell described above was modelled to operate as an enhancer of the antenna shown in Fig. 5. The MTS loading was created by fitting the triangular patch

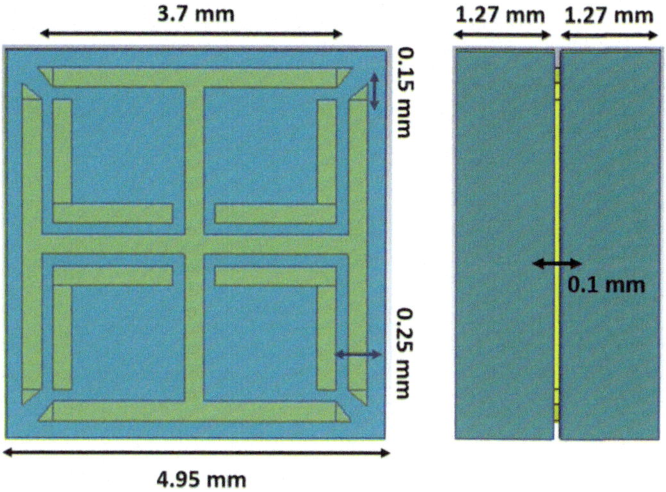

Fig. 3 MTS unit cell design comprising a Jerusalem Cross-based copper lattice embedded between two high-dielectric substrates (left). Side-view of the MTS unit cells, showing the two Rogers high-dielectric substrates embedding the copper lattice (right)

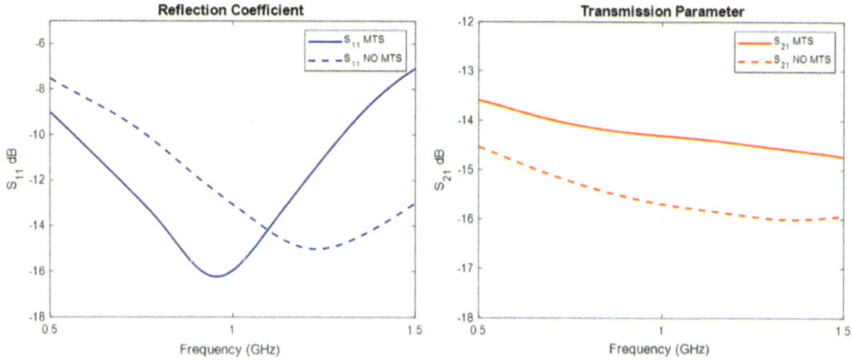

Fig. 4 Reflection coefficient (left) and transmission parameter (right) comparison with (solid lines) and without (dashed lines) MTS

antenna's surface with a suitable number of unit cells and glued on the antenna's radiating element. Thus, keeping the unit cell design and size fixed, a 5×6 unit cell MTS was generated and radar and tomographic measurements were performed with the microwave multi-port system shown in Fig. 1 [42]. Triangular antenna's photos both with and without MTS coating are depicted in Fig. 6.

Before proceeding with the measurements, the experimental setup of Fig. 1 was simulated in CST Microwave Studio to check the functionality and performance of the MTS and the DBIM-TwIST and Huygens principle-based algorithm.

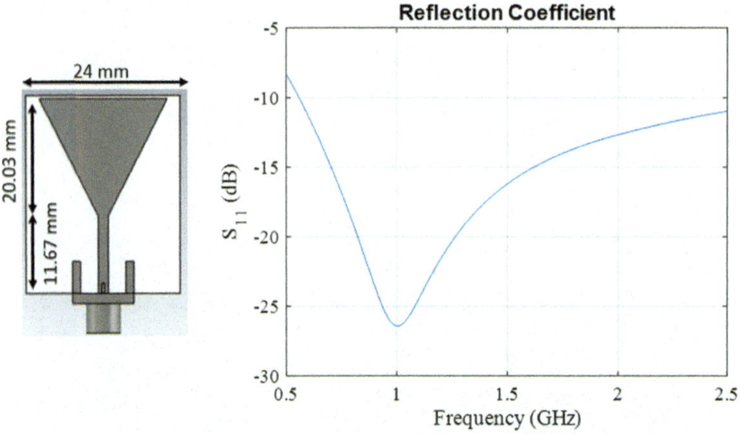

Fig. 5 Geometry and reflection coefficient of the triangular patch antenna used with the MTS

Fig. 6 Triangular antenna without (left) and with MTS coating (right)

2.3.3 Impact of MTS on Antenna Performance

Our CST simulation model featured an array of eight antennas immersed in a 90% glycerol-water mixture and closely fixed on the elliptical head phantom made of an external plastic layer ($\epsilon = 2.56$; $\tan\delta = 0.0119$) and average brain tissue inside ($\epsilon = 45.8$; $\sigma = 0.76$ S/m).

Figure 7 shows the total E-field distribution on the surface of the head phantom at 1.25 GHz for antenna 1 (T1) transmitting with and without MTS. To isolate the MTS effect, a new "negative control" (NC) scenario, where a substrate without the MTS-defining metallic elements is placed on the antennas, was also simulated. The plots show that the MTS does not focus the E-field onto a tighter beam, but on the contrary, it widens the E-field distribution inside the head model, resulting in better signal coverage.

To demonstrate the MTS's impact on the radiated field, the E-field value at two points belonging to different imaging planes and located in the centre of the head was calculated. For all the array configurations, the E-field intensity values in those

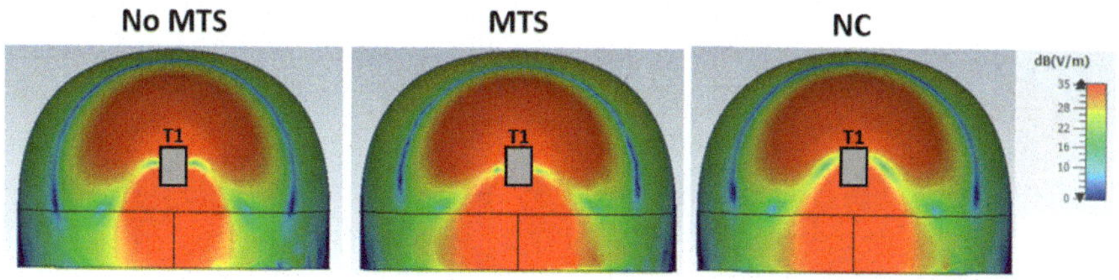

Fig. 7 E-field distribution on the surface of the head phantom for the "no MTS", "MTS", "NC" scenarios when antenna 1 (T1) is transmitting. The MTS produces a wider E-field distribution

Table 1 Differences between E-field intensity values at 2 different points for "no MTS", "MTS" and "NC" configurations

	Point1 (0, 60, 0) mm (V/m)	Point2 (0, 43, 0) mm (V/m)
ΔE (no MTS)	0.41	0.65
ΔE (MTS)	0.50	0.81
ΔE (NC)	0.47	0.76

points were computed for a model of the head made of average brain only ("no target" scenario) and for the same model including a blood-mimicking target ("with target" scenario). Then, the difference between the E-field intensity values calculated in the two scenarios (ΔE) was estimated. Values of the scattered fields for the "no MTS", "MTS", and "NC" configurations are shown in Table 1. These values suggest that the field scattered from the target inclusion has a higher intensity when the MTS is present. This might be the reason why the target response shown in Sect. 2.4.2 is stronger in the presence of the MTS, leading to better target detection.

Moreover, the E-field distributions on the transmitting antennas' substrate for the "no MTS", "MTS" and "NC" scenarios (with and without the head model) shown in Fig. 8 indicate that a stronger field is excited by the transmitter's substrate when the MTS is present. However, it is important to notice that this higher-field intensity on the antenna's substrate covered by the MTS sheet may be due to the fact that the monopole's radiating part is not in direct contact with our lossy coupling medium.

We also examined the impact of the MTS on receive mode. The simulation results suggest that the MTS seems to act as an "amplifier" of the received field. To show this enhancement, different antenna arrays ("no MTS", "MTS", "NC") immersed in matching medium only were considered. The first array includes a triangular patch transmitter (antenna 1) and 7 MTS-enhanced receivers. The second array consists of the same transmitter and 7 negative-control receivers. The E-field was plotted on the substrate of the receiver opposite to the transmitter (with and without MTS and with NC). Figure 9 shows that the E-field on receiver 5 is significantly enhanced when the MTS covers the antenna's substrate. Thus, considering the same source of excitation, the MTS leads to an improvement of the received field.

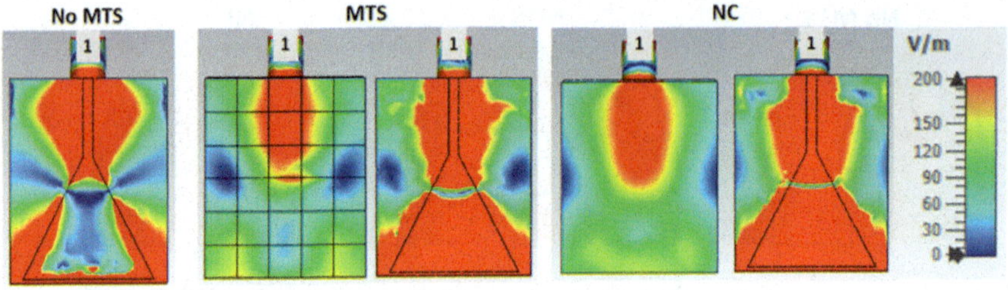

Fig. 8 E-field plotted on the transmitter's substrate for the "no MTS", "MTS", "NC" scenarios with the head model. A higher-intensity E-field is transmitted when the MTS is present

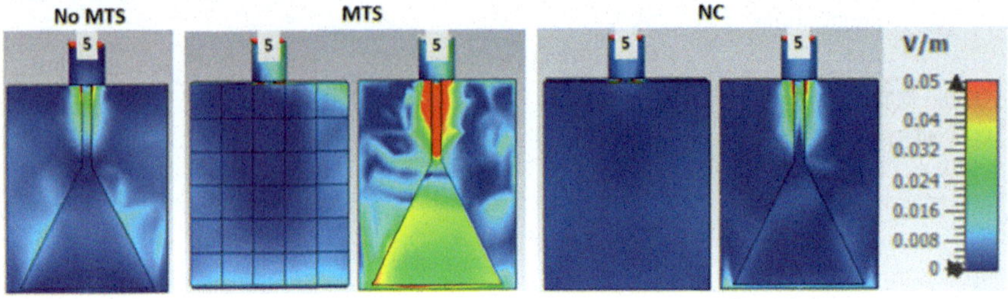

Fig. 9 E-field plotted on the substrate of receiver 5 (with and without MTS and with NC loading) when antenna 1 is transmitting. A higher-intensity E-field is received when the MTS is present

2.4 Experimental Validation

2.4.1 Methodology

To investigate whether the proposed MTS could improve the quality of the images reconstructed through the radar and tomographic algorithms, a gelatin-oil mixture was produced to mimic the average brain and blood tissues. The concentration of each material used in preparing the mixtures is presented in Table 2. This experimental procedure described in this section is reported in [30].

Table 2 Quantities of materials used for 100 ml of tissue mimicking phantoms

	Average brain	Blood
Gelatin powder	11 g	16 g
Kerosene	13 ml	–
Propanol	2.5 ml	1.5 ml
Safflower oil	13 ml	–
Surfactant	4 ml	–
Water	60 ml	80 ml

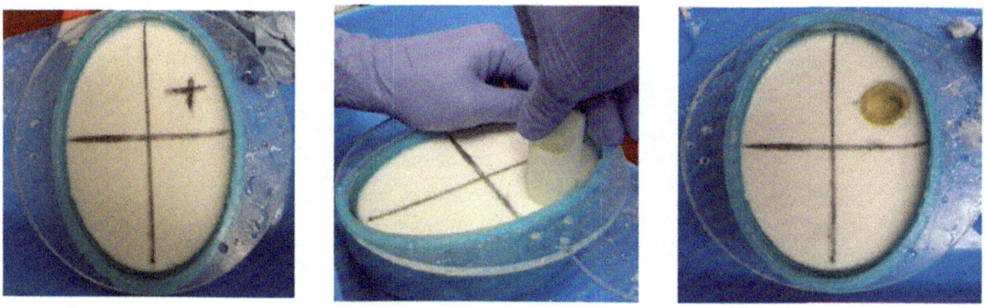

Fig. 10 Elliptical brain phantom before (left) and after (right) inserting the eccentric target. The target was positioned at two different locations in the two constructed phantoms

The preparation process started with preparing a water-gelatine solution, which was mixed until the gelatine particles were fully dissolved, resulting in a transparent mixture. Next, propranolol was added to remove the air bubbles from the surface. When the mixture reached 70°C, a 50% kerosene-safflower oil solution was added to the existing mixture and stirred at the same temperature, until the oil particles were fully dissolved. The mixture was stirred until the temperature dropped to 50°C, at which point surfactant was added. As a final step, the prepared mixture (at 35°C) was poured inside a thin elliptical plastic mould, as shown in Fig. 10, and was left to cool and solidify until the next day, when measurements were preformed.

Before inserting the target into the average brain mixture, the "no target" scenario (only average brain) was measured using both tomography and radar antenna configurations, with both "without MTS" and "with MTS" antenna types. After performing all the "no target" measurements, the "with target" measurements were carried out, by placing a 30 mm diameter cylindrically shaped target mimicking the blood inside the average brain solution. As shown in Fig. 10, the target was placed at a distance of 30 mm horizontally to the right and 25 mm vertically downwards from the centre of the phantom [30, –25], at an angular position of ≈320°C relative to the position of the first transmitting antenna. Next, all the "with target" measurements for both tomography and radar configurations, both with and without MTS were performed.

2.4.2 Results

All images presented in this section have been obtained by subtracting the signals from the phantom before and after inserting the target.

Figure 11 shows the DBIM-TwIST results using data from the CST model for three different frequencies corresponding to the phantom without and with MTS. An improvement in the target localization, with less artefacts can be obtained when using MTS.

Moving on to the radar simulation results, Fig. 12 shows the normalized and adjusted reconstructed intensity images with and without MTS, respectively. The image adjustment procedure allows us to have a better and clearer visualization of

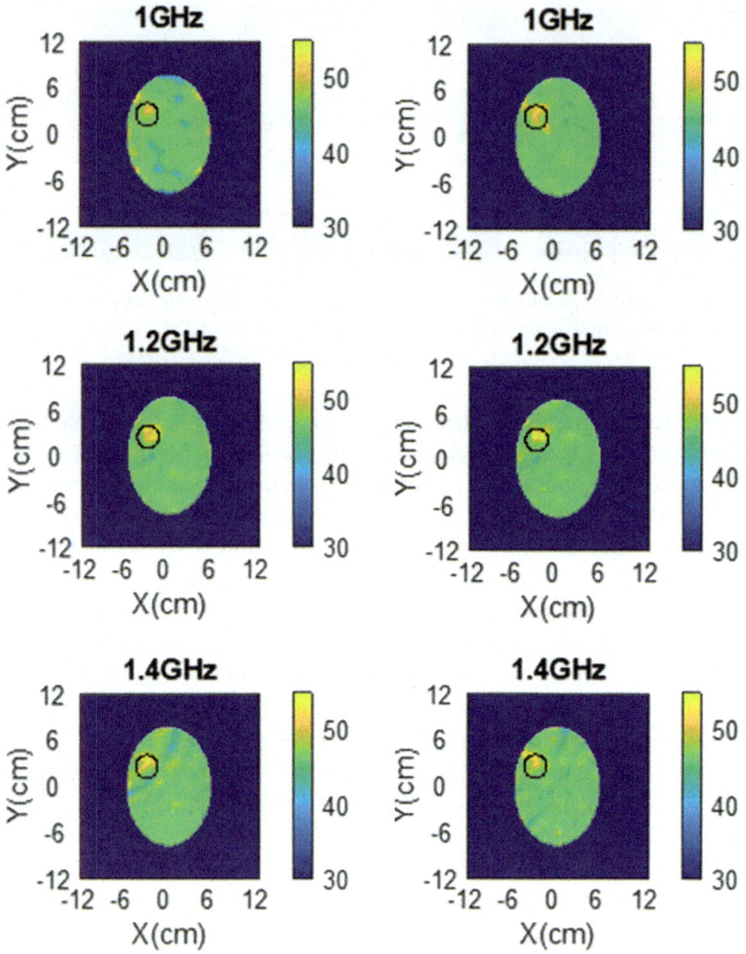

Fig. 11 Reconstructed dielectric constant at three different frequencies using data from the CST model and antennas with (left) and without the MTS (right)

the target and the region of interest, and is done here by converting intensity values lower than 0.65–0, and re-ranging values higher than 0.65, from 0 to 1. In both cases, the target can be localized in its approximate position without artefacts. In comparison, no detection is achieved without the use of MTS on the antennas. When the MTS is not applied on the antennas, the algorithm detects artefacts, which appear as peaks when the image is normalized and adjusted.

To demonstrate the positive impact of the MTS on radar-based detection, a negative control scenario was simulated in CST. Although the electric field distribution plotted in Fig. 7 for this case is very similar to the case of the MTS, the resulting reconstruction in Fig. 13 shows a stronger ghost target close to the real one, which

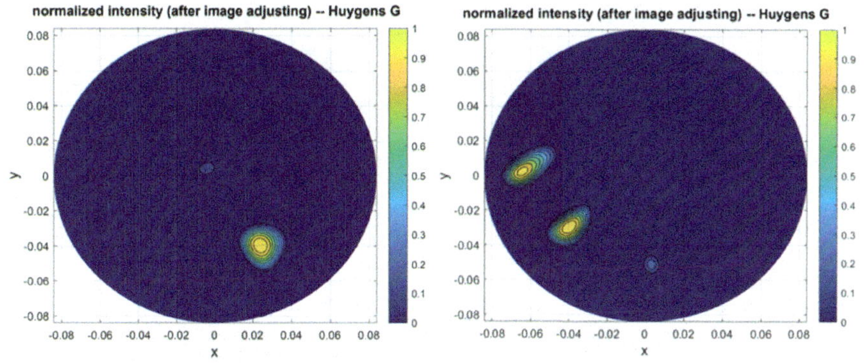

Fig. 12 Radar-based images after normalization and adjustment with (left) and without MTS (right) using data from the CST model. Axes units are meters

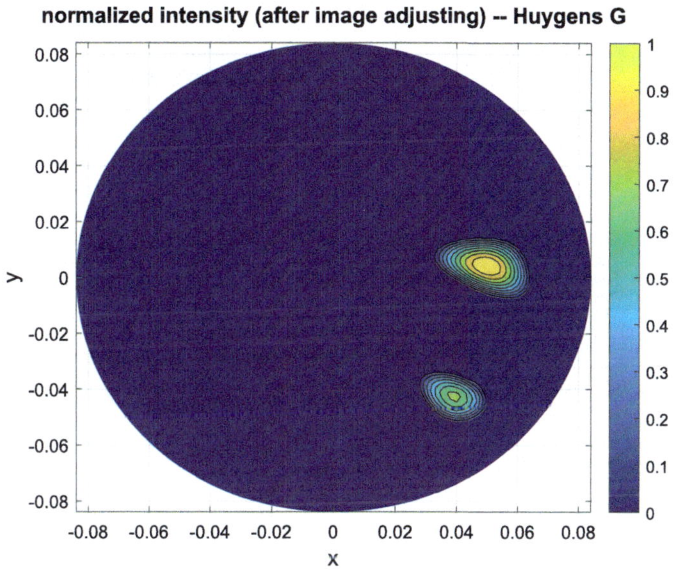

Fig. 13 Radar-based image after normalization and adjustment, using data from the CST model and the "negative control" (NC) scenario. Axes units are meters

obstructs accurate detection. Indeed, the NC-antennas have worse receiving performance (see Fig. 8). This example suggests that using the MTS can enhance detection by virtue of its transmitting and receiving properties, which increases the target response relative to other signal artefacts.

Focusing now on the reconstruction results from the measurements, Fig. 14 shows the reconstructed images using DBIM-TwIST for 1 GHz, 1.2 GHz and 1.4 GHz, with and without MTS. These results show a more clear improvement in target detection when using the MTS than the improvement observed with the simulated data.

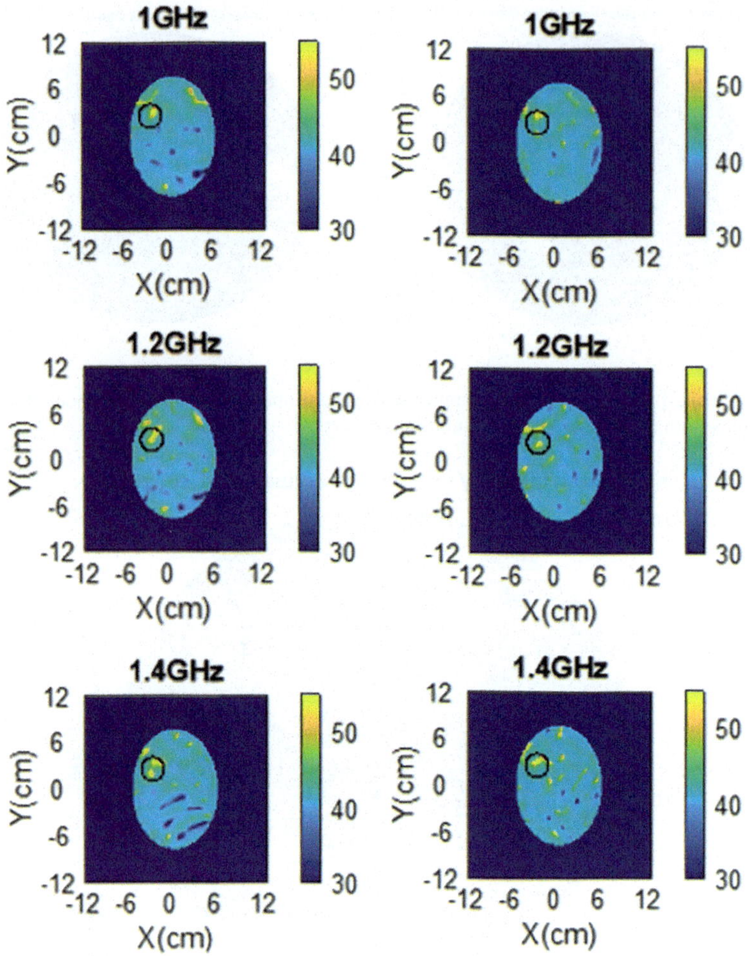

Fig. 14 Reconstructed dielectric constant at three different frequencies using experimental data and antennas with (left) and without (right) the MTS

Referring to radar measurements, Fig. 15 shows the S-parameter comparison for the recorded frequency band (0.5–2 GHz) without and with MTS, when the triangular antennas are closest (S21), midway or 90° apart (S71) and farthest to each other (S13-1), respectively. From the figure, signal propagation improvement over the whole frequency band can be observed when using the MTS. Investigating the recorded frequency range while processing the measured data indicated a frequency range of 0.6–0.8 GHz as the optimum frequency band for image reconstruction when using the Huygens algorithm. An average gain of 4.8 dB when using the MTS is achieved in this frequency range.

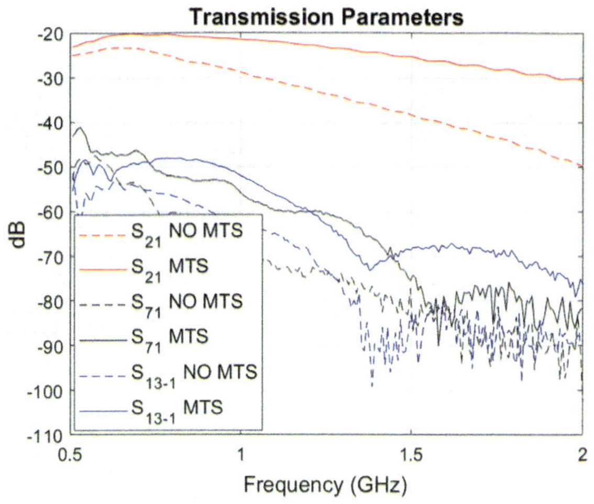

Fig. 15 Magnitude (dB) plots for different antenna distances within frequency band of 0.5–2 GHz. Solid lines represent MTS measurement while dashed lines represent "no MTS" measurement

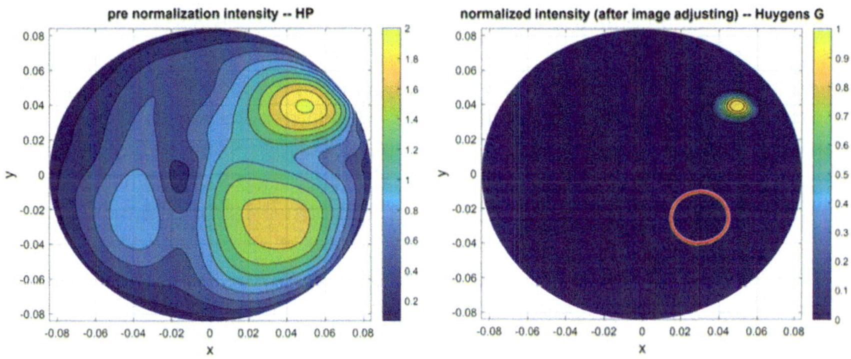

Fig. 16 Radar-based images using experimental data and antennas without the MTS before normalizing (left) and after normalization and thresholding adjustment (right). Axes units are meters

Figure 16 shows intensity reconstruction images without the use of MTS obtained through Huygens algorithm, before and after image normalization and adjusting, respectively. Similarly, Fig. 17 shows reconstruction images of the same phantom when placing the MTS on the antennas. We can observe that, when using the MTS, the target is detected and localized in its approximate position without noticeable artefacts. The actual location of the target is shown with a red circle.

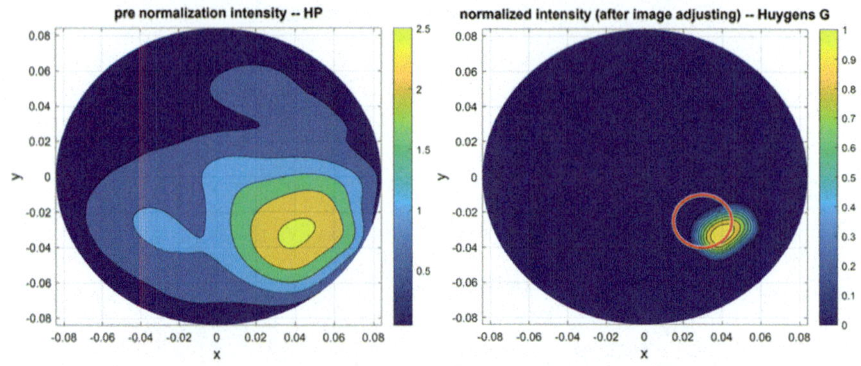

Fig. 17 Radar-based images using experimental data and antennas with the MTS before normalizing (left) and after normalization and thresholding adjustment (right). Axes units are meters

2.5 *Discussion*

This section has presented preliminary numerical and experimental results which demonstrate the potential of a novel MTS structure to enhance an MWI systems for brain imaging. The proposed MTS is capable of improving transmission through a simple model of the human head (Fig. 2). Simulation and experimental studies were performed to assess the impact of a 5×6 unit cell MTS on a printed monopole antenna. The investigation verifies that both the radar and tomographic algorithms described in Sect. 2.2 can detect and localize a haemorrhagic stroke-mimicking target in its approximate location (if the background signal in the absence of the target is known). Both simulations and measurements were carried out with the custom-made MWI system shown in Fig. 1. The results presented in Sect. 2.4.2 shows that the MTS permits a considerable performance enhancement, which results in the improvement in the detection and localization of the target with both radar and tomographic approaches.

3 Enhancing Cancer Treatment Monitoring Through Metamaterial Technology

The demand for new and personalized cancer therapies is a challenge involving numerous research fields. Many procedures and drugs are now available and many more are being studied. In this section, we investigate the feasibility of enhancing liver's ablation monitoring through MTS technology. In particular, a MTS film is proposed as a hardware advancement of the microwave imaging system for thermal ablation monitoring developed by CNR-IREA (Naples, Italy) and Sapienza University of Rome (Rome, Italy). More details on the systems are given in Chap. "An Initial Assessment of a Microwave Imaging System to Monitor Microwave Ablation Treatments" of this book.

3.1 Liver Cancer: Prognosis and Treatments

Estimates from the year 2000 indicate that liver cancer is the fifth most common malignancy in men and the eighth in women worldwide. As a result of the peak of infection with hepatitis B and C viruses in the 1950s–1980s, the incidence of primary liver cancer (PLC) is now increasing and will likely increase for decades in several developed countries [43]. Clinical studies have shown that PLC can be surgically treated only if detected at a presymptomatic stage [44]. However, only 5–15% of patients are early diagnosed and eligible for surgical removal [45]. Other treatment options for more advanced stages include trans-arterial chemoembolization (TACE) and chemiotherapy with oral dosing of sorafenib, which is the most accepted treatment for late-stage cases. However, fewer than one-third of patients benefit from this treatment, as patients may have toxicity issues and develop drug resistance after a few months of treatment [45].

Microwave thermal ablation (MTA) is a recent promising cancer treatment able to destroy liver cancerous cells with a minimally invasive approach. During this treatment, a microwave antenna is placed directly into the tumour area with imaging guidance. Then, electromagnetic (EM) microwaves are emitted in the tissue to produce heat and induce cellular death via coagulation necrosis [46].

During MTA, temperature monitoring of the ablation region is needed to avoid damaging healthy cells. For this purpose, several imaging modalities have been used, such as computed tomography (CT) and magnetic resonance imaging (MRI). Currently, MRI is the most used technique for real time temperature monitoring, as it provides high-resolution temperature's distributions and it is more suitable than CT for long time exposures [47].

An emerging imaging method which might be used for thermal ablation monitoring is MWI. The basic principle of this technique is to record the field back-scattered during the ablation treatment thorough an array of antennas positioned externally to the patient's body. The variations of the recorded data between different time instants are then processed through an inverse scattering algorithm and an image of the changes occurring in the electromagnetic properties of the heated area is produced [48]. MWI is a valid alternative to the current imaging modalities as it is a non-ionising low-cost technology which could be made compact and portable and thus compatible with MTA devices. For this reason, numerous research groups have put efforts in the development of new MWI devices for real-time monitoring during thermal ablation in different anatomical regions such as liver [49–51], breast [52], and brain [53].

3.2 Microwave Imaging System to Monitor Thermal Ablation of Liver Tumours

The microwave imaging system for thermal ablation monitoring developed by CNR-IREA (Naples, Italy) and Sapienza University of Rome (Rome, Italy) comprises

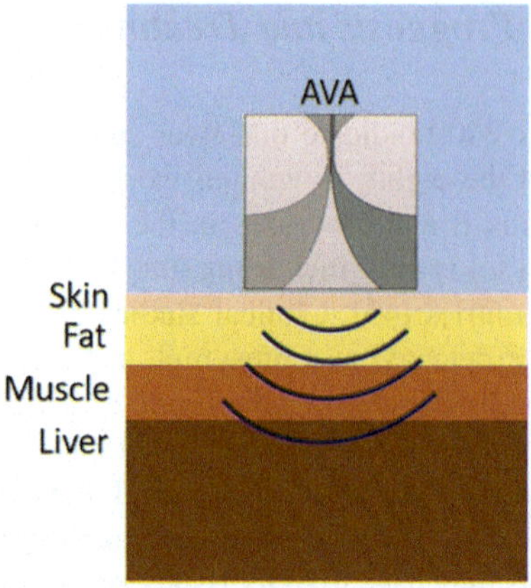

Fig. 18 Schematic design of the microwave thermal ablation monitoring system reported in [50]. The liver phantom includes a first thin layer of skin (pink), then fat (in yellow) and muscle (orange). The last layer (brown) represents the liver. The operating frequency and matching medium are identified in order to improve the match between the probing field and the human abdomen. The antipodal Vivaldi antenna (AVA) is designed and optimized to operate in the identified conditions

compact antipodal Vivaldi antennas (AVAs) operating in the 500 MHz-5 GHz frequency range. The proposed antennas are immersed in a liquid matching medium with a permittivity value close to 23, chosen such that the largest possible portion of the EM power enters the abdomen and provide an adequate backscattered signal after interacting with the liver (Fig. 18) [50].

3.2.1 Antipodal Vivaldi Antenna

The AVA developed by CNR-IREA and Sapienza University of Rome is shown in Fig. 19. As this antenna was initially designed to operate in air, the connector's pin was covered with an epoxy resin of permittivity equal to 4 and dimensions of 4.76 $\times$ 5.75 $\times$ 2.77 mm in order to overcome mismatching issues [50].

The reflection coefficient of the antenna tested on an abdomen phantom consisting of a stack of tissue slabs is shown in Fig. 20. The phantom block includes layers of skin ($thickness_s$ = 2.3), fat ($thickness_f$ = 12.2 mm) and muscle ($thickness_m$ = 20.2 mm). The last layer represents the liver ($thickness_l$ = 80 mm) and contains a target mimicking the ablated area (ϵ = 26.67 and σ = 1.26 S/m). Dimensions and position of the ablated liver are reported in Fig. 20. For this simulation, the frequency dielectric properties of the tissues were taken from [41] and fixed at central frequency 2.5 GHz.

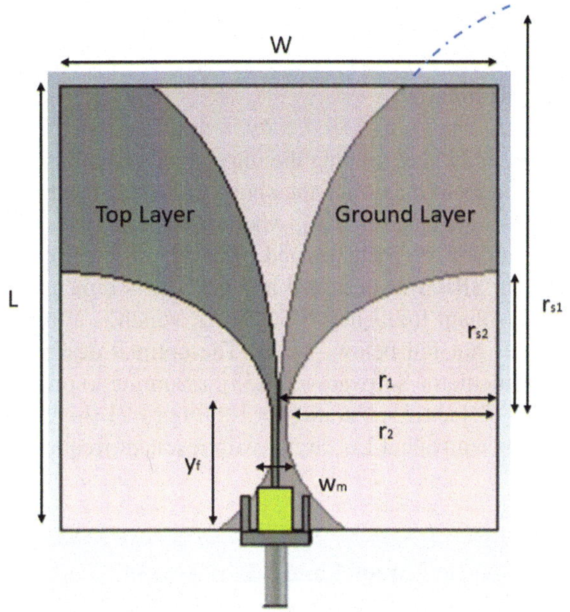

Fig. 19 Antipodal Vivaldi antenna (AVA). W = 60 mm, L = 60 mm, w_m = 0.9 mm, y_f = 14.5 mm, r_1 = 30 mm, r_2 = 29.1 mm, r_{s1} = 50.1 mm, r_{s2} = 20.37 mm. The SMA connector's pin is covered with epoxy resin on the back side of the antenna [50]

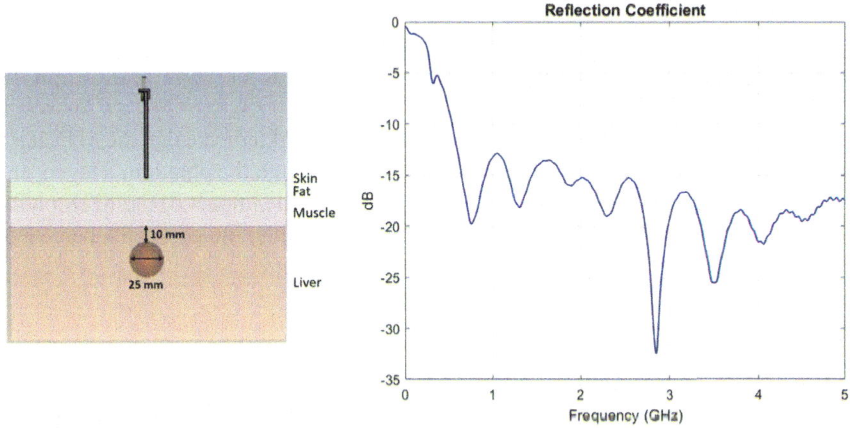

Fig. 20 Abdomen phantom with target mimicking the ablated liver tissue (left) and reflection coefficient of the AVA in front of the abdomen phantom containing an ablated area

3.3 Metasurface Design

3.3.1 Split Ring Resonator

The purpose of this study is enhancing the impedance matching at low frequencies through a MTS in order to maximize the signal backscattered by the ablated area. A similar study was conducted in [54], where a MM slab based on a SRR array was designed to improve the focusing and penetration depth in human biological tissue. To this end, a SRR was designed and the SRR's dimensions were tailored in order to set its resonant frequency at 500 MHz, which is a frequency where the S_{11} of the AVA does not fall below –10dB. The optimal design parameters were found using SRR Calculator Software tool [55], assuming to print the resonator on a 0.1 mm thick high-dielectric substrate ($\epsilon = 10$). Figure 21 shows the SRR's design and Fig. 21 shows its equivalent LC circuit with resonant frequency $f_0 = 500$ MHz (Fig. 22).

3.3.2 Full Metasurface Structure and 1-Port System

After designing the resonator element, an MTS structure made of three SRRs was tested on the abdomen phantom. This structure was centred on the skin slab in order to cover great part of the phantom's surface, as shown in the left panel of Fig. 23.

To assess the MTS impact on the AVA's near-field characteristics, the reflection coefficient of the antenna was calculated with and without MTS, see Fig. 23 right panel. The results show that the AVA's performance is greatly enhanced at lower frequencies. In particular, a deep resonance (–34 dB) at 500 MHz is shown in the presence of the MTS.

Furthermore, the E-field distributions shown in Fig. 24 indicate that the MTS has a focusing effect, allowing better field's penetration through the phantom's layers and increasing the E-field intensity in the target's region. The impact of the MTS on the AVA's near-field radiation properties might be attributed to the unit cell's behaviour in correspondence of its resonant frequency.

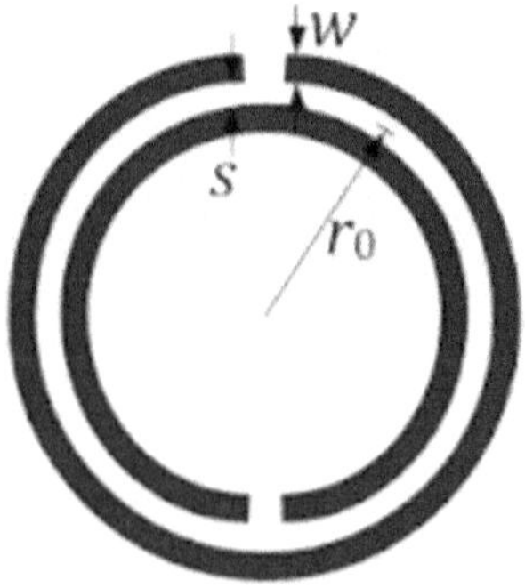

Fig. 21 Split-Ring Resonator (SRR). w = 1 mm, s = 1 mm, $r_o = 23.4$ mm (outer radius)

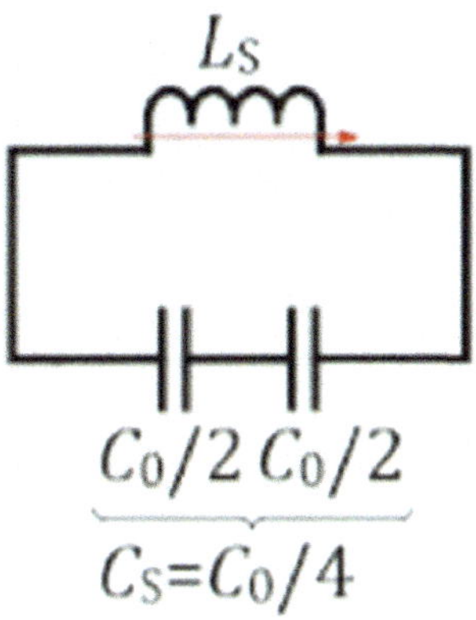

Fig. 22 Equivalent LC circuit. $C_0 = 2\pi r_0 C_{pul}$ is the total capacitance between the two rings, where C_{pul} is the per-unit-length capacitance of a CPS line. The resulting series capacitance C_S of the equivalent circuit is given by two capacitors of $C_0/2$ in series, that is $C_S = C_0/4$. The series inductance can be approximated by that of a single ring of width w and a radius the averaged between the two rings. The resonant frequency of the SRR is given by: $f_0 = 1/2\pi\ \sqrt{LsCs}$

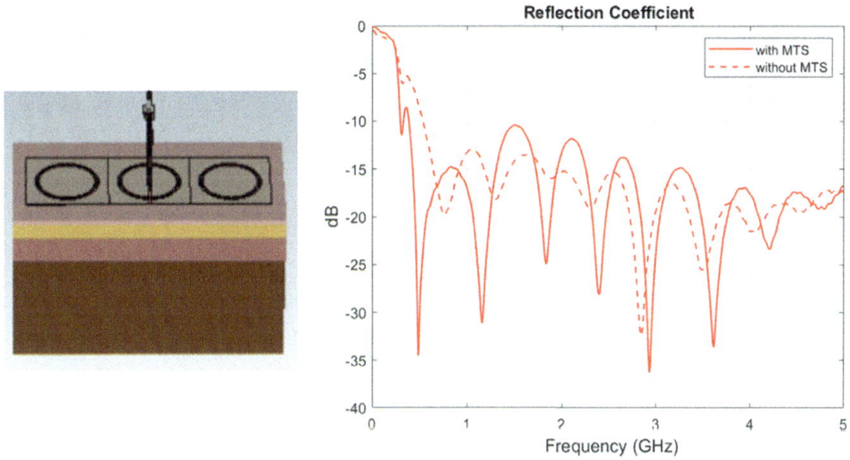

Fig. 23 Full MTS structure placed on abdomen phantom (left) and reflection coefficient of the AVA with and without MTS. The phantom is not including the ablated area

3.3.3 Multiple-Port System

In order to investigate whether the proposed MTS could enhance the "weak" signal backscattered by the ablated area, preliminary simulations of a multiple-port system were carried out. This system is shown in Fig. 25 and includes five AVAs placed at a distance of 23 mm, as reported in [51]. The EM wave propagation was studied for an abdomen phantom including a target mimicking the ablated area ("with target" scenario) and for a phantom without inclusions ("no target" scenario). The S-Parameters for reflection and transmission were calculated over the 0.5–2.5 GHz

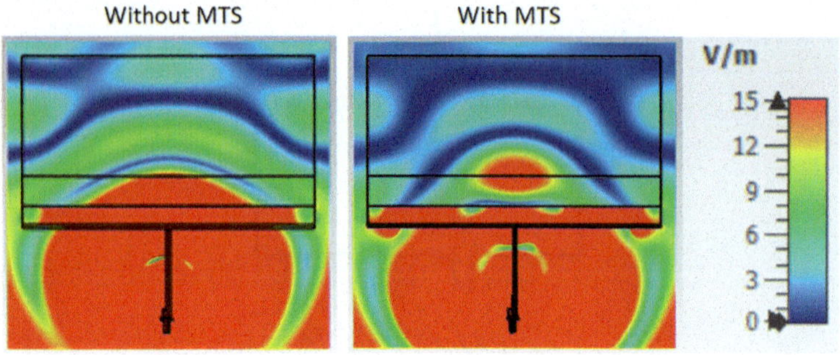

Fig. 24 E-Field plotted on a transverse section of the phantom without MTS

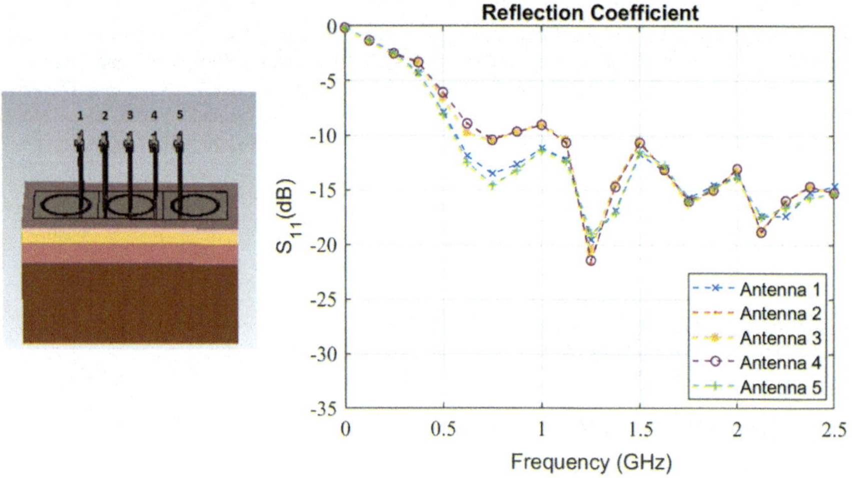

Fig. 25 5-Port setup and reflection coefficient for the antennas measured in the "no target" configuration without MTS

frequency range, with and without MTS. Then, the signal difference "with target–no target" (dB) was calculated at five equally spaced frequencies in the 0.5–2.5 GHz frequency range and plotted in function of the receivers' location.

The right panel of Fig. 25 shows the reflection parameters of the antennas calculated for the "no target" scenario, without MTS. Plots of the antennas' reflection coefficient calculated in the presence of the MTS are shown in Fig. 26. As expected, the antennas exhibit deeper resonances when radiating in front of the MTS.

In addition, a reduction of the coupling effects between the antennas was observed in the presence of the MTS. Figure 27 shows the transmission parameters between antenna 1 and antennas 2, 3 and 4, calculated with and without MTS.

Furthermore, the graphs below show that the differential signal "with target–no target" (dB) is overall enhanced when the antennas radiate in front of the MTS. This suggests that the MTS is capable of enhancing "weak" signals, which may be useful for the considered applications (Figs. 28, 29, 30, 31, 32).

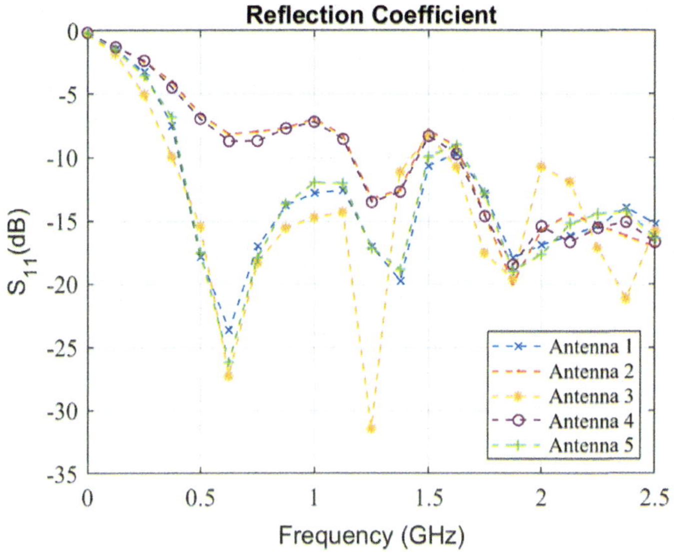

Fig. 26 Reflection coefficient for the antennas measured in the "no target" configuration measured with MTS

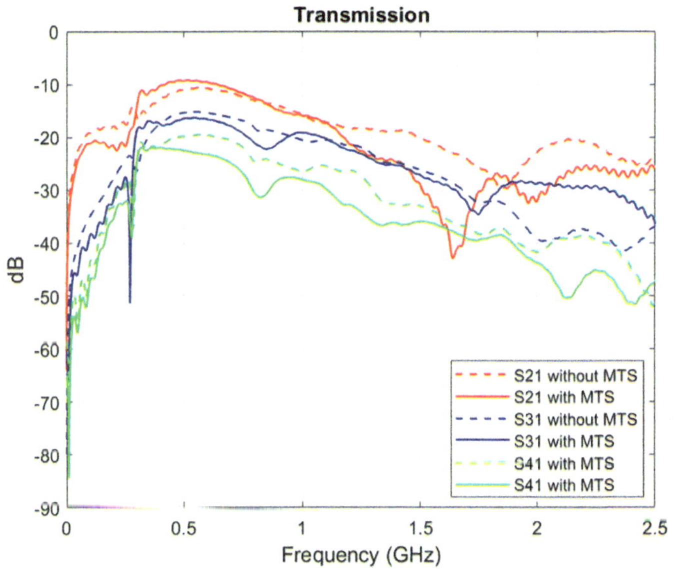

Fig. 27 S_{21}, S_{31} and S_{41} parameters calculated with and without MTS for the setup shown in Fig. 25a. The transmission parameters are reduced due to the presence of the MTS

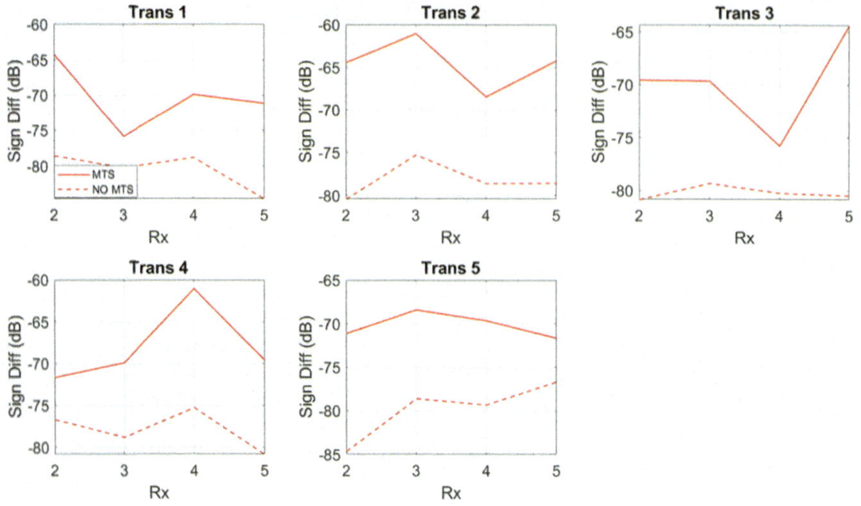

Fig. 28 Signal difference "with target–no target" (dB) plotted in function of the receivers' location, calculated at 0.5 GHz

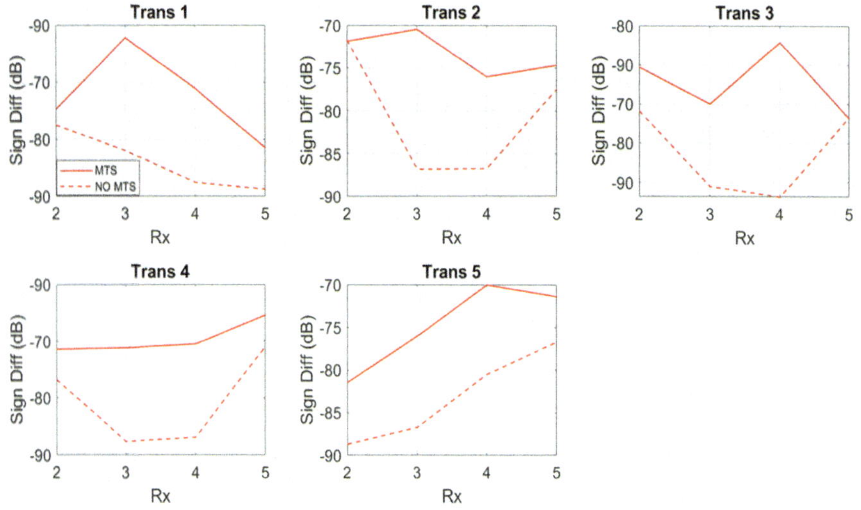

Fig. 29 Signal difference "with target–no target" (dB) plotted in function of the receivers' location, calculated at 1 GHz

3.4 Discussion

The feasibility study presented in this Section is a first step towards the realisation of an enhanced MWI system for temperature monitoring during microwave thermal ablation (MTA). The main goal was to improve the near-field performance of the AVA

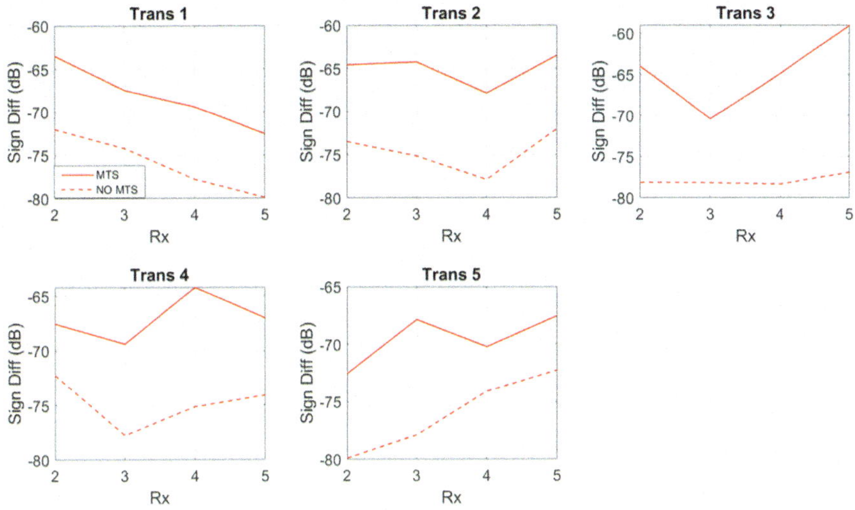

Fig. 30 Signal difference "with target–no target" (dB) plotted in function of the receivers' location, calculated at 1.5 GHz

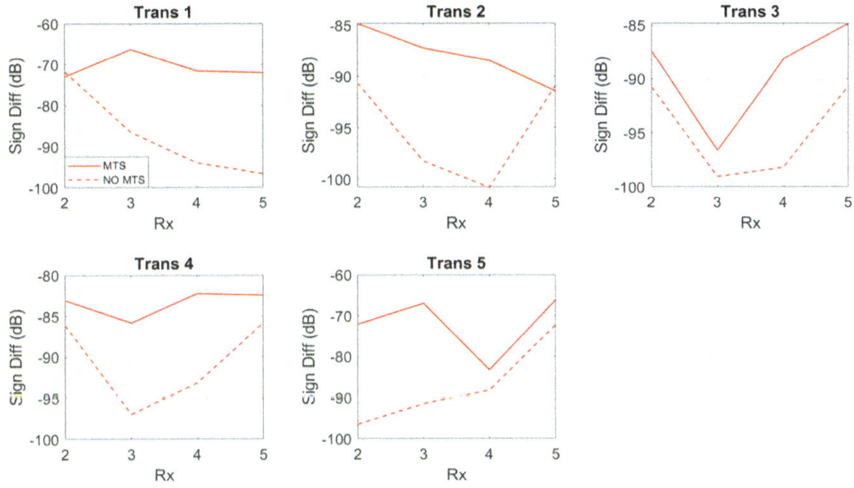

Fig. 31 Signal difference "with target–no target" (dB) plotted in function of the receivers' location, calculated at 2 GHz

for thermal ablation monitoring developed by CNR-IREA and Sapienza University of Rome. To this end, an MTS structure capable of reducing the reflection coefficient of the antenna at low frequencies and focusing the E-field inside an abdomen phantom model was designed. Then, a multiple-port system was studied. The simulation results show that the signal backscattered by a target mimicking an ablated area is overall increased in the presence of the MTS. This indicates that the MTS is capable of

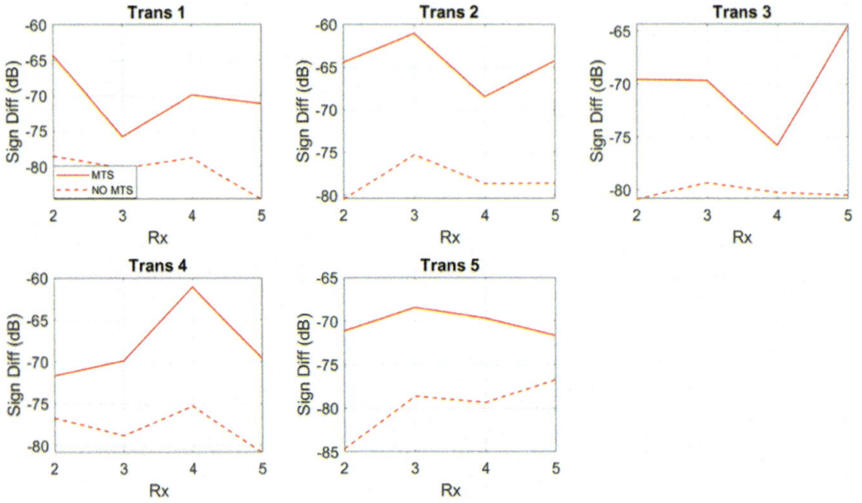

Fig. 32 Signal difference "with target–no target" (dB) plotted in function of the receivers' location, calculated at 2.5 GHz

enhancing "weak" signals imaging in MWI ablation monitoring. In conclusion, the results presented in this Section suggest that the proposed MTS might improve the near-field characteristics of compact antennas, thereby providing a powerful tool towards the development of compact and ergonomic imaging systems.

4 Conclusions

This chapter has presented hardware advances towards the development of innovative MWI prototypes. In particular, the feasibility of enhancing brain stroke detection using MTS technology has been extensively investigated through full-wave simulations and experiments. Furthermore, several numerical studies have been carried out to investigate the impact of employing MTSs on thermal ablation monitoting. To this aim, a new MTS film has been proposed to focus the E-field inside a liver phantom. The results suggest that the proposed MTS films might improve small-size antennas' near-field characteristic and hence, be a powerful element towards the development of portable, compact and ergonomic imaging systems. Thus, MTSs can greatly benefit MWI, providing a significant hardware advance towards the realisation of new systems with the desired clinical accuracy.

Acknowledgement This research was supported by the EMERALD project funded from the European Union's Horizon 2020 under the Marie Skłodowska-Curie grant agreement No. 764479.

References

1. M. Puentes Vargas, *Planar Metamaterial Based Microwave Sensor Arrays for Biomedical Analysis and Treatment* (Springer International Publishing, Switzerland, 2014)
2. K. Aydin, I. Bulu, K. Guven, M. Kafesaki, C.M. Soukoulis, E. Ozbay, Investigation of magnetic resonances for different split-ring resonator parameters and designs. New J. Phys. **7**, 1–15 (2005)
3. M.J. Freire, R. Marques, L. Jelinek, Experimental demonstration of a μ=-1 metamaterial lens for magnetic resonance imaging. Appl. Phys. Lett **93**(23), 1–4 (2008)
4. M.A. Lopez, M.J. Freire, J.M. Algarin, V.C. Behr, P.M. Jakob, R. Marqús, Nonlinear split-ring metamaterial slabs for magnetic resonance imaging. Appl. Phys. Lett. **98**(13), 1–3 (2011)
5. A.P. Slobozhanyuk, A.N. Poddubny, A.J. Raaijmakers, C.A. Van Den Berg, A.V. Kozachenko, I.A. Dubrovina, I.V. Melchakova, Y.S. Kivshar, P.A. Belov, Enhancement of magnetic resonance imaging with metasurfaces. Adv. Mater. **28**(9), 1832–1838 (2016)
6. L.L. Spada, F. Bilotti, L. Vegni, Metamaterial biosensor for cancer detection, in *IEEE Sensors Proceedings* (Limerik, Ireland, 2011)
7. S. Sugumaran, M. Faizal, M. Noor, C. Shekar, D. Schreurs, Biosensors and Bioelectronics Nanostructured materials with plasmonic nanobiosensors for early cancer detection: a past and future prospect. Biosens. Bioelectron. **100**, 361–373 (2018)
8. D. Isakov, C.J. Stevens, F. Castles, P.S. Grant, A split ring resonator dielectric probe for near-field dielectric imaging. Sci. Rep. **7**(1), 1–9 (2017)
9. S. Mukherjee, X. Shi, L. Udpa, S. Udpa, Y. Deng, P. Chahal, Design of a split-ring resonator sensor for near-field microwave imaging. IEEE Sensors J. **18**(17), 7066–7076 (2018)
10. S. Mukherjee, Z. Su, L. Udpa, S. Udpa, A. Tamburrino, Enhancement of microwave imaging using a metamaterial lens. IEEE Sensors J. **20**, 1–11 (2019)
11. O. Yurduseven, V.R. Gowda, J.N. Gollub, D.R. Smith, Printed aperiodic cavity for computational and microwave imaging. IEEE Microw. Wireless Compon. Lett. **26**(5), 367–369 (2016)
12. J.N. Gollub, O. Yurduseven, K.P. Trofatter, D. Arnitz, M.F. Imani, T. Sleasman, M. Boyarsky, A. Rose, H. Odabasi, T. Zvolensky, G. Lipworth, D. Brady, D.L. Marks, M.S. Reynolds, D.R. Smith, Large metasurface aperture for millimeter wave computational imaging at the human-scale. Sci. Rep. **7**, 1–10 (2017)
13. J.N. Mait, G.W. Euliss, R.A. Athale, Computational imaging. Adv. Opt. Photonics **10**(2), 409 (2018)
14. O. Yurduseven, D. Smith, A reconfigurable millimeter-wave spotlight metasurface aperture integrated with a frequency-diverse microwave imager for security screening. Proc. SPIE **10634**, 1–9 (2018)
15. M. Pastorino, *Microwave Imaging* (Wiley, Hoboken, New Jersey, U.S.A., 2010)
16. Q. Yang, J. Gu, D. Wang, X. Zhang, Z. Tian, C. Ouyang, R. Singh, J. Han, W. Zhang, Efficient flat metasurface lens for terahertz imaging. Opt. Express **22**(21), 25931 (2014)
17. G. Wang, J. Fang, X. Dong, Resolution of near-field microwave target detection and imaging by using flat LHM lens. IEEE Trans. Anten. Propag. **55**, 3534–3541 (2007)
18. H. Cano-Garcia, P. Kosmas, E. Kallos, Demonstration of enhancing the transmission of 60 GHz waves through biological tissue using thin metamaterial antireflection coatings, in *10th International Congress on Advanced Electromagnetic Materials in Microwaves and Optics, METAMATERIALS 2016* (Heraklion, Crete, Greece, 2016)
19. World Health Organization, *The World Health Report 2002: Reducing Risks, Promoting Healthy Life* (2002)
20. S.V. Semenov, D.R. Corfield, Microwave tomography for brain imaging: feasibility assessment for stroke detection. Int. J. Antennas Propag. **2008**, 1–8 (2008)
21. M. Hopfer, R. Planas, A. Hamidipour, T. Henriksson, S. Semenov, Electromagnetic tomography for detection, differentiation, and monitoring of brain stroke: a virtual data and human head phantom study. IEEE Trans. Antennas Propag. **59**(5), 86–97 (2017)
22. R. Scapaticci, L. Di Donato, I. Catapano, L. Crocco, A feasibility study on microwave imaging for brain stroke monitoring. Prog. Electromagn. Res., pp. 305–324 (2014)

23. S.V. Semenov, R. Planas, M. Hopfer, A. Hamidipour, A. Vasilenko, E. Stoegmann, E. Auff, Electromagnetic tomography for brain imaging: initial assessment for stroke detection, in *IEEE Biomedical Circuits and Systems Conference: Engineering for Healthy Minds and Able Bodies (BioCAS)* (Atlanta, Georgia, U.S.A., 2015)
24. P.M. Meaney, F. Shubitidze, M.W. Fanning, M. Kmiec, N.R. Epstein, K.D. Paulsen, Surface wave multipath signals in near-field microwave imaging. Int. J. Biomed. Imag. **2012**, 1–11 (2012)
25. C.J. Fox, P.M. Meaney, F. Shubitidze, L. Potwin, K.D. Paulsen, Characterization of an implicitly resistively-loaded monopole antenna in lossy liquid media. Int. J. Antennas Propag. **2008**, 1–9 (2008)
26. H. Cano-Garcia, P. Kosmas, E. Kallos, Enhancing electromagnetic transmission through biological tissues at millimeter waves using subwavelength metamaterial antireflection coatings, in *9th International Congress on Advanced Electromagnetic Materials in Microwaves and Optics, METAMATERIALS 2015* (Oxford, UK, 2015)
27. H.T. Chen, J. Zhou, J.F. O'Hara, F. Chen, A.K. Azad, A.J. Taylor, Antireflection coating using metamaterials and identification of its mechanism. Phys. Rev. Lett. **105**(7), 1–4 (2010)
28. H.T. Chen, J. Zhou, J. O'Hara, A. Taylor, A numerical investigation of metamaterial antireflection coatings. Terahertz Sci. Technol. **3**(2), 66–73 (2010)
29. S. Ahsan, Z. Guo, Z. Miao, I. Sotiriou, M. Koutsoupidou, E. Kallos, G. Palikaras, P. Kosmas, Multiple-frequency microwave tomography system. Sensors **18**, 1–13 (2018)
30. N. Ghavami, E. Razzicchia, O. Karadima, P. Lu, W. Guo, I. Sotiriou, E. Kallos, G. Palikaras, P. Kosmas, The use of metasurfaces to enhance microwave imaging: experimental validation for tomographic and radar-based algorithms. IEEE Open J. Antennas Propag. **3**, 89–100 (2022)
31. M. Lazebnik, E.L. Madsen, G.R. Frank, S.C. Hagness, Tissue-mimicking phantom materials for narrowband and ultrawideband microwave applications. Phys. Med. Biol. **50**(18), 4245–4258 (2005)
32. J.D. Shea, P. Kosmas, S.C. Hagness, B.D. Van Veen, Three-dimensional microwave imaging of realistic numerical breast phantoms via a multiple-frequency inverse scattering technique. Med. Phys. **37**(8), 4210–4226 (2010)
33. J.M. Bioucas-Dias, M.A. Figueiredo, A new TwIST: two-step iterative shrinkage/thresholding algorithms for image restoration. IEEE Trans. Image Process. **16**(12), 2992–3004 (2007)
34. Z. Miao, P. Kosmas, Microwave breast imaging based on an optimized two-step iterative shrinkage/thresholding method, in *9th European Conference on Antennas and Propagation (EuCAP)* (Lisbon, Portugal, 2015)
35. N. Ghavami, G. Tiberi, D.J. Edwards, A. Monorchio, UWB microwave imaging of objects with canonical shape. IEEE Trans. Antennas Propag. **60**(1), 231–239 (2012)
36. N. Ghavami, P.P. Smith, G. Tiberi, D. Edwards, I. Craddock, Non-iterative beamforming based on Huygens principle for multistatic ultrawide band radar: application to breast imaging. IET Microw. Antennas Propag. **9**(12), 1233–1240 (2015)
37. N. Ghavami, G. Tiberi, M. Ghavami, S. Dudley, M. Lane, Huygens principle based UWB microwave imaging method for skin cancer detection, in *International Symposium on Communication Systems, Networks and Digital Signal Processing (CSNDSP)* (Prague, Czech Republic, 2016)
38. E. Razzicchia, I. Sotiriou, H. Cano-Garcia, E. Kallos, G. Palikaras, P. Kosmas, Feasibility study of enhancing microwave brain imaging using metamaterials. Sensors **19**(24), 1–17 (2019)
39. S. Ahsan, M. Koutsoupidou, E. Razzicchia, I. Sotiriou, P. Kosmas, Advances towards the development of a brain microwave imaging scanner, in *13th European Conference on Antennas and Propagation (EuCAP)* (Krakow, Poland, 2019)
40. CST Microwave Studio, *Workflow and Solver Overview* (CST—Computer Simulation Technology AG, Darmstadt, Germany, 2016), p.2016
41. D. Andreuccetti, An internet resource for the calculation of the dielectric properties of body tissues in the frequency range 10 Hz-100 GHz. [Online]. Available: http://niremf.ifac.cnr.it/tissprop/

42. N. Ghavami, E. Razzicchia, O. Karadima, P. Lu, W. Guo, I. Sotiriou, E. Kallos, G. Palikaras, P. Kosmas, The use of metasurfaces to enhance microwave imaging: experimental validation for tomographic and radar-based algorithms. IEEE Open J. Antennas Propag December, 1–1 (2021)
43. X.F. Bosch, J. Ribes, M. Díaz, R. Cléries, Primary liver cancer: worldwide incidence and trends. Gastroenterology **127**(5 SUPPL. 1), 5–16 (2004)
44. B. Yang, B. Zhang, Y. Xu, W. Wang, Y. Shen, A. Zhang, Z. Xu, Prospective study of early detection for primary liver cancer. J. Cancer Res. Clin. Oncol. **123**(6), 357–360 (1997)
45. D. Anwanwan, S.K. Singh, S. Singh, V. Saikam, R. Singh, Challenges in liver cancer and possible treatment approaches. Biochim. Biophys. Acta Rev. Cancer **1873**(1), 188314 (2020)
46. C.J. Simon, D.E. Dupuy, W.W. Mayo-Smith, Microwave ablation: principles and applications. Radiographics **25**(25 SUPPL. 1), 69–83 (2005)
47. B. Quesson, J.A. De Zwart, C.T. Moonen, Magnetic resonance temperature imaging for guidance of thermotherapy. J. Magn. Reson. Imaging **12**(4), 525–533 (2000)
48. R. Scapaticci, V. Lopresto, R. Pinto, M. Cavagnaro, L. Crocco, Monitoring thermal ablation via microwave tomography: an Ex Vivo experimental assessment. Diagnostics **8**(4), 81 (2018)
49. R. Scapaticci, G.G. Bellizzi, M. Cavagnaro, V. Lopresto, L. Crocco, Exploiting microwave imaging methods for real-time monitoring of thermal ablation. Int. J. Antennas Propag., **2017** (2017)
50. M. Wang, L. Crocco, M. Cavagnaro, On the design of a microwave imaging system to monitor thermal ablation of liver tumors. IEEE J. Electromagn. RF Microw. Med. Biol. **5**(3), 231–237 (2021)
51. M. Wang, R. Scapaticci, M. Cavagnaro, L. Crocco, Towards a microwave imaging system for continuous monitoring of liver tumor ablation: design and in silico validation of an experimental setup. Diagnostics, 11(5) (2021)
52. K. Kanazawa, K. Noritake, Y. Takaishi, S. Kidera, Microwave imaging algorithm based on waveform reconstruction for microwave ablation treatment. IEEE Trans. Antennas Propag. **68**(7), 5613–5625 (2020)
53. G. Chen, J. Stang, M. Haynes, E. Leuthardt, M. Moghaddam, Real-time three-dimensional microwave monitoring of interstitial thermal therapy. IEEE Trans. Biomed. Eng. **65**(3), 528–538 (2018)
54. V. Portosi, A.M. Loconsole, F. Prudenzano, A split ring resonator-based metamaterial for microwave impedance matching with biological tissue. Appl. Sci. (Switzerland) **10**(19), 1–14 (2020)
55. Split Ring Resonator (SRR) Calculator. [Online]. Available: https://srrcalculator.blogspot.com/p/calculator.html

Numerical Modeling of Complex 3D Electromagnetic Scenarios for Medical Microwave Imaging

Tushar Singh, Branislav Ninkovic, Mladjen Stevanetic, Miodrag Tasic, Marija Nikolic Stevanovic, and Branko Kolundzija

Abstract The core of a microwave imaging (MWI) simulation environment is general-purpose three-dimensional (3D) electromagnetic (EM) software. For this purpose, we use the solver WIPL-D Pro. In addition to numerical simulation, the software, developed by the authors of this chapter, provides the following functionalities: (a) efficient editors for importing and processing STL and Voxel files describing realistic human phantoms, (b) triangular to quadrilateral mesher that converts STL files into WIPL-D input data files, and (c) library of human phantoms with a variety of tissues, and (d) library of antennas and antenna arrays templates. The software also includes the STL decimator that reduces the number of triangles with a controllable deviation from the original STL model. Antennas in the library include matching medium and are optimized for high-speed simulation using principles of "smart 3D EM modeling". The performance of the proposed 3D EM simulation environment is tested on the stroke-detection scenario comprising a helmet-like antenna array placed on a realistic human head model.

T. Singh (✉) · B. Ninkovic · M. Stevanetic · M. Tasic · B. Kolundzija
WIPL-D, Gandijeva 7, Belgrade, Serbia
e-mail: tushar.singh@wipl-d.com

B. Ninkovic
e-mail: branislav.ninkovic@wipl-d.com

M. Stevanetic
e-mail: mladjen.stevanetic@wipl-d.com

M. Tasic
e-mail: tasic@etf.bg.ac.rs

B. Kolundzija
e-mail: branko.kolundzija@wipl-d.com; kol@etf.bg.ac.rs

M. Tasic · M. Nikolic Stevanovic · B. Kolundzija
School of Electrical Engineering, University of Belgrade, Bulevar Kralja Aleksandra 73, Belgrade, Serbia
e-mail: mnikolic@etf.bg.ac.rs

F. Vipiana and L. Crocco (eds.), *Electromagnetic Imaging for a Novel Generation of Medical Devices*, Lecture Notes in Bioengineering,
https://doi.org/10.1007/978-3-031-28666-7_4

Keywords Antenna arrays · Convergence of numerical methods · Medical microwave imaging · Antenna probes · Numerical simulation · Phantoms

1 Introduction

In many practical applications, detecting and analyzing visually inaccessible objects/structures, such as distant aircraft or vessels, defects in buildings, or buried objects, is necessary. Most often of interest is the location of the unknown target, its shape, and its composition. The nature of these examined objects is quite diverse: in the building construction, it could be a crack in the wall or reinforcement; in the excavation work, it is usually an examination of an archaeological site or a hidden mine; in medicine, an altered tissue. The problems mentioned above belong to the class of inverse scattering problems and can be solved using microwave imaging (MWI) algorithms. MWI modalities for medical diagnostic purposes are known as medical microwave imaging (MMWI).

MMWI focuses on tissue changes inside human organs, which may indicate the occurrence of tumors, cysts, strokes, etc. Nowadays, clinicians rely on Magnetic Resonance Imaging (MRI), Computed Tomography (CT), X-Ray, and Positron Emission Tomography (PET) [1, 2]. CT and MRI are the gold standards as they provide high-resolution images with well-resolved tissues. However, they are inappropriate for bedside monitoring and unavailable in remote places [3]. Moreover, CT uses ionizing radiation.

Recent investigations have turned the attention to MMWI to overcome some of the shortcomings of previously mentioned techniques. MMWI is envisioned primarily as a complementary screening tool for medical applications [4–7]. The advantages of MMWI are its non-invasiveness, low cost, utilization of nonionizing radiation, and ease of portability. The essential modules of MMWI are (1) reliable numerical models of the human body (phantoms) for testing (with the accurate distribution of electrical parameters), (2) precise measuring equipment, (3) accurate reconstruction algorithms, and (4) efficient 3D EM solvers.

The core of an MMWI measurement system is an antenna array placed around the body part of interest to measure self and mutual *s*-parameters. MMWI algorithms use these data to reconstruct tissue inhomogeneity in the domain of interest (quantitative approach) or estimate the size and position of tissue changes compared with the previous patient condition (qualitative approach). In addition, most MMWI algorithms require computation of EM field inside or around the body part of interest or the equivalent numerical model (phantom) mimicking that part.

The chapter is organized as follows: Sect. 2 introduces a library of different human phantoms. Section 3 presents the library of frequency-dependent human tissues. Section 4 presents the antennas and antenna arrays suitable for creating MMWI measurement systems. Section 5 describes a complete numerical scenario composed of a numerical phantom and measurement system. Section 6 elaborates on different MWI algorithms, focusing on truncated singular value decomposition (TSVD) and linear sampling method (LSM). The focus of Sect. 7 is on the WIPL-D Pro, EM Solver, which serves as a base for MMWI upgrades. In Sect. 8, we elaborate on accuracy assessment and self-convergence strategies.

2 Anthropomorphic Phantom and Phantom Libraries

The creation of realistic human phantoms is a very demanding task. The most realistic human phantoms have been derived from MRI or CT scans [8] and cross-section images [8–12]. Only a few human models are available, mainly in voxel [8–10, 12], STL [8, 10], or some other CAD format [8, 13]. WIPL-D has developed a user interface that enables users to load and modify phantoms defined using STL and voxel files through STL editor and Voxel editor modules, respectively [14, 15].

2.1 Voxel Model

Online repositories comprise the databases of anatomically realistic numerical, mainly breast and head phantoms derived from MRI images. In this analysis, we will consider a breast phantom. To obtain the breast phantom, we utilize an online repository [15]. The breast data is defined using three different text files, named "breastinfo.txt," "mtype.txt," and "pval.txt," which all have to be loaded to define the breast phantom fully. An example of a breast phantom stored in voxel format is shown in Fig. 1a. The model is opened using the WIPL-D Voxel Editor, shown in Fig. 1b. Voxel Editor requires usage of cubic voxels of constant shape. However, dimensions of a cubic voxel can be different along different axes.

Apart from the antenna system, the voxel files are simulation ready. Due to the high complexity/resolution of voxel-based phantoms, mechanisms have been developed that enable simplifications while maintaining the accuracy of the obtained results.

2.2 STL Phantom

Currently, there are not many open-access realistic human phantoms defined in STL format. Through cooperation with NEVA (Bio) Electromagnetics [8], WIPL-D obtained access to anatomically accurate male and female models for electromagnetic simulations. Although the models differ in complexity (the male model is more complex regarding the number of tissues and the number of triangles per tissue), they both represent a complete and accurate human organism. The models consist of many STL files that can be combined into a single STL project. Each file represents a closed structure corresponding to some tissue or organ (muscles, bones, nerves, blood vessels, skin, fat, etc.). A glimpse of one such triangular-based human model

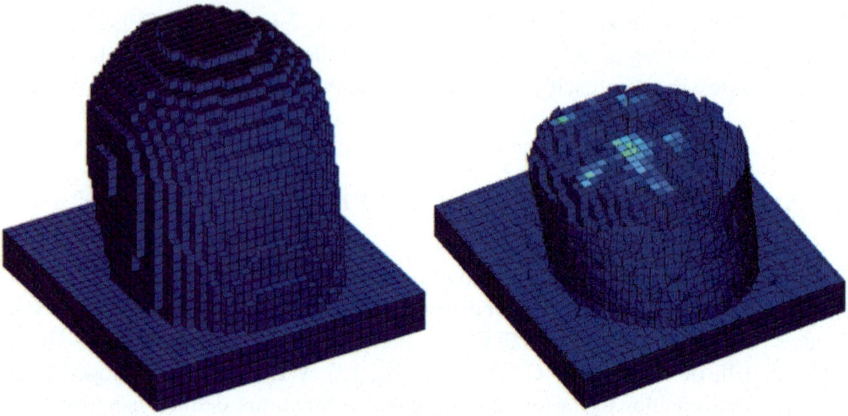

(a)

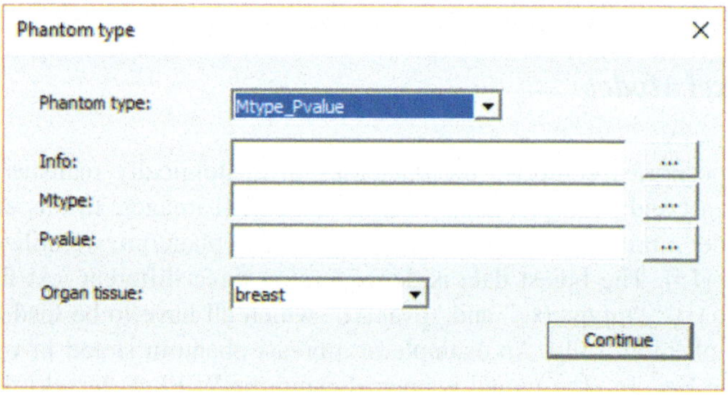

(b)

Fig. 1 **a** Voxel model of a breast. **b** WIPL-D voxel editor

(in this case, it is "Static VHP-Female model v2.2 of NEVA Electromagnetics") is shown in Fig. 2.

Different EM simulations intended for understanding, analysis, or testing medical devices can be performed on these human models or some of their parts. A stroke detection scenario, for example, requires only tissues located inside the head (e.g., skin, fat, mucous membrane, brain, and skull). Combining only the tissues significant for a particular scenario into an STL project simplifies the measurement setting, reduces the number of unknowns, and reduces simulation time.

Thus, when designing a measurement scenario using STL files, the first step should be choosing the phantom and significant tissues. Some selected tissues may entirely be in the zone of interest, while others may cover the whole body. Skin and fat layers are well-known examples of tissues stretching over the entire body. Such extended tissues require cropping and filling tools that are integral to the WIPL-D developed STL editor.

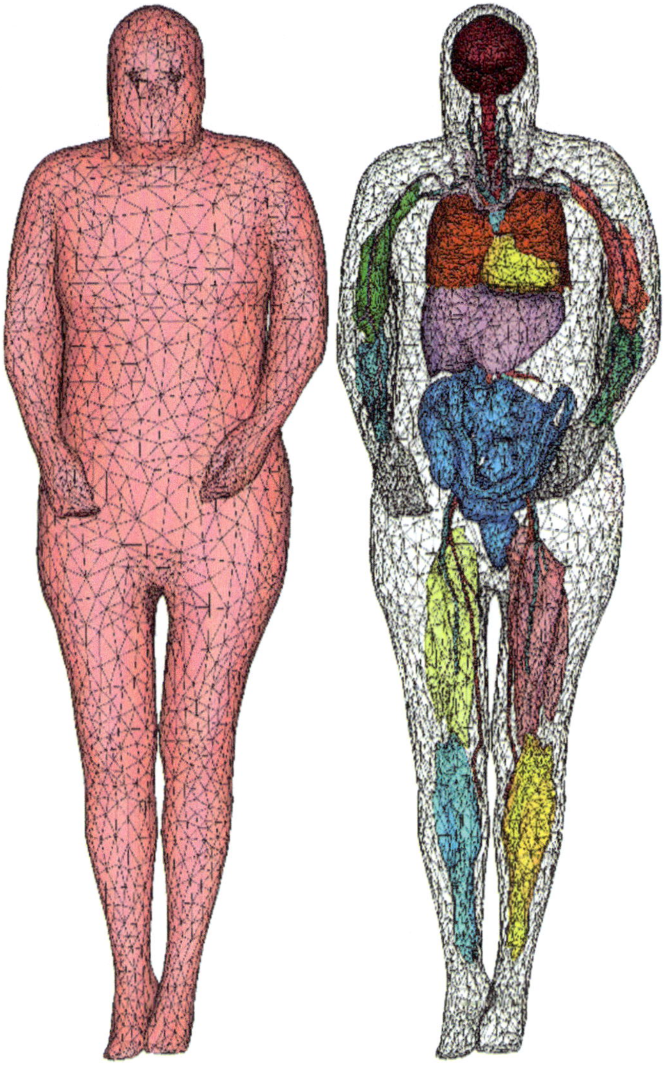

Fig. 2 NEVA EM developed an anatomical human model for electromagnetic simulations

3 Tissue Electrical Properties Library

Defining the electromagnetic properties of human tissues is essential for MMWI applications. In the WIPL-D user interface, setting the tissue properties depends on the phantom type. While voxel-based phantoms have the inbuilt definition of electrical parameters for each voxel, STL files require the user to define them for each tissue. The guidelines for defining the parameters, along with the homogenization process, are given in the following sub-section.

3.1 Electrical Parameters Definition for Voxel Model

The description of dielectric parameters of tissues in the utilized breast model is given in [15–17]. The repositories describe the total number of the voxel along each axis ("breastinfo.txt"), tissue types ("mtype.txt"), and lastly, the value of the dielectric constant for each voxel ("pval.txt"). The complex permittivity is defined using the single pole Cole–Cole model [18, 19], whose general form is

$$\varepsilon(\omega) = \varepsilon'(\omega) - \mathrm{j}\varepsilon''(\omega) = \varepsilon_\infty + \sum_{p=1}^{n} \frac{\Delta\varepsilon_p}{1 + \left(\mathrm{j}\omega\tau_p\right)^{(1-\alpha_p)}} + \frac{\sigma_\mathrm{s}}{\mathrm{j}\omega\varepsilon_0}, \tag{1}$$

where ω is the angular frequency, ε' and ε'' are the frequency-dependent real and negative imaginary part of the complex permittivity, respectively. Further, ε_∞ is the permittivity at the highest frequency, σ_s is the static ionic conductivity, p is the index of the Cole–Cole pole, n is the total number of poles, $\Delta\varepsilon_p$, τ_p, and α_p are the permittivity change, relaxation time, and exponent parameter ($0 \leq \alpha_p \leq 1$), respectively, of the p pole.

3.1.1 Homogenization in the Voxel Model

The voxel module allows the user to decrease the complexity of a phantom through homogenization [20]. In the process of homogenization, the voxels are grouped into the cubes of the size $n \times n \times n$, where n represents the number of voxels along each axis of the cube. Then, an effective dielectric constant is assigned to the selected group of voxels.

Let us consider a breast that is, together with the surrounding space, defined with a grid of 258 × 253 × 251 voxels, where each voxel is a cube of a side length of 0.5 mm. The electromagnetic properties of each voxel depend on the tissue to which it belongs. Also, within each tissue, the permittivity varies within a specific range. To illustrate this, we show in Fig. 3a a group of 8 × 8 × 8 voxels from the considered breast model. Figure 3b shows the full breast model obtained for $n = 8$.

Each color refers to one value of the complex permittivity magnitude, going from blue for the lowest value to red for the highest value. Direct simulation of this model would take an inadmissible long time. Through homogenization, a large heterogonous cube is replaced by a single homogeneous cube, which occupies the same space. The effective permittivity of the homogeneous cube is obtained using various mixing formulas initially developed for physical mixtures.

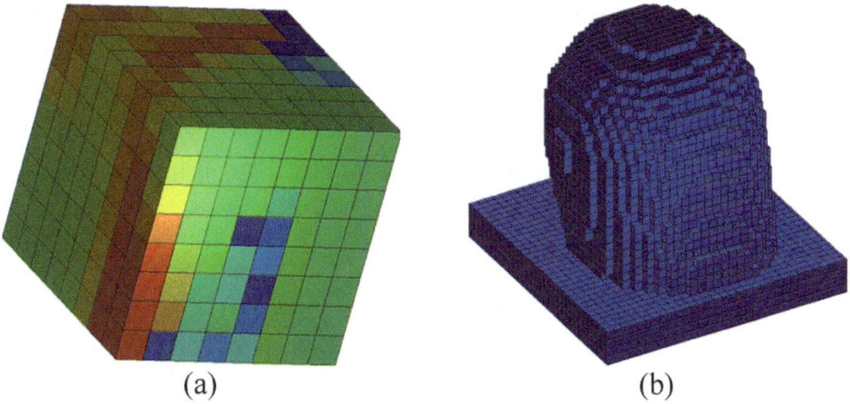

Fig. 3 Voxel-based breast phantom obtained for $n = 8$. **a** Heterogeneous single cube. **b** The whole breast model

By increasing n, the complexity of the homogenized model decreases. It is of interest to determine the maximum n, which produces the model that negligibly deviates from the initial model (in terms of the scattering parameters and near field). If this value is found, the requirements for computer resources are reduced, and simulation time is safely shortened.

Standard Averaging Method

The simplest way to determine the effective permittivity is provided by the standard averaging procedure [20]

$$\varepsilon_{\mathrm{avg}} = \frac{1}{N}\sum_{i=1}^{N} \varepsilon_i, \quad (2)$$

where N defines the total number of voxels in the cube, and ε_i is the permittivity of the ith voxel, $i = 1, \ldots, N$.

Lichtenecker Mixing Formula

Lichtenecker's logarithmic mixing formula was established for determining the effective permittivity of biological materials such as human blood [21].

$$\varepsilon_{\mathrm{avg}} = \exp\left(\frac{1}{N}\sum_{i=1}^{N} \ln(\varepsilon_i)\right), \quad (3)$$

where N and ε_i have the same meaning as above.

Looyenga Mixing Formula

Finally, the Looyenga equation is considered to be the most reliable formulation for predicting the effective permittivity of a mixture [22, 23]

$$\varepsilon_{\text{avg}} = \left(\frac{1}{N} \sum_{i=1}^{N} \varepsilon_i^{1/M} \right)^M, \tag{4}$$

where $M = 1, 2, 3, \ldots$ is a dimensionless integer factor.

In the context of human-tissue homogenization, the results obtained by the standard averaging procedure were significantly poorer than those obtained with the Lichtenecker mixing formula and, particularly, Looyenga mixing formula [24].

3.2 *Electrical Parameters Definition for STL Model*

STL files contain only geometrical description of tissues without their electromagnetic properties. Thus, after importing and (possibly) modifying STL files, the next important step is to assign permittivity data to appropriate tissues. By binding tissue descriptions to the appropriate groups of plates, the complete characterization of human organs is performed.

The most utilized database with electrical properties of human tissues is related to the work of Gabriel et al., described in [19, 25]. In [12], the Cole–Cole formula for frequency-dependent tissue properties has been implemented for several tissues and made available online to the broad research community. WIPL-D platform offers a library of frequency-dependent human tissue parameters based on [12, 19, 25]. Figure 4 shows the user interface of the developed library of materials. The library includes both human tissue materials of Cole–Cole type and different materials used in antenna and filter design and scattering problems.

Figure 5 illustrates a five-tissue head model, obtained using the STL editor and materials library. Different colors refer to different organs with their specific electromagnetic parameters.

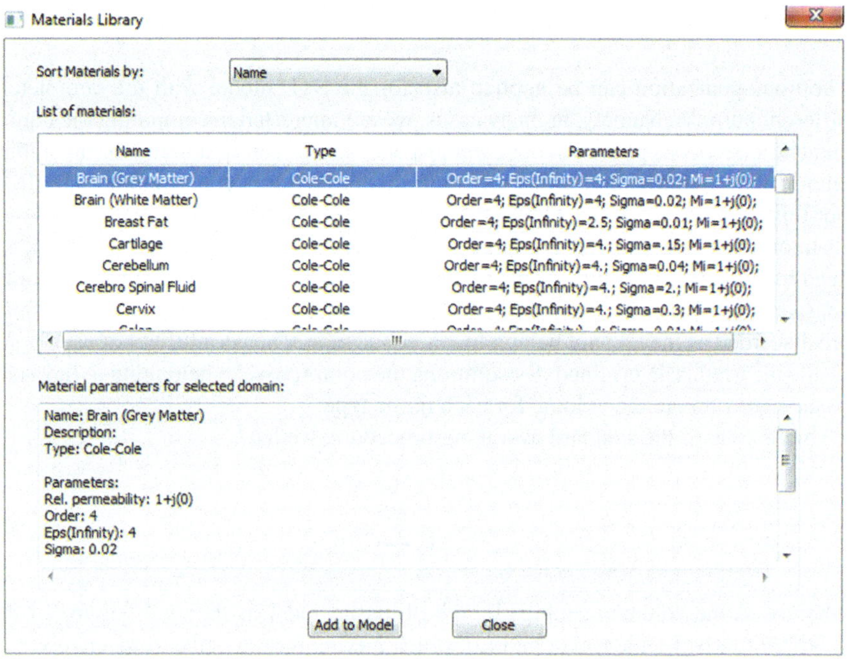

Fig. 4 Materials library is introduced to WIPL-D software, providing users with numerous frequency-dependent material types

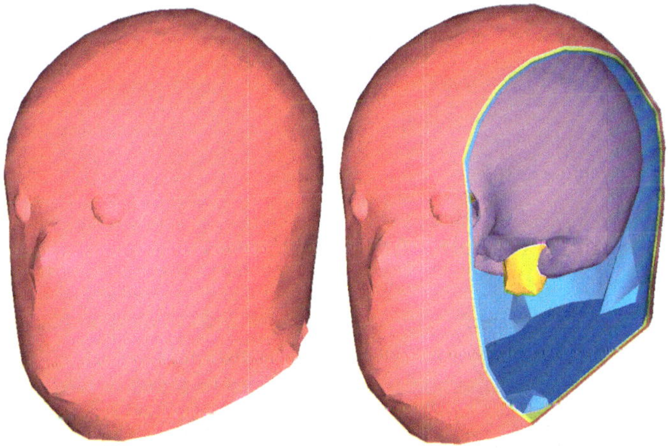

Fig. 5 Head phantom resulting from methodical approach with multiple crops and fillings, after all adequate domain parameters were assigned to tissues

3.2.1 Homogenization in STL Model

The homogenization can be applied also for the STL model with the completely different purpose. Namely, in many cases, we are interested in computing the equivalent homogeneous phantom. For example, such model can be utilized for the initialization of the iterative algorithms such as the distorted Born iterative method (DBIM). For this purpose, we utilize the averaging formulas (2)–(4).

In order to determine the complex permittivity of the homogeneous phantom, we need to compute the volumes occupied by each tissue. A uniform 3D grid of points is assumed inside and around the head for that purpose. The points outside the head are discarded as they do not belong to any tissue type. The permittivity of each point inside the head is determined. By summing the points (voxels) belonging to the same tissue, we compute the volume for each tissue type.

In the case of the standard averaging procedure, we have

$$\varepsilon_{\mathrm{avg}} = \frac{1}{V}\sum_{i=1}^{N} \varepsilon_i v_i, \tag{5}$$

where v_i is the volume occupied by the ith tissue and V is the total volume of the head. Similarly, Lichtenecker and Looyenga mixing formula give, respectively

$$\varepsilon_{\mathrm{avg}} = \exp\left(\frac{1}{V}\sum_{i=1}^{N} v_i \ln(\varepsilon_i)\right), \tag{6}$$

$$\varepsilon_{\mathrm{avg}} = \left(\frac{1}{V}\sum_{i=1}^{N} v_i \varepsilon_i^{1/M}\right)^{M}. \tag{7}$$

4 Measurement System (Antenna Probes)

MMWI measurement systems typically utilize an array of densely packed antennas in the vicinity of the body. The survey of antennas utilized for various medical purposes (including hyperthermia and ablation) can be found in [26]. This research resulted in a library of predefined antennas, which is included in the WIPL-D Pro software package. The antennas in the library are designed to work around 1 GHz. However, they are defined using frequency-dependent symbols. By changing the central frequency, the dimensions of the antenna change accordingly.

The antennas are fed by coaxial lines whose characteristic impedance is $Z_{\mathrm{c}} = 50\ \Omega$. They contain a spacer made of matching media. The spacer enhances the penetration of the radiated field into the human tissues/organs and improves the communications between the antennas in the array, which leads to better MWI results.

In addition, it is very important that the antenna impedance is properly matched to its feeder.

Antennas adapted to MMWI computation are:

- Microstrip Trapezoidal Patch (MTP) antenna
- Refined Microstrip Trapezoidal Patch (MTPR) antenna
- Pentagonal Arm Dipole (PAD) antenna
- Open waveguide (OWG) antenna
- Hexagon Dipole Brick (HDB) antenna.

These five antennas can be seen in Fig. 6.

All antennas introduced to the library are of original design. The inspiration for their creation was found in [27, 28].

In this chapter, we use the MTPR antenna to develop the MMWI measurement system. Figure 7 shows the s_{11} parameter of a single MTPR antenna in a vacuum, calculated in the frequency range 0.8–1.2 GHz. To improve the matching around 1 GHz, the antenna is optimized using the WIPL-D optimization tool. The s_{11} parameter of the optimized MTPR antenna is also shown in Fig. 7. The antenna is further optimized in the presence of the head phantom. As shown in Fig. 7, the s_{11} parameter is below -10 dB at all frequencies of interest. Thus, the antenna is ready for integrating in a MMWI measurement system.

5 Simplification of Phantoms and Complete Numerical Scenarios

The NEVA phantom models are defined using the triangular mesh stored in STL format. The original mesh is very dense with fine details in the areas of eyes, nostrils, ears, veins, white matter, arteries, etc. The EM simulation for dense mesh structures requires high computation cost, leading to a considerable simulation time. The WIPL-D simulation platform offers a newly developed module, STL editor, with decimation and meshing tool [29]. This tool simplifies the structure in a controlled manner and converts the triangular mesh into quadrilateral mesh. In this way, computational time is significantly reduced while maintaining high accuracy.

Once the electrical properties are assigned to respective tissue types, neighboring triangles from STL model are merged into quadrilaterals if they share material properties and if the angle between their unit surface normals is less than five degrees. Such small-angle tolerance does not allow all triangles to be merged. Usually, around 20–30% of the triangles are left unmerged, resulting in a mesh with quadrilateral and triangular plates. Further, each quadrilateral and triangle are subdivided into four and three quadrilaterals, respectively, to obtain pure quadrilateral mesh. In this way, the initial number of quadrilaterals is greater than the initial number of triangles a little more than two times.

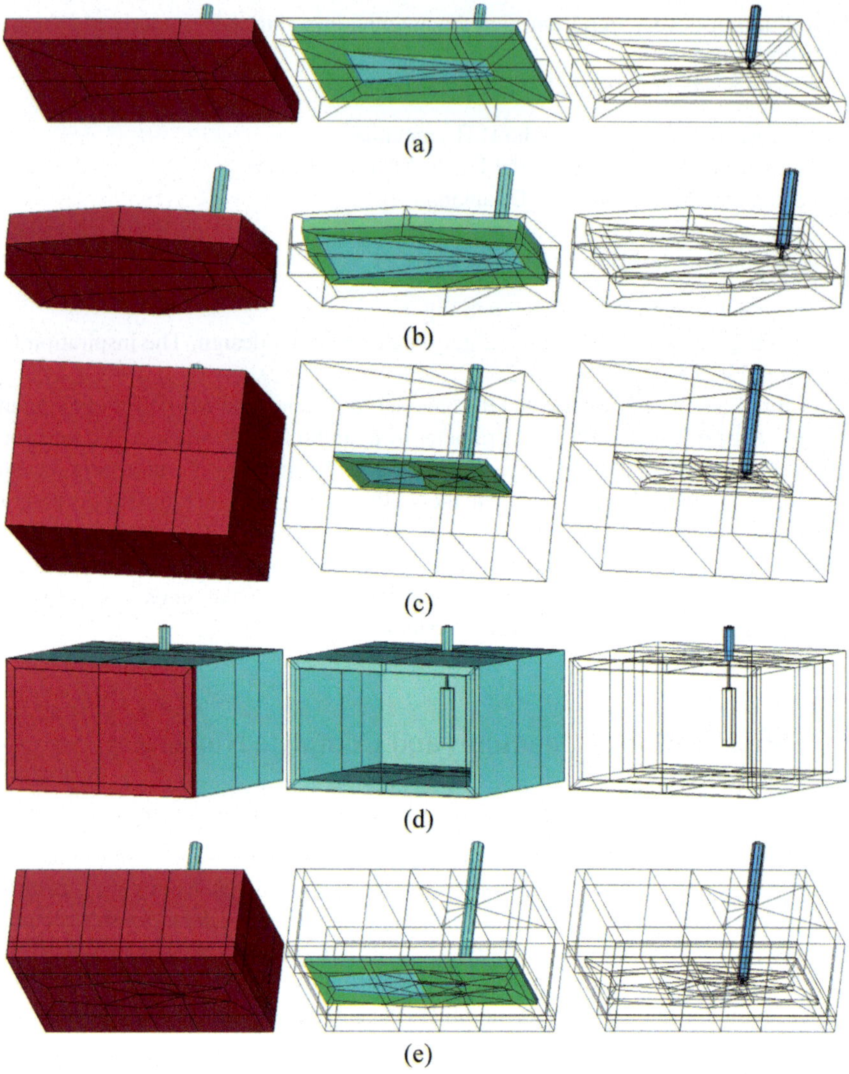

Fig. 6 Antenna Library for medical applications. **a** Microstrip trapezoidal patch antenna (MTP), **b** refined microstrip trapezoidal patch antenna (MTPR), **c** pentagonal arm dipole antenna (PAD), **d** open waveguide antenna (OWG), and **e** hexagon dipole brick antenna (HDB)

Also, the initial number of triangles can be reduced using the vertex collapse technique. The maximum Euclidian distance, σ, between the nodes from the initial mesh and newly created triangles has to be lower than the maximum allowed value, σ_{max}. To measure the discrepancy between the simplified model and the initial model, we compute the average distance for all nodes, which we call the average deviation,

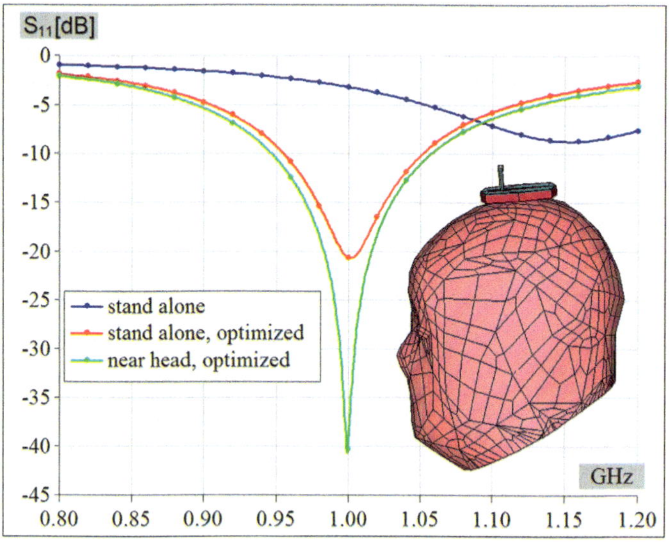

Fig. 7 The s_{11} parameters of MTPR antenna in a vacuum, optimized MTPR antenna, and MTPR antenna optimized in the presence of the head phantom

σ_{av}. (We note that σ defined here has nothing to do with material conductivity defined in (1).).

In meshing of mixed quadrilateral and triangular mesh, all new nodes are glued to the initial triangular mesh. In this way, the average deviation of the initial nodes from the final quadrilateral mesh is reduced compared with the deviation from the mixed mesh.

The application of the decimation and meshing tool is illustrated in Fig. 8. Figure 8a shows the original triangular mesh for the outer surface of the phantom, scull, and brain, whereas Fig. 8b–e show meshes obtained using different values of maximum deviation, σ_{max}.

In addition to the female head phantom, other phantoms are developed on the WIPL-D platform. The NEVA male model given in Fig. 9 distinguishes for the complexity of the mesh. Figure 10 shows a torso phantom with all essential internal organs.

5.1 *Complete Numerical Scenario with STL Phantom*

The antenna systems play a key role in efficient and reliable simulations of MMWI scenarios. The antennas need to be as compact as possible, as close to the body's surface as possible, and to cover the maximum surface around the body part of interest. In order to design such a system around the NEVA head, a semi-automatic procedure is developed. In the first step, we determine the position and semi-axes

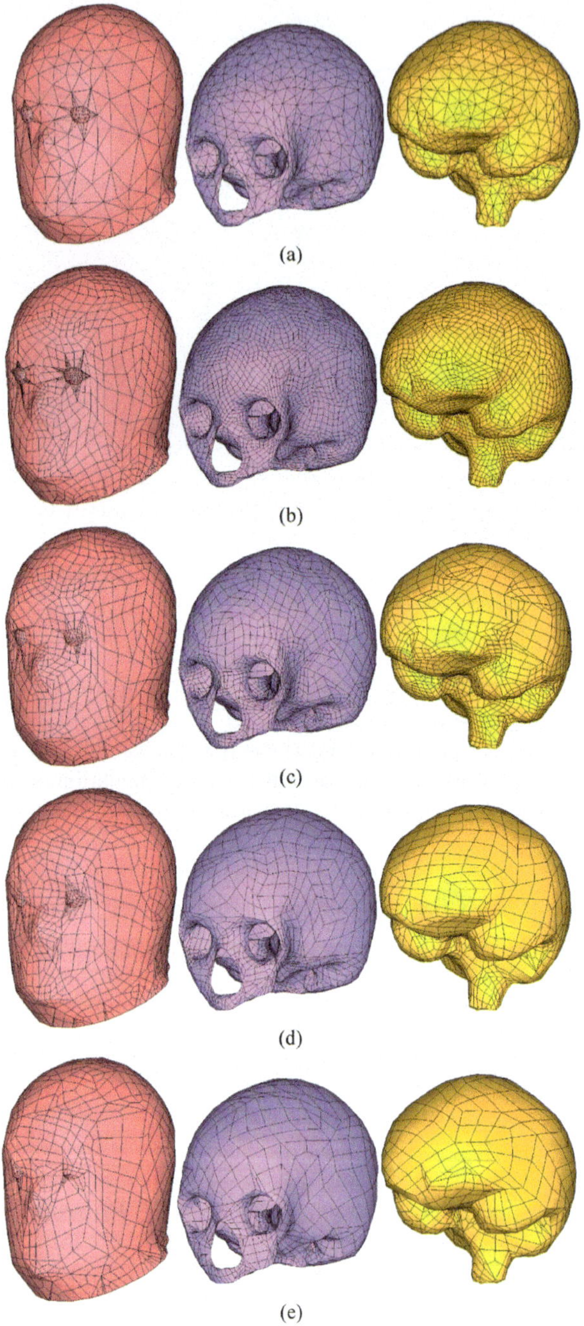

Fig. 8 Female head phantom (skin, skull, and brain) **a** Original STL model. **b** WIPL-D model without decimation. **c** WIPL-D model decimated with $\sigma_{max} = 1$ mm. **d** WIPL-D model decimated with $\sigma_{max} = 2$ mm. **e** WIPL-D model decimated with $\sigma_{max} = 3$ mm

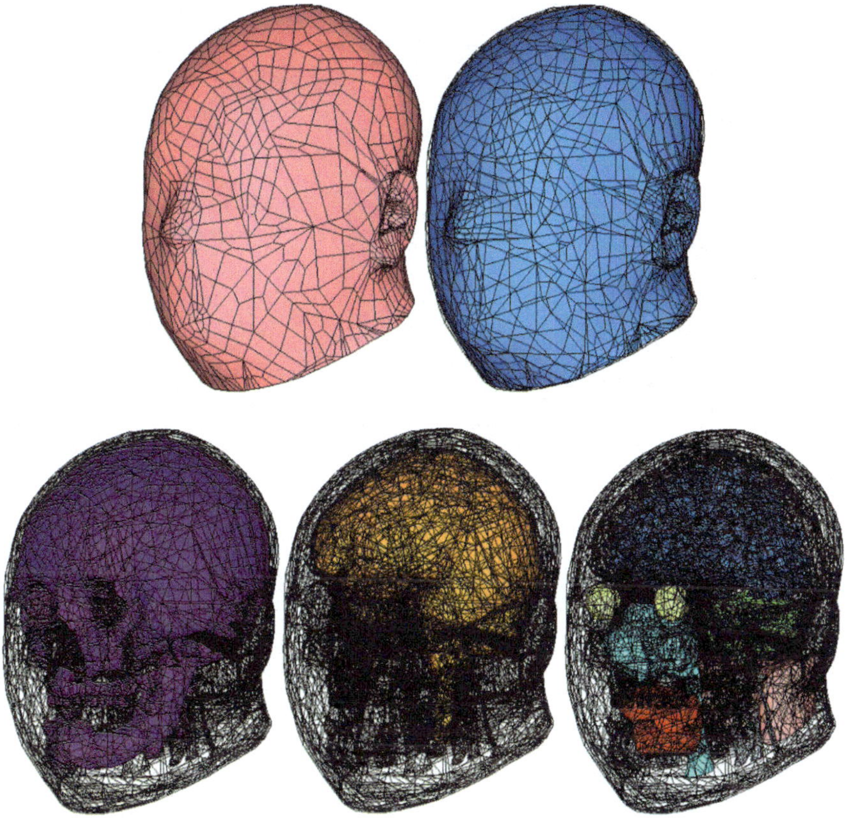

Fig. 9 Complex male head phantom

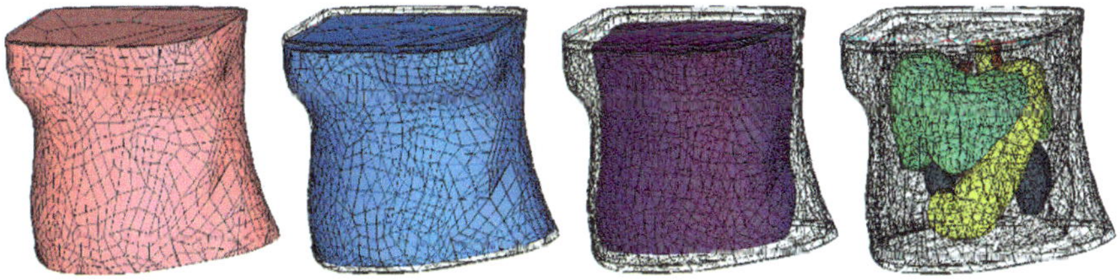

Fig. 10 Torso phantom

of a half ellipsoid that closely approximates the upper part of the head, where the stroke possibly occurs, as shown in Fig. 11. Other input data for the array design is a rectangular or octagonal footprint of the antenna that will be used as array element, as shown in Fig. 12. In the next step, octagon patterns are symmetrically placed around the half-ellipsoidal shape with the help of the semi-automatic algorithm (Fig. 13). The number of antennas is set to 21, as this is the maximum number of patterns

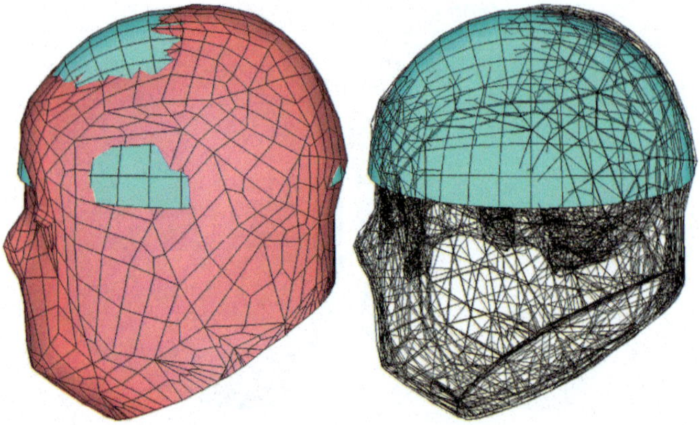

Fig. 11 Upper half of the head and the approximating half ellipsoid

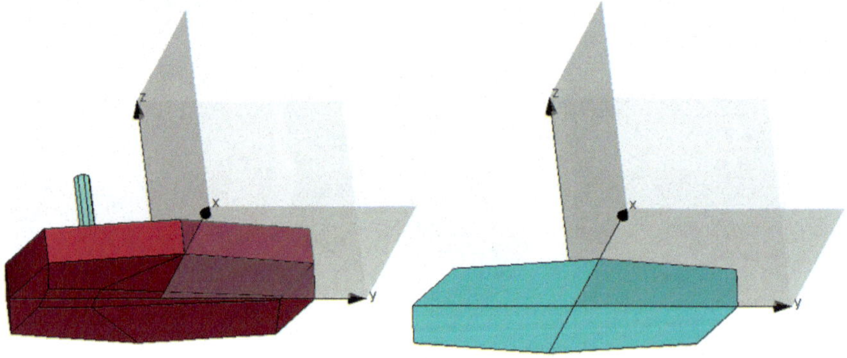

Fig. 12 MTPR antenna and the octagon pattern with the same footprint

covering the entire ellipsoidal shape's surface. Figure 14 shows the corresponding antenna system placed on the head.

5.2 Complete Numerical Scenario Using Voxel Phantom

In this section, we explain how do develop a complete scenario for voxel-based phantoms. Figure 15 shows two microstrip patch antennas surrounding the breast phantom simplified using the homogenization tool for $n = 16, 20, 24, 32$. The number of unknowns associated with each phantom is given in the figure's caption.

Figure 16 shows the results for s_{11} and s_{21}, obtained using different homogenization levels. By reducing the voxel size, the results converge, particularly for $n <$

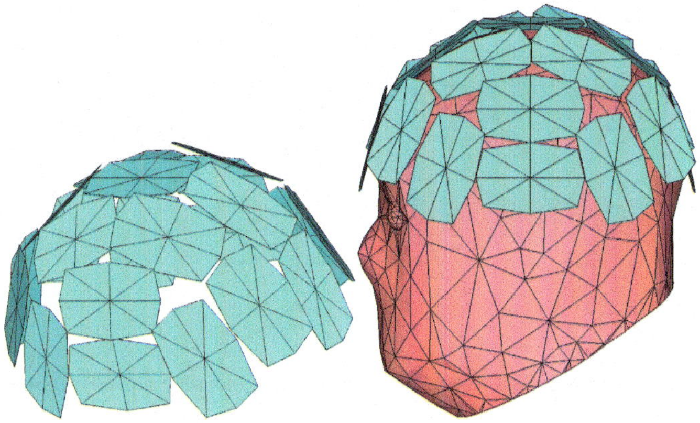

Fig. 13 Semi-automatic helmet around the NEVA head

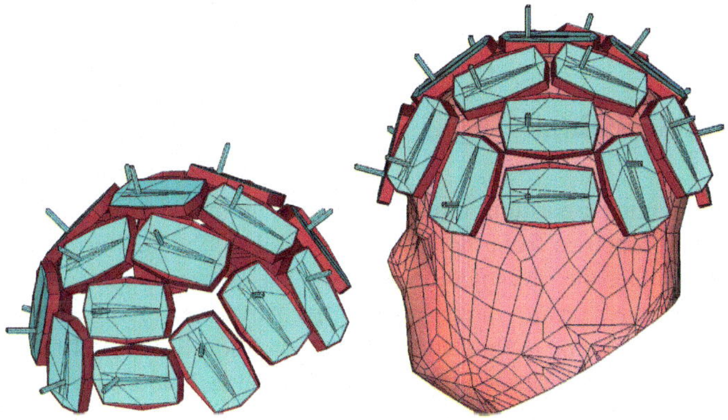

Fig. 14 MTPR Semi-automatic helmet placement around the NEVA head

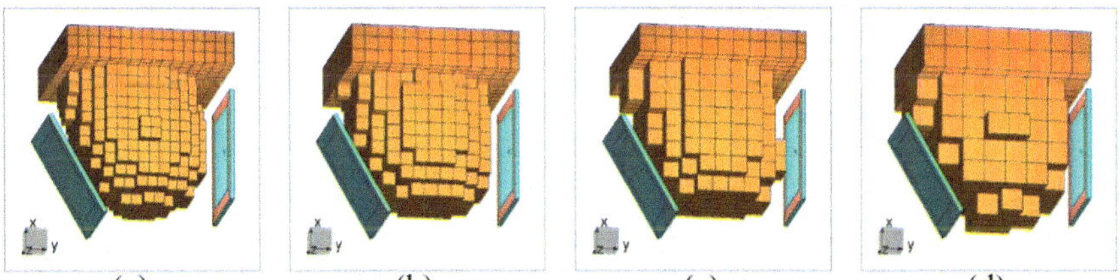

(a) (b) (c) (d)

Fig. 15 Homogenization of voxel-based breast model for **a** $n = 16$ (13,792 unknowns), **b** $n = 20$ (8406 unknowns), **c** $n = 24$ (5274 unknowns), and **d** $n = 32$ (2768 unknowns)

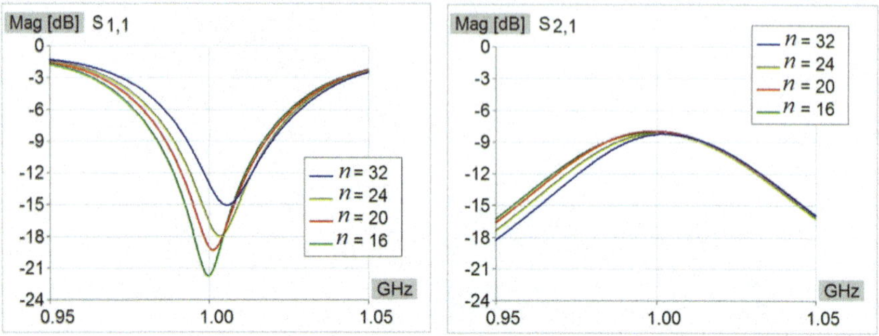

Fig. 16 Results for *s*-parameters when different homogenizations are applied to the breast phantom

32. Nevertheless, from the engineering point of view, the accuracy of all models is acceptable.

6 MWI Algorithms

Microwave imaging algorithms aim to restore the unknown target's shape, location, and electrical properties from the scattered field data obtained using measurement probes surrounding the investigation domain. The starting point for most of these methods is the inverse scattering equation (ISE). However, solving inverse scattering problems is difficult due to their ill-posedness and non-linearity. To mitigating the ill-posedness, it is necessary to utilize regularization techniques.

MWI algorithms are classified as qualitative and quantitative, based on the solution obtained using ISE. Qualitative algorithms determine the shape and position of the examined object without its composition. These algorithms are characterized by simple implementation and short execution time. Examples of qualitative algorithms are the truncated singular value decomposition (TSVD) [30] and the linear sampling method (LSM) [31]. Quantitative techniques attempt to reconstruct the complex permittivity in the whole domain of interest. They are time-consuming, iterative, and prone to false solutions. Examples are the inexact Newton methods [32] and the distorted Born iterative method (DBIM) [33, 34].

Here, we utilize TSVD and LSM algorithms to reconstruct the shape and location of the target. The 3D EM simulation is performed on the NEVA head and 21 antenna helmet scenario. The obtained data are the *s*-parameter matrix estimated for two scenarios, one with a spherical stroke and one without a stroke. In order to obtain the electric field inside the observation domain, the near-field calculation is performed. In the case of a weak scatterer, the incident field is considered the same in both scenarios (Born approximation).

The results are obtained in the form of a pixelated image, as shown in Fig. 17. The pixels are normalized to unity for several vertical cuts parallel to the *yOz* plane,

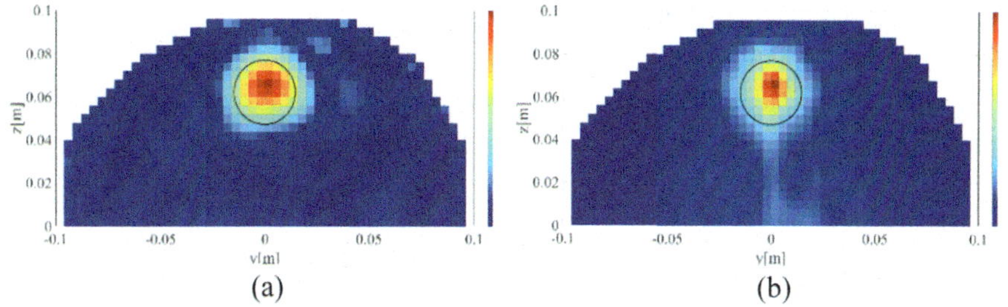

Fig. 17 Microwave imaging results in the plane containing the center of stroke (0, 0, 62 mm) obtained using **a** TSVD and **b** LSM algorithms

including the cut containing the center of the stroke. The actual shape of the spherical stroke is represented by a black circle. In both cases, the shape of the stroke is well estimated.

7 EM Solver

WIPL-D Pro, a general-purpose 3D EM solver, is a crucial element of the proposed MMWI system design. It is based on the Method of Moment applied to the Surface Integral Equation (MoM/SIE). Solvers based on MoM/SIE determine unknown equivalent currents over boundary surface building elements (e.g., triangles and quadrilaterals).

Since boundary surfaces made of triangles are utilized in STL models of human tissues, a preferred method for the 3D EM modeling is MoM/SIE with the triangular mesh. However, the quadrilateral mesh is advantageous in terms of efficiency over the triangular mesh [35]. Therefore, we transform the triangular mesh into a quadrilateral one, allowing us to use the WIPL-D Pro 3D EM simulation platform.

As for voxel files, it is worth noting that methods such as MoM/VIE (MoM applied to Volume Integral Equation), Finite Element Method (FEM), Finite Integration Technique (FIT), and Finite Difference Time Domain (FDTD), all determine unknown fields inside the volume-building elements (e.g., tetrahedrons and hexahedrons). Thus, they can utilize voxel-based data in its original form. For MoM/SIE based solvers, such as WIPL-D Pro, preprocessing is necessary. In the case of the WIPL-D Pro solver, homogenization is utilized, as explained previously. Such preprocessing of voxel data ensures high accuracy and efficiency in working with voxel-based models.

Here are some of the principal features of the WIPL-D Pro solver:

(a) Unknown (electric and magnetic) currents over domain boundaries are determined by solving EFIE (Electric Field Integral Equation) and PMCHWT (Poggio, Miller, Chang, Harrington, Wu, Tsai) boundary integral equations.

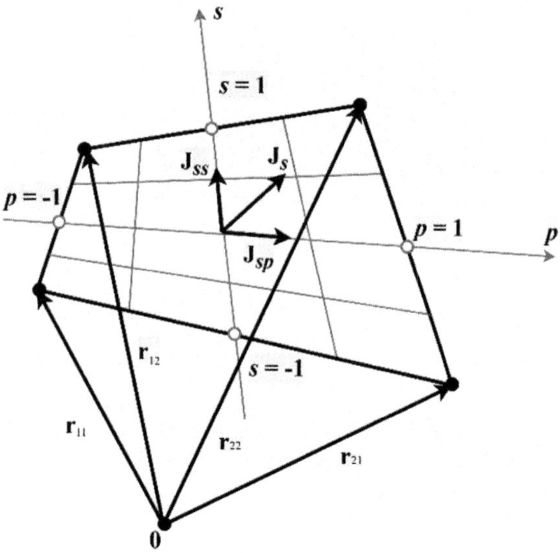

Fig. 18 A sketch of a bilinear surface and decomposition of surface current in the local parametric *ps*-coordinate system

(b) Geometry of boundary surfaces is modelled using quadrilateral patches (bilinear surfaces). A bilinear surface is uniquely defined by four arbitrary vertices, as presented in Fig. 18. The equation of a quadrilateral, using the local *s*-coordinate system, can be written in two alternative forms:

$$\mathbf{r}(p, s) = \frac{1}{4}\sum_{i=1}^{2}\sum_{j=1}^{2}\mathbf{r}_{ij}\left[1 + (-1)^{i} p\right]\left[1 + (-1)^{j} s\right], -1 \le p, s < 1, \quad \text{(8a)}$$

$$\mathbf{r}(p, s) = \mathbf{r}_{\mathrm{c}} + \mathbf{r}_{p} p + \mathbf{r}_{s} s + \mathbf{r}_{ps} ps, -1 \le p, s < 1, \quad \text{(8b)}$$

where $\mathbf{r}_{ij}$, $i, j = 1{,}2$, are the position vectors of its vertices. The vectors, $\mathbf{r}_p$, $\mathbf{r}_s$, and $\mathbf{r}_{ps}$ determine the center of the quadrilateral, semi *p*-axis, semi *s*-axis, and a deviation of the quadrilateral from rhombus, respectively.

(c) The current distribution over a bilinear surface is decomposed into two components in the local *ps*-coordinate system, as shown in Fig. 18. The higher-order approximation of the *s*-component can be written as

$$\mathbf{J}_{ss}(p, s) = \frac{\mathbf{a}_s}{\left|\mathbf{a}_p \times \mathbf{a}_s\right|}\sum_{j=0}^{n_p-1}\sum_{k=0}^{n_s} a_{jk} P_j(p) S_k(s), \mathbf{a}_s = \frac{\partial \mathbf{r}}{\partial s}, \mathbf{a}_p = \frac{\partial \mathbf{r}}{\partial p} \quad \text{(9)}$$

where n_{p} and n_s are the expansion orders in the *p* and *s*-coordinates, respectively, a_{jk} are the unknown coefficients, and $P_j(p)$ and $S_k(s)$ are the expansion functions in the directions perpendicular and parallel to the current flow, respectively. The same formula is valid for approximating the *p*-component of the current with indices p and s interchanged. The lowest expansion orders along the p and s-coordinates are $n_p = 1$ and $n_s = 1$. These settings reduce the approximation to those of rooftop bases. Maximum applicable expansion orders for "classic" higher-order bases, which provide a well-conditioned matrix and stable solution, are limited to $n_{\max} = 8$. This limit is extended for one order of magnitude (tested up to $n_{\max} = 64$) by maximal orthogonalization of P and S functions, which results in max-ortho basis functions [36].

The classic form of P and S expansion functions, which should be coupled at interconnections of bilinear surfaces in order to satisfy the continuity of normal current components, is given as

$$P_j(p) = p^j,\ S_k(s) = \begin{cases} 1 - s & k = 0 \\ 1 + s & k = 1 \\ s^k - 1 & k = 2, 4, 6... \\ s^k - s & k = 3, 5, 7... \end{cases} \tag{10a, b}$$

(d) Galerkin test procedure is applied to convert a system of integral equations into a determined system of linear equations. The singularity extraction and the singularity cancellation techniques provide a highly accurate MoM system matrix (near to machine precision) [37]. Conditionality of the matrix is improved by matrix equilibration so that for max-ortho bases, the matrix condition number practically does not depend on expansion order [38]. The system is solved by the direct method (LU decomposition) so that accuracy of results depends only on the basis functions to compute unknown currents.

The choice of expansion orders affects the accuracy of the analysis. In general, expansion orders related to the electrical size of patches achieve acceptable accuracy. For each patch, we determine the maximum physical lengths in p and s-coordinates, respectively, as:

$$l_{p\max} = \max(\ |\mathbf{r}_{21} - \mathbf{r}_{11}|,\ |\mathbf{r}_{22} - \mathbf{r}_{12}|\ ), \tag{11a}$$

$$l_{s\max} = \max(\ |\mathbf{r}_{12} - \mathbf{r}_{11}|,\ |\mathbf{r}_{22} - \mathbf{r}_{21}|\ ). \tag{11b}$$

In order to provide control over the expansion orders, along with the operating frequency (f_0), a parameter called reference frequency (f_{ref}) is introduced. The reference frequency determines the expansion orders along p and s-axes in the following way:

$$n_p = 1 + \left\lfloor \left| \frac{3l_{p\max}}{\lambda_{\text{ref}}} - \frac{1}{6} \right| \right\rfloor,\ n_s = 1 + \left\lfloor \left| \frac{3l_{s\max}}{\lambda_{\text{ref}}} - \frac{1}{6} \right| \right\rfloor, \tag{12a, b}$$

where $\lfloor \cdot \rfloor$ represents the floor operation, and λ_{ref} is the wavelength at the reference frequency f_{ref}. From (12), it is clear that if $l_{p\max}$, $l_{s\max} < \lambda_{\text{ref}}/6$, the expansion orders are set to $n_p = n_s = 1$, which is equivalent to rooftop bases. For each further increase of $l_{p\,\max}$, $l_{s\,\max}$ for $\lambda_{\text{ref}}/3$, the expansion orders are increased by one. In the solver, the default value for f_{ref} is f_0.

For classical basis functions, the maximum patch size (MPS) parameter, whose expansion order is limited to $n_{\max} = 8$, is MPS $= 2.5\lambda_{\text{ref}}$. If the electrical size of the patch is larger than the given MPS value, for at least one of the two coordinates, it is automatically subdivided into a minimum number of patches with the appropriate electrical size. By default, MPS is set to $2\lambda_{\text{ref}}$, enabling the solver to use classic higher-order bases to their full potential. In order to use only rooftop bases ($n_p = n_s = 1$), MPS should be set to $\lambda_{\text{ref}}/6$. In the general case, increasing the number of unknown coefficients improves the accuracy of the analysis/simulation. For higher-order expansions, this is achieved by increasing the f_{ref}, while for rooftop basis functions this is done by decreasing the MPS, leading to additional divisions.

8 Accuracy Assessment and Validation: Self Convergence Strategies

We proposed several self-convergence strategies to verify the accuracy of EM simulations for MWI scenarios. The quality of three data types is assessed. Primary, or regular, EM data are the electric field inside the head (**E**) and *s*-parameters (s_{ij}, $i, j = 1, 2, \ldots, N$). Secondary, or differential EM data comprise the differential electric field ($\Delta\mathbf{E}$) and differential *s*-parameters ($\Delta s_{ij}\; i, j = 1, 2, \ldots, N$). Differential data is obtained as the difference of the primary EM data computed when the stroke is present and when it is absent. Tertiary data are associated with inverse scattering operator, as we will detail later.

We measure the overall accuracy of primary and secondary data using the relative mean absolute (RMA) error:

$$E_{\text{RMA}} = \frac{\sum_{k=1}^{M}\left|\mathbf{E}_k - \mathbf{E}_k^{\text{ref}}\right|}{\sum_{k=1}^{M}\left|\mathbf{E}_k^{\text{ref}}\right|},\; s_{\text{RMA}} = \frac{\sum_{i=1}^{N}\sum_{j=1}^{N}\left|s_{ij} - s_{ij}^{\text{ref}}\right|}{\sum_{i=1}^{N}\sum_{j=1}^{N}\left|s_{ij}^{\text{ref}}\right|}, \tag{13a, b}$$

where $\mathbf{E}_k$ is the electric field in the kth of M sample point inside the head and "ref" denotes the reference values. In formulas for RMA error of differential quantities, $\Delta\mathbf{E}/\Delta s_{ij}$ are used instead of $\mathbf{E}/s_{ij}$.

We define the (residual) r_{ij}-parameters, which occur due to numerical treatment and/or approximations applied while computing the inverse scattering equation, as

$$r_{ij} = \Delta s_{ij} + \frac{\mathrm{j}\omega}{2}\int_{v} \Delta\varepsilon(\mathbf{r}')S_{ij}(\mathbf{r}')\,\mathrm{d}v i, j = 1, \ldots, N \tag{14a}$$

$$S_{ij}(\mathbf{r}') = \mathbf{E}_{i1}(\mathbf{r}') \cdot \mathbf{E}_{j2}(\mathbf{r}'), i, j = 1, \ldots, N, \tag{14b}$$

where ω is the angular frequency, $\Delta\varepsilon(\mathbf{r}')$ is the difference between the permittivity of the stroke and the head, $S_{ij}(\mathbf{r}')$ is the exact scattering operator, $\mathbf{E}_{i1}$ ($\mathbf{E}_{j2}$) is the electric field vector inside the head without (with) stroke, obtained using a unit amplitude excitation at the *ith*(*jth*) port when other $N-1$ ports are matched, and v is the volume occupied by the stroke.

The total field $\mathbf{E}_{j2}(\mathbf{r}')$ in the stroke is unknown, making the inverse problem nonlinear. When observed target, in our case the blood clot, is a weak scatterer (in most realistic examples, this is the case), it is possible to apply the Born approximation by replacing $\mathbf{E}_{j2}(\mathbf{r}')$ with $\mathbf{E}_{j1}(\mathbf{r}')$ in (14b). Thus, we define the approximate scattering operator as

$$S_{ij}(\mathbf{r}') \approx \mathbf{E}_{i1}(\mathbf{r}') \cdot \mathbf{E}_{j1}(\mathbf{r}'), i, j = 1, ..., N. \tag{15}$$

The exact and approximate r-parameters and scattering operators represent the tertiary EM data utilized in MMWI applications. Namely, the goal of MMWI is to find such $\Delta\varepsilon(\mathbf{r}')$ that minimizes Δr_{ij}. Thus, to estimate the quality of the obtained 3D EM data, we define the RMA residual as

$$R_{\mathrm{RMA}} = \frac{\sum_{i=1}^{N}\sum_{j=1}^{N}|r_{ij}|}{\sum_{i=1}^{N}\sum_{j=1}^{N}\Delta s_{ij}}, \tag{16}$$

The integration in (14a) is performed numerically, using the exact values of $\Delta\varepsilon(\mathbf{r}')$, which are different from zero inside the stroke.

We have developed a self-convergence study as an iterative procedure in which the number of unknown coefficients gradually increases. The procedure continues until the difference between the results of two consecutive iterations becomes smaller than a previously specified threshold. The three refinement techniques assure that the number of unknown coefficients increases:

(1) *h*-refinement, achieved by increasing the number of patches while keeping the lowest expansion orders for all patches
(2) *p*-refinement, achieved by increasing the expansion orders over patches while keeping the mesh unchanged
(3) *hp*-refinement, a combination of *h* and *p*-refinements.

As defined in (12a, b), the number of unknowns depends on the reference frequency (f_{ref}), whose default value is equal to the operating frequency (f_0). However, if we want to increase the number of unknowns further, we should increase the reference frequency (f_{ref}) or decrease the maximum patch size (MPS).

To demonstrate the effectiveness of the proposed self-convergence test, we utilized 5-tissue head phantom (with and without the stroke), obtained with $\sigma_{\max} = 3$ mm. For the refinement technique, we chose *hp*-refinement. Hence, MPS is set to the default value of $2\lambda_{\mathrm{ref}}$, where λ_{ref} is the wavelength at the reference frequency. The number of unknowns for full MMWI scenario, shown in Fig. 14, obtained for chosen settings is about 50,000.

The convergence tests can be accelerated having in mind that:

(a) Step-by-step refinement of any distinctive part (phantom, antenna array), while keeping the default parameters for the rest of the scenario, gives a clear insight into the influence of that part on the overall convergence rate. (For the stroke, we utilize a very refined model since it requires small number of unknowns.)
(b) Determining the impact of the phantom refinement on the overall convergence rate for **E** and Δ**E** can be facilitated with the plane wave excitation instead of the full antenna system.
(c) The impact of the antenna system refinement on the overall convergence rate for *s*-parameters can be observed by reducing the antenna array to a single antenna. For example, the results for *s*-parameters of a single MTPR antenna, shown in Fig. 7, are obtained using *hp*-refinement and $f_{\text{ref}} = 2f_0$. It is found that these results practically do not change with further increase of reference frequency. Hence, such a local setting is used for all antenna elements in a complete scenario simulation.

8.1 Self-Convergence Testing for Electric Field

Convergence rates of **E** and Δ**E** are estimated using head phantoms excited broadside by the vertically polarized plane wave of the unit electric field. Figure 19 shows RMA errors of **E** and Δ**E**, obtained by *hp*-refinement, versus the number of unknowns. Convergence is completed once RMA errors (E_{RMA} and ΔE_{RMA}) are minimized to 0.03%.

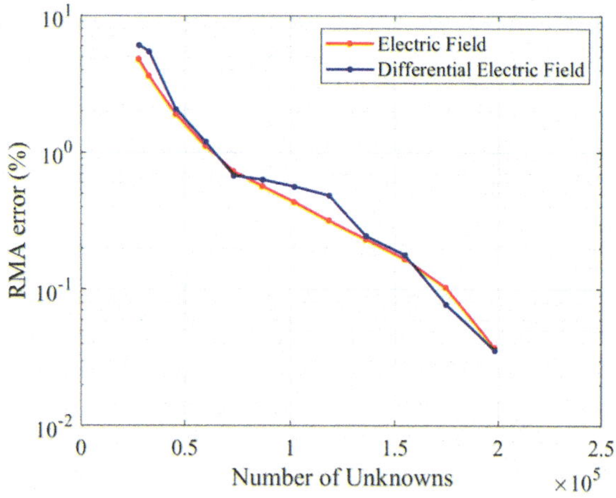

Fig. 19 RMA errors of electric field and differential electric field versus the number of unknowns used for phantom

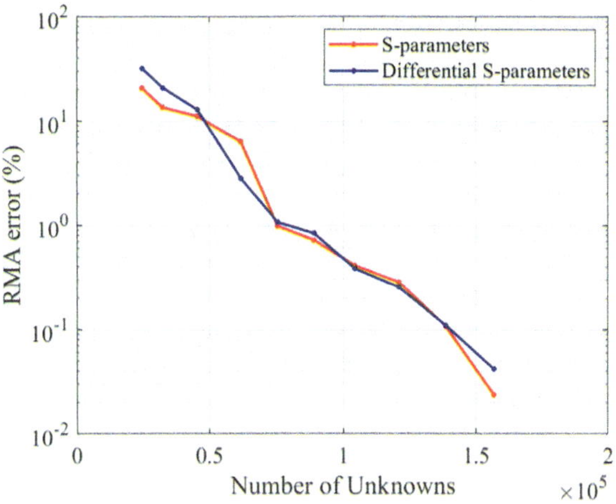

Fig. 20 RMA errors of *s*-parameters and differential *s*-parameters versus number of unknowns used for phantom

8.2 Self-Convergence Testing for Scattering Parameters

The convergence rates of *s* and differential *s*-parameters are estimated using head phantoms excited by the 21 MTPR antenna helmet. Corresponding RMA errors obtained through *hp*-refinement versus the number of unknowns are shown in Fig. 20. Local settings are applied to the antennas, so the refinement is only applied to the phantom. Thus, the number of unknowns is estimated only for the phantom. RMA errors (s_{RMA} and Δs_{RMA}) are decreased to 0.03% in a controlled manner. Figures 19 and 20 show that RMA errors of regular quantities (**E** and *s*) and RMA errors of differential quantities (Δ**E** and Δs) are almost the same in magnitude. In contrast, the differential quantities are at least one order of magnitude smaller than the regular ones, as is shown in Fig. 21. Figure 21a shows the magnitude of the electric field inside the head with stroke in the symmetry plane, ranging from 0 to 500 mV/m. Figure 21b shows the magnitude of the differential electric field in the same plane, ranging from 0 to 50 mV/m.

The self-convergence test of the whole scenario can also be performed considering the output of MMWI algorithms. In this case, the iterative procedure finishes when, visually, the images obtained in two consecutive iterations are practically the same.

8.3 Self-Convergence Testing for Residual Parameters

RMA residuals of the exact and approximate ISE for the full-wave scenario versus the number of unknowns are shown in Fig. 22. The simulations were performed using default settings for the antennas and the head ($\sigma_{\max} = 3$ mm). In the *hp*-refinement,

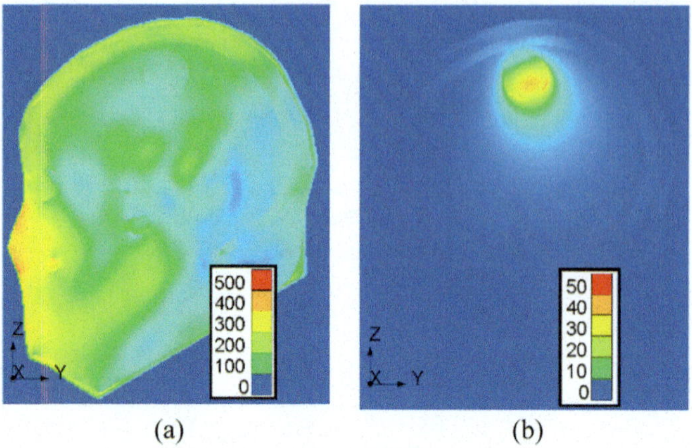

Fig. 21 **a** Magnitude of electric field and **b** differential electric field in the symmetry plane of the head excited broadside by a plane wave of unit electric field

the reference frequency was increased from $f_{\text{ref}} = f_0$ to $f_{\text{ref}} = 9f_0$, in steps of $\Delta f = 0.5f_0$.

Figure 22 shows that with the increase in the number of unknowns, both exact and approximate RMA residuals practically remain constant. RMA residual is constantly below 1% for the exact ISE, and somewhat below 20% for the approximate ISE. This 1% of the exact RMA residual appears due to inadequate EM modeling, while the difference of 19% between two residuals is due to the Born approximation applied in the approximate ISE.

Images obtained using standard TSVD and LSM methods [30, 31], shown in Fig. 17, do not change with the change of reference frequency setting. This can be explained by the fact that in all these cases, the RMA residual computed with exact

Fig. 22 RMA residual versus total number of unknowns

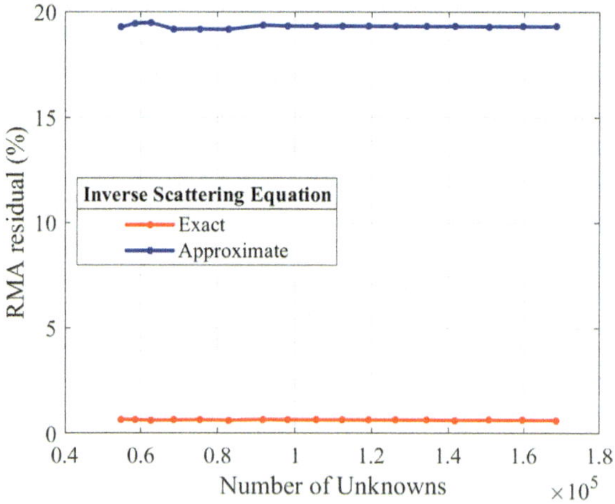

ISE is much smaller than the one computed with approximate ISE, i.e.,

$$R_{\text{RMA}}^{\text{exact}} \ll R_{\text{RMA}}^{\text{approx}} \tag{17}$$

Generally, we found that if inequality (17) is satisfied, tertiary EM data are validated for approximate ISE based MMWI.

Lastly, we should consider the error introduced with the decimation procedure. As stated above, all convergence procedures are performed on a phantom obtained with meshing that included maximum deviation of $\sigma_{\text{max}} = 3$ mm. Here, we consider various levels of decimation, ranging from $\sigma_{\text{max}} = 0$ mm to $\sigma_{\text{max}} = 3$ mm, with the step of $\Delta\sigma_{\text{max}} = 0.5$ mm. The reference frequency was increased from $f_{\text{ref}} = f_0$ to $f_{\text{ref}} = 9f_0$, in steps of $\Delta f = f_0$. The results obtained with $\sigma_{\text{max}} = 0$ mm and $f_{\text{ref}} = 10f_0$ are adopted as the reference results. Figure 23 shows the corresponding RMA errors versus the number of unknowns for the phantom. Due to distortions introduced by decimation, after reaching a certain threshold value, RMA error cannot be decreased by further refinement. With the increase of the maximum allowed deviation, the threshold also increases. Figure 24 shows the threshold values versus the number of unknowns for $0 \leq \sigma_{\text{max}} \leq 5$ mm scenarios. Similar convergence curves are observed for the electric field. A threshold can be explained by the fact that the decimation slightly modifies the tissue boundaries. Thus, the near field, mostly around these boundaries, as well as *s*-parameters of the antenna system, are slightly changed. However, tissue shapes are mostly identical for the head with and without stroke, regardless of the applied decimation. Therefore, the corresponding convergence curves for differential quantities (for different decimation) have the same shape (without threshold) as those in Figs. 19 and 20. The same is true for residual curves demonstrated in Fig. 22.

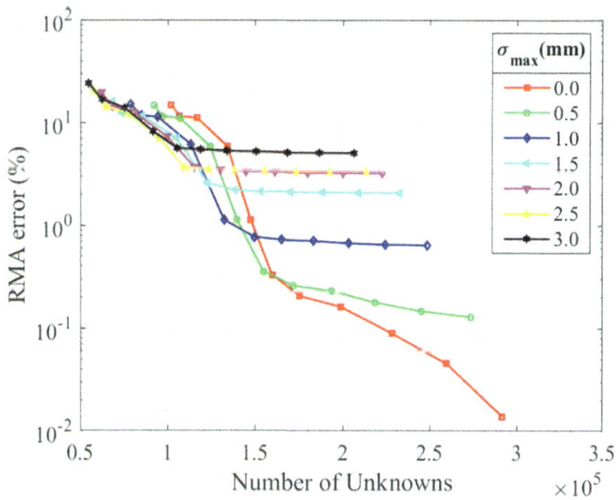

Fig. 23 RMA error of *s*-parameters versus number of unknowns needed for head for different levels of maximum allowed deviations, σ_{max}

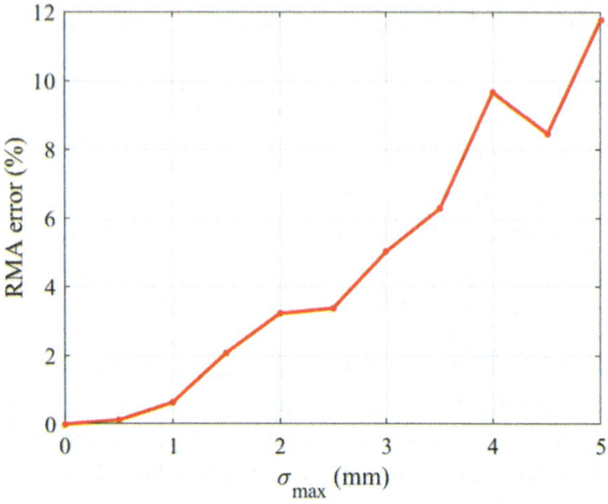

Fig. 24 Minimal RMA error of *s*-parameters versus maximum allowed deviation

9 Conclusions

In this chapter, we described the electromagnetic modeling of microwave medical scenarios using the EM simulation environment WIPL-D. Several tools and strategies have been developed to facilitate this complex task. Among others, we have developed procedures for importing and simplifying realistic human phantoms defined in the STL and voxel format. Particular attention has been paid to algorithms for reducing the number of triangles from the initial STL model in order to preserve the accuracy of EM simulation. Moreover, we have developed libraries that contain antenna models and arrays particularly suitable for medical MWI scenarios. In addition, the library encompasses realistic human phantoms with multiple tissues, including their electromagnetic properties. We have also defined convergence tests, which provide information and control to the user over the accuracy of the obtained results. This chapter was mainly devoted to the analysis of STL-defined models. In future work, we will focus our attention on EM modeling of voxel-based models.

Acknowledgements This research was supported by the EMERALD project funded from the European Union's Horizon 2020 under the Marie Skłodowska-Curie grant agreement No. 764479.

References

1. G. Zhu, A. Bialkowski, L. Guo, B. Mohammed, A. Abbosh, Stroke classification in simulated electromagnetic imaging using graph approaches. IEEE J. Electromagn. RF Microw. Med. Biol. **5**(1), 46–53 (2021). https://doi.org/10.1109/JERM.2020.2995329
2. A.S.M. Alqadami, A. Trakic, A.E. Stancombe, B. Mohammed, K. Bialkowski, A. Abbosh, Flexible electromagnetic cap for head imaging. IEEE Trans. Biomed. Circuits Syst. **14**(5), 1097–1107 (2020). https://doi.org/10.1109/TBCAS.2020.3025341

3. Z. Wang, E.G. Lim, Y. Tang, M. Leach, Medical applications of microwave imaging. Sci. World J. **2014**, 7 (2014). https://doi.org/10.1155/2014/147016
4. R. Chandra, H. Zhou, I. Balasingham, R.M. Narayanan, On the opportunities and challenges in microwave medical sensing and imaging. IEEE Trans. Biomed. Eng. **62**(7), 1667–1682 (2015). https://doi.org/10.1109/TBME.2015.2432137
5. M. Pastorino, A. Randazzo, Microwave imaging methods and applications, in *Artech House* (2018)
6. L. Crocco, I. Karanasiou, M. James, R. Conceição, *Emerging Electromagnetic Technologies for Brain Diseases Diagnostics, Monitoring and Therapy* (Springer, Cham, Switzerland, 2018)
7. R. Conceição, J. Mohr, M. O'Halloran, *An Introduction to Microwave Imaging for Breast Cancer Detection* (Springer, Switzerland, 2018)
8. NEVA Electromagnetics. https://www.nevaelectromagnetics.com/
9. IT'IS—The Virtual Family. https://itis.swiss/virtual-population/virtualpopulation/overview/
10. The Zubal Phantom. http://noodle.med.yale.edu/zubal/
11. S.N. Makarov, G.M. Noetscher, A. Nazarian, *Low-Frequency Electromagnetic Modeling for Electrical and Biological Systems Using Matlab* (John Wiley & Sons, Inc., 2016)
12. NLM Visible Human Project. https://www.nlm.nih.gov/databases/download/vhp.html
13. Anatomium. www.anatomium.com
14. WIPL-D software suite (WIPL-D Pro v17 & Pro CAD 2020), WIPL-D d.o.o, Belgrade, 2020
15. E. Zastrow, S.K. Davis, M. Lazebnik, et al., *Database of 3D Grid-Based Numerical Breast Phantoms for Use in Computational Electromagnetics Simulations* (University of Wisconsin-Madison, Madison (WI), c2014) [cited 2014 Sep 21]. Available from https://uwcem.ece.wisc.edu/phantomRepository.html
16. M. Lazebnik, L. McCartney, D. Popovic, C.B. Watkins, M.J. Lindstrom, J. Harter, S. Sewall, A. Magliocco, J.H. Booske, M. Okoniewski, S.C. Hagness, A large-scale study of the ultrawideband microwave dielectric properties of normal breast tissue obtained from reduction surgeries. Phys. Med. Biol. **52**, 2637–2656 (2007)
17. M. Lazebnik, D. Popovic, L. McCartney, C.B. Watkins, M.J. Lindstrom, J. Harter, S. Sewall, T. Ogilvie, A. Magliocco, T.M. Breslin, W. Temple, D. Mew, J.H. Booske, M. Okoniewski, S.C. Hagness, A large-scale study of the ultrawideband microwave dielectric properties of normal, benign, and malignant breast tissues obtained from cancer surgeries. Phys. Med. Biol. (in press)
18. M. Lazebnik, M. Okoniewski, J.H. Booske, S.C. Hagness, Highly accurate Debye models for normal and malignant breast tissue dielectric properties at microwave frequencies. IEEE Microw. Wirel. Components Lett. (2007) (in press)
19. S. Gabriel, R.W. Lau, C. Gabriel, The dielectric properties of biological tissues: III. Parametric models for the dielectric spectrum of tissues. Phys. Med. Biol. **41**, 2271–2293 (1996)
20. T. Singh, M. Stevanetic, M. Stevanovic, B. Kolundzija, Homogenization of voxel models using material mixing formulas, in *14th European Conference on Antennas and Propagation*, 2020, pp. 1–4. https://doi.org/10.23919/EuCAP48036.2020.9135081
21. R. Simpkin, Derivation of Lichtenecker's logarithmic mixture formula from Maxwell's equations. IEEE Trans. Microw. Theory Techniques. **58**(3) (2010)
22. S.O. Nelson, Measurement and computation of powdered mixture permittivities, in *Proceedings of the 17th IEEE Instrumentation and Measurement Technology Conference*. https://doi.org/10.1109/IMTC.2000.848661
23. K.K. Kärkkäinen, A.H. Sihvola, K.I. Nikoskinen, Effective permittivity of mixtures: numerical validation by the FDTD method. IEEE Trans. Geosci. Remote Sens. **38**(3), 1303 (2000)
24. T. Singh, B. Ninkovic, M. Tasic, M.N. Stevanovic, B. Kolundzija, New method for calculation of average electric properties of reference head phantom in microwave imaging, in *2022 16th European Conference on Antennas and Propagation (EuCAP)*, 2022, pp. 1–5. https://doi.org/10.23919/EuCAP53622.2022.9769058
25. C. Gabriel, Compilation of the Dielectric Properties of Body Tissues at RF and Microwave Frequencies, Final Report, Physics Department King's College London, Feb 1992
26. T. Singh, M.N. Stevanovic, B. Kolundzija, Survey and classification of antennas for medical applications, in *2019 13th European Conference on Antennas and Propagation (EuCAP)*, 2019, pp. 1–5

27. S. Semenov, M. Hopfer, R. Planas, A. Hamidipour, T. Henriksson, Electromagnetic tomography for brain imaging: 3D reconstruction of stroke in a human head phantom. IEEE Conf. Antenna Measur. Appl. (CAMA) **2016**, 1–4 (2016). https://doi.org/10.1109/CAMA.2016.7815752
28. D.O. Rodriguez-Duarte, J.A.T. Vasquez, R. Scapaticci, L. Crocco, F. Vipiana, Brick-shaped antenna module for microwave brain imaging systems. IEEE Antennas Wirel. Propag. Lett. **19**(12), 2057–2061 (2020). https://doi.org/10.1109/LAWP.2020.3022161
29. T. Singh, S. Abedi, B. Ninkovic, M.N. Stevanovic, N. Joachimowz, H. Roussel, B. Kolundzija, Smart simplification of anthropomorphic head phantom aimed for microwave imaging, in *2021 15th European Conference on Antennas and Propagation (EuCAP)*, Düsseldorf, Germany, 2021, pp. 1–4
30. J.D. Shea, B.D. Van Veen, S.C. Hagness, A TSVD analysis of microwave inverse scattering for breast imaging. IEEE Trans. Biomed. Eng. **59**(4), 936–945 (2012). https://doi.org/10.1109/TBME.2011.2176727
31. L. Crocco, I. Catapano, L. Di Donato, T. Isernia, The linear sampling method as a way to quantitative inverse scattering. IEEE Trans. Antennas Propag. **60**(4), 1844–1853 (2012). https://doi.org/10.1109/TAP.2012.2186250
32. G. Bozza, C. Estatico, M. Pastorino, A. Randazzo, An inexact newton method for microwave reconstruction of strong scatterers. IEEE Antennas Wirel. Propag. Lett. **5**, 61–64 (2006)
33. L. Guo, N. Nguyen-Trong, A. Al-Saffar, A. Stancombe, K. Bialkowski, A. Abbosh, Calibrated frequency-division distorted born iterative tomography for real-life head imaging, in *IEEE Transactions on Medical Imaging*. https://doi.org/10.1109/TMI.2021.3132000
34. L. Guo, M. Khosravi-Farsani, A. Stancombe, K. Bialkowski, A. Abbosh, Adaptive clustering distorted born iterative method for microwave brain tomography with stroke detection and classification. IEEE Trans. Biomed. Eng. **69**(4), 1512–1523 (2022)
35. B.M. Kolundzija, A.R. Djordjevic, *Electromagnetic Modeling of Composite Metallic and Dielectric Structures* (Artech House, Boston, MA, 2002)
36. M.M. Kostić, B.M. Kolundžija, Maximally orthogonalized higher order bases over generalized wires, quadrilaterals, and hexahedra. IEEE Trans. Antennas Propag. **61**(6), 3135–3148 (2013). https://doi.org/10.1109/TAP.2013.2249036
37. A.J. Krneta, B.M. Kolundzija, Evaluation of potential and impedance integrals in analysis of axially symmetric metallic structures to prescribed accuracy up to machine precision. IEEE Trans. Antennas Propag. **65**(5), 2526–2539 (2017). https://doi.org/10.1109/TAP.2017.2673760
38. B.M. Kolundzija, M.M. Kostic, Matrix equilibration in method of moment solutions of surface integral equations. Radio Sci. **49**(12), 1265–1276 (2014). https://doi.org/10.1002/2014RS005536

Assessment and Validation of 2-D and 3-D DBIM-TwIST Algorithm for Brain Stroke Detection and Differentiation

Olympia Karadima, Pan Lu, Ioannis Sotiriou, and Panagiotis Kosmas

Abstract Microwave Imaging (MWI) methods rely on the existing different dielectric properties of human tissue, due to the different water content. By solving an inverse, ill-posed electromagnetic (EM) scattering problem, a map of the spatial distribution of the dielectric properties of the region of interest is reconstructed, allowing to locate and detect a target with unknown properties. The use of non-ionized radiation, the portability and the low cost of a microwave imaging system are the main motivations that have established MWI as a promising way of imaging in medical conditions like brain stroke detection, where early detection and continuous imaging is vital and not all patients have quick access to the traditional imaging devices. MWI methods are challenged by the high heterogeneity of the human body as well as the increased computational cost that arises from the non-linearity of the problem and the non-uniqueness of the solution. This chapter focuses on incorporating the distorted Born iterative method with the two-step iterative shrinkage thresholding (DBIM-TwIST) algorithm into an experimental microwave imaging prototype for brain stroke detection and classification.

O. Karadima (✉) · P. Lu · I. Sotiriou · P. Kosmas
Faculty of Natural and Mathematical Sciences, King's College London, Strand, London, WC2R 2LS, UK
e-mail: olympia.karadima@kcl.ac.uk

P. Lu
e-mail: pan.lu@kcl.ac.uk

I. Sotiriou
e-mail: ioannis.sotiriou@kcl.ac.uk

P. Kosmas
e-mail: panagiotis.kosmas@kcl.ac.uk

F. Vipiana and L. Crocco (eds.), *Electromagnetic Imaging for a Novel Generation of Medical Devices*, Lecture Notes in Bioengineering,
https://doi.org/10.1007/978-3-031-28666-7_5

1 Introduction to Microwave Medical Imaging

Medical imaging is a primary mechanism for visualizing the tissues and the structures of the body for detecting any abnormality or malignancy [1]. The main imaging techniques which are widely used include X-rays, computed tomography (CT), magnetic resonance imaging (MRI), ultrasound (US) and positron emission tomography (PET) [2]. Although these methods provide accurate clinical data with a high resolution, they suffer from high operating costs and long examination times. For example, MRI offers the optimal resolution of images, but it is an expensive procedure. X-rays and CT can produce images of high resolution yet they cannot image soft tissues [3]. Also, they have been accused for side effects like nausea, hair loss, skin and breathing complications on top of an increased cancer risk due to the ionized radiation that is used in this type of imaging [4]. PET scans can produce good images of the soft tissues, but they suffer from poor spatial resolution which can lead to a false diagnosis [5]. Moreover, the above mentioned imaging methods and devices are usually not portable, creating delays in the diagnosis window as they require the patient to be admitted to the hospital [2]. Therefore, the use of the above imaging methods is unsuitable for critical conditions when a fast and continuous diagnosis is vital [6].

Furthermore, the increased associated costs can prevent their operation in underserved communities, maintaining and increasing the imbalance of health care access worldwide [7]. These challenges motivated the development of alternative approaches, which aim to be fast, safe, portable and cost-effective. MWI technology incorporates the applications and the techniques that use non-ionized EM radiation, within the range of 300 MHz–300 GHz, with the aim of sensing and imaging a dielectric difference located in an investigation domain of interest [8]. This frequency range includes both microwave and millimeter wave bands, but they can both be grouped under MWI, as they use electromagnetic radiation. The interaction of EM radiation with human tissues has been widely studied in the past and gained a lot of interest due to the prospective advantages it can offer [9]. Many studies investigated and measured the dielectric properties of biological tissues at the microwave frequencies range [10]. For example, the outcomes of these studies showed that there is a difference between the dielectric properties of a healthy and a malignant tissue, due to the different water content changes when a malignancy occurs in the tissue [11]. The dielectric properties (permittivity and conductivity) of an area in the brain affected by a stroke also, differ from the ones of healthy brain tissue [3].

MWI relies on the electric difference between the healthy and diseased tissues. This difference is called dielectric contrast and can be detected with MWI methods [12]. In MWI, the region of interest (ROI) is surrounded by measurement probes that radiate EM signals. The signals are received by the same or a different probe and the scattered or reflected signals are processed with various techniques to locate, shape and reconstruct an existing dielectric contrast [13]. A two-dimensional (2-D) or a three-dimensional (3-D) dielectric map is formed, that visualizes the permittivity and the conductivity of different tissues or the location of a strong scatterer inside the ROI. The use of non-ionized radiation, the low cost of an MWI system and the portability are the main motivations that have established MWI as a promising and non-invasive alternative to the current medical imaging techniques [2, 14, 15].

The remainder of the chapter is structured as follows. Section 2 discusses the background theory of MWI for brain stroke imaging and the adopted algorithm (DBIM-TwIST). Section 3 presents the experimental procedure towards the validation of the 2-D DBIM-TwIST algorithm for brain stroke detection and classification using a brain phantom. Section 4 validates the 3-D version of DBIM-TwIST first for simulated realistic scenarios. Then it presents a novel multi-layer head phantom and validates experimentally both the 2-D and 3-D versions of the DBIM-TwIST for different stroke scenarios.

2 Microwave Imaging for Brain Stroke

Brain stroke is a very serious condition that occurs when a blood vessel, is either blocked (ischemic) or burst (haemorrhagic), causing damage in the brain tissue. Treatment depends on the type of the stroke, the cause and the part of the brain that is affected. However, in any case of stroke, it is vital for the survival of the patient to detect successfully and as soon as possible, the type of the stroke, as the wrong or delayed treatment could have fatal results [6]. As a consequence, this section focuses on the brain stroke detection by validating the performance of existing algorithms for numerical and experimental results.

Specifically for the problem of brain stroke detection and classification, the most progressive systems that include human trials are the ones proposed by [3, 16, 17]. From a theoretical point of view, [18, 19] studied the optimization of the characteristics of a MWI apparatus for brain imaging, suggesting that the working frequency should be between 0.5–2.0 GHz, with the optimal number of antennas being in the order of 24.

The proposed techniques for brain imaging follow two main paths that include microwave imaging (MWI) or microwave radar algorithms [20]. Radar methods depict the geometrical characteristics of an area using scattered signals [20–22]. Alternately, MWI algorithms use the scattered fields to reconstruct the dielectric properties of a region of interest and identify a target with unknown dielectric properties by solving an inverse scattering electromagnetic problem [23, 24]. A new hybrid imaging method was proposed at [25], using information from a radar-based algorithm to locate the target and then apply MWI to reconstruct the dielectric properties of the head, gaining promising preliminary results. The research in this field is continuous and recently, machine learning approaches have been proposed for brain stroke classification as they are quick and robust, providing accurate results [17, 26–29]. Howbeit, to provide precise results, a large data set is needed to be trained. Moreover, the lack of images could act as a disadvantage for clinical applications.

MWI algorithms for brain imaging face two main challenges. First, the increased complexity of the head in combination with the high dielectric contrast of the individual tissues, as well as the need for a robust algorithm that can overcome the non-linearity and ill-posedness of the arisen EM problem [30]. For the first one, anatomical models and phantoms that mimic the structure and the dielectric properties of the brain have been developed [31, 32]. While the dielectric properties of the human tissues have been widely reported [33, 34], there is a lack of knowledge regarding the dielectric properties of ischemic tissues. To the best of our knowledge, [35] is the only study that measured the dielectric properties of ischemic infected areas ex-vivo. In this paper it was reported that the properties of ischemic tissues vary from − 10 to − 25% in comparison with the dielectric properties of healthy brain tissue.

MWI methods suffer from the non-linearity and ill-posedness of the inverse problem, that increases the complexity of the solution [36]. Several solutions have been proposed in the literature that include among others, the contrast source inversion (CSI) approach [37] and the linear inversion with truncated singular value decomposition (TSVD) [15, 38]. Nevertheless, the increased complexity of the brain imaging requires regularization of the problem and often adjusting the algorithm parameters for each case. As a consequence, the distorted born iterative method with the inverse solver two-step iterative shrinkage/thresholding (DBIM-TwIST) have been proposed in [39] where an inverse scattering problem is solved at each iteration. MWI algorithms reconstruct images from experimental or numerical 3-D data, with representations of the set-ups simulated by finite difference time domain (FDTD) or finite element method (FEM)-based solvers.

Initially, research towards this direction was focused on 2-D algorithms as they are easier to model and solve. On the other hand, 3-D algorithms are associated with very high computational cost as the ill-posedness of the inverse problem is increasing [24]. Some 3-D methods for MWI applications in medical imaging have been presented in [24, 40–42] but mainly for breast imaging. For the problem of head imaging, [43, 44] used scattering signals from 3-D head phantoms applying FDTD and FEM, respectively, while [15] suggested incorporating FEM with TSVD. Recent work by [45] presented a 3-D version of the DBIM, combined with the fast iterative shrinkage/thresholding algorithm (FISTA) for microwave head imaging, with some promising results for realistic cases.

2.1 Inverse Scattering Problems

Inverse scattering problems aim at determining characteristics of an unknown object according to how it scatters the incoming radiation from field measurements in the accessible region [46].

A popular approach to solve the scattering problem is by using volume integral equations where the unknowns in the problem are expressed in terms of the flowing current volume in the inhomogeneity [47]. The current volume is composed by the conduction current and the displacement current induced by the total electric field. Then, an integral equation is formed from which the total field can be solved [48].

Next, the mathematical formulation of EM scattering is analysed. Considering Maxwell's equations theory, it is assumed that the electric field satisfies the following equation:

$$\nabla \times \mu^{-1}\nabla \times \vec{E}(\vec{r}) - \omega^2\epsilon\vec{E}(\vec{r}) = \iota\omega\vec{J}(\vec{r}) \tag{1}$$

where μ and ϵ are functions of position inside the inhomogeneous region V. By defining with ϵ_b and μ_b the electric properties of the background medium and by subtracting the term $\nabla \times \mu_b^{-1}\nabla \times \vec{E}(\vec{r}) - \omega^2\epsilon_b\vec{E}(\vec{r})$ from both sides of the equation, the resulted equation is obtained:

$$\begin{aligned}\nabla \times (\mu^{-1} - \mu_b^{-1})\nabla \times \vec{E}(\vec{r}) - \omega^2(\epsilon - \epsilon_b)\vec{E}(\vec{r})\\ = \iota\omega\vec{J}(\vec{r}) - \nabla \times \mu_b^{-1}\nabla \times \vec{E}(\vec{r}) - \omega^2\epsilon_b\vec{E}(\vec{r})\end{aligned} \tag{2}$$

The dyadic Green's function in the absence of the scatterer is formed as follows:

$$\nabla \times \mu_b^{-1}\nabla \times \vec{G}(\vec{r},\vec{r}') - \omega\epsilon_b\vec{G}(\vec{r},\vec{r}') = \mu_b^{-1}\vec{I}\delta(\vec{r} - \vec{r}') \tag{3}$$

By combining Eqs. 2 and 3, the integral equation is formulated:

$$\begin{aligned}\vec{E}(\vec{r}) = \iota\omega\int_V dr'\vec{G}(\vec{r},\vec{r}')\cdot\mu_b\vec{J}(\vec{r}') + \omega^2\int_V dr'\vec{G}(\vec{r},\vec{r}')\cdot(\epsilon - \epsilon_b)\vec{E}(\vec{r}')\\ - \int_V dr'\vec{G}(\vec{r},\vec{r}')\cdot\mu_b\nabla' + \left(\frac{1}{\mu} - \frac{1}{\mu_b}\right)\nabla' \times \vec{E}(\vec{r}')\end{aligned} \tag{4}$$

The first term of the Eq. 4 is defined as the incident field, $\vec{E}_{inc}$, i.e., the field in the absence of the scatterer, therefore Eq. 4 becomes:

$$\begin{aligned}\vec{E}(\vec{r}) = \vec{E}_{inc}(\vec{r}) + \omega^2\int_V dr'\vec{G}(\vec{r},\vec{r}')\cdot(\epsilon - \epsilon_b)\vec{E}(\vec{r}')\\ - \int_V dr'\vec{G}(\vec{r},\vec{r}')\cdot\mu_b\nabla' + \left(\frac{1}{\mu} - \frac{1}{\mu_b}\right)\nabla' \times \vec{E}(\vec{r}')\end{aligned} \tag{5}$$

By substituting the complex permittivity, the two integrals of the above equation can be combined. Furthermore, by considering non-magnetic materials, $\mu = \mu_b = 1$. Hence, Eq. 5 is simplified:

$$\vec{E}(\vec{r}) = \vec{E}_{inc}(\vec{r}) + \int_V dr' \vec{G}(\vec{r}, \vec{r}') \cdot O(\vec{r}') \vec{E}(\vec{r}') \tag{6}$$

where $O(\vec{r}') = \omega^2(\epsilon - \epsilon_b)$ over a finite domain V. In an inverse scattering problem, the only available information includes the data that are outside the scatterer. As a result, the total field is connected to the object through the volume integral Eq. 6.

The integral in Eq. 6 includes information on the scatterer represented by contrast function $O(\vec{r}) = \kappa^2(\vec{r}) - \kappa_b^2$, where $\kappa^2(\vec{r}) = \omega^2\mu(\vec{r})\epsilon(\vec{r})$ is an inhomogeneous medium over a finite domain V. The scattered field is non-linear in relation to $O(\vec{r})$, as $\vec{E}(\vec{r})$ is a function of $O(\vec{r})$.

The solution of the volume integral equation can be solved numerically. In this study, we employ the Born approximation which is presented in the next section.

2.2 Distorted Born Iterative Method

When the scatterer is small compared to the wavelength or when is small, or when the contrast $O(\vec{r}) = \kappa^2(\vec{r}) - \kappa_b^2$ is weak, so the second term on the right of the Eq. 6 is very small in comparison with the first term. Accordingly, $\vec{E}(\vec{r}) \approx \vec{E}_{inc}(\vec{r})$ and Eq. 6 is approximated as follows:

$$\vec{E}(\vec{r}) = \vec{E}_{inc}(\vec{r}) + \int_V dr' \vec{G}(\vec{r}, \vec{r}') \cdot O(\vec{r}') \vec{E}_{inc}(\vec{r}') \tag{7}$$

This formulation is known as the first-order Born approximation. Born approximation considers a linear integral equation, however for inverse scattering EM problems the integral equation is non-linear. This is due to the non-linear relationship between the scattered fields and the inhomogeneity as a result of the multiple scatterers. Non-linearity increases as the body grows larger relative to the wavelength or as the contrasts of the inhomogeneity become larger [47]. A reliable method for resolving the inverse scattering problem includes an optimization approach. However, when the operating frequency is high, the non-linearity increases in the inverse problem and this could lead the algorithm to a local minima.

As a result of the non-linearity, consecutive linear approximations are applied to solve the non-linear equation [48]. Some popular inverse scattering methods include the Gauss-Newton optimization algorithms [49], [50] and the conjugate gradient techniques [51]. This chapter uses the Gauss-Newton solution of the inverse EM scattering problem that is more commonly known as DBIM [52]. More specifically, the contrast function $O(\vec{r}')$ is the difference between the complex relative permittivity of the background and object. Exploiting the Born approximation, the linear integral equation that relates the continuous spatial distribution of dielectric properties within V to the scattered electric field is given by:

$$\vec{E}(\vec{r}) = \vec{E}_{inc}(\vec{r}) + \int_V dr' \vec{G}(\vec{r}, \vec{r}', \epsilon_b) \cdot [\kappa^2(\vec{r}') - \kappa_b^2] \vec{E}_{inc}(\vec{r}') \tag{8}$$

As the measurement data are available only outside the scatterer in the inverse scattering problem, the only item available is:

$$\vec{E}_{sca}(\vec{r}) = \vec{E}(\vec{r}) - \vec{E}_{inc}(\vec{r}), r \in S \tag{9}$$

where $\vec{r} \in S$ and S is the surface outside V. As E_{sca} holds information on the scatterer, Eq. 8 can be composed as follows:

$$\vec{E}_{sca}(\vec{r}) = \int_V dr' \vec{G}(\vec{r}, \vec{r}', \epsilon_b) \cdot [\kappa^2(\vec{r}') - \kappa_b^2] \vec{E}_{inc}(\vec{r}') \tag{10}$$

The resulting equation is linear in $\kappa^2(\vec{r}') - \kappa_b^2$. However, multiple data are required to reconstruct κ^2 efficiently. Hence, an iterative method is applied in which the new κ^2 is used to estimate κ_b^2.

Next, an updated $\vec{G}(\vec{r}, \vec{r}', \epsilon_b)$ is computed until a convergence is reached, as there are no analytical Green's functions available for the heterogeneous background profiles estimated at each iteration of the DBIM. For a point source at a specific antenna feed point r in V, the Green's function estimates the field received at a given antenna at r_m [24].

The inhomogeneous Green's function $\vec{G}(\vec{r}, \vec{r}', \epsilon_b)$ is related to the internal electric field. As a result of the differential equations controlling the Green's function and electric field, Green's function is acquired as follows:

$$\vec{G}(\vec{r}, \vec{r}', \epsilon_b) \approx \frac{i}{\omega\mu} \vec{E}_a(\vec{r}, \vec{r}', \epsilon_b) \tag{11}$$

where $\vec{E}_a(\vec{r}, \vec{r}', \epsilon_b)$ is the total electric field at the receivers, created by a 2-D point source inside V filled by the background medium.

By employing the reciprocity theorem, Eq. 11 can by given as [24]:

$$\vec{G}(\vec{r}, \vec{r}', \epsilon_b) \approx \frac{i}{\omega\mu} \vec{E}_b(\vec{r}', \vec{r}, \epsilon_b) \tag{12}$$

The 2-D version of DBIM is a simplified version of the 3-D formulation. For the 3-D EM inverse scattering problem, an $x - y$ horizontal plane is considered. Then, the 3-D inverse scattering problem is approximated, considering that only electric fields at the z-polarized antennas are significant [24]. Therefore, E_{xi} and E_{yi} are assumed to be zero and Eqs. 10 and 12 are applied for the z-directed tensor elements of the electric field and the dyadic Green's function, respectively. In principal, 3-D algorithms should include all three E-field components in the integral equation of the scattered field, but previous work [53] has shown that the scalar approximation does not affect the performance significantly. The validity of the scalar approximation has been also recently demonstrated for the head imaging problem by our group in [45].

The integral equation can then be linearized by discretizing Eq. 10 combined with Eq. 12 for all Tx-Rx pairs. The linearization of Eq. 10 results in the following equation:

$$A(\omega)o = b(\omega) \tag{13}$$

where A is an $M - by - K$ propagation matrix, with M being the number of transmit-receive pairs in the antenna array and K the number of elements in the discretization inside the reconstruction domain, while $b(\omega)$ is the $M - by - 1$ vector of the scattered fields recorded at the receivers. The $K - by - 1$ vector o is the unknown dielectric properties contrast function.

Solving the above equation derives into a discrete approximation $\hat{o}$ for the true value of the contrast function o. The approximate contrast function $\hat{o}$ is added iteratively to the background profile. This is then used to calculate the new background electric field and Green's function by employing FDTD simulations. The iterative implementation of this sequence of computations shapes the basis of the DBIM.

The DBIM algorithm starts with an initial guess ϵ_{b0} of the background profile. Hence, at the i-th iteration, the background electric field and the Green's function are calculated for the background profile ϵ_{bi}. The discretised equation is given by:

$$A_i(\omega)o_{i+1} = b_i(\omega) \tag{14}$$

The background profile is updated as:

$$\epsilon_{bi+1} = \epsilon_{bi} + \hat{o}_{i+1} \tag{15}$$

In this chapter, the FDTD simulations compute the fields in the entire region of interest, including both the background fields at the antennas and the reconstruction domain V. The resulted problem is still ill-posed and thus, the TwIST solver is used [54]. TwIST breaks the matrix into a two-step iterative equation where the next solution depends both on the current as well as the previous solution. The advantages of the shrinkage/thresholding adaptive operators of TwIST over other ill-posed solvers have been analyzed at [39].

The resulted contrast function is added to the background profile which is updated at each iteration until the residual scattering is minimized and consequently the dielectric properties of the region of interest (ROI) are reconstructed.

To apprehend the frequency-depended behaviour of the materials, we apply the single-pole Debye model for the complex permittivity at each frequency as follows:

$$\epsilon_s(\omega) = \epsilon_\infty + \frac{\Delta\epsilon}{1 + j\omega\tau} + j\frac{\sigma_s}{\omega\epsilon_0} \tag{16}$$

where ϵ_∞, $\Delta\epsilon$, σ_s and τ are the Debye parameters which are updated at each iteration.

3 Experimental Validation of Microwave Imaging with the DBIM-TwIST Algorithm for Brain Stroke Detection and Classification

3.1 Phantoms Preparation and Characterization

Experimental testing of MWI prototypes is essential to validate their capability for clinical applications. Therefore, phantoms that imitate the dielectric and structure properties of the human head and brain provide efficient, easy to fabricate, and low-cost alternatives for experimental validation in the pre-clinical stage. Gelatine emulsions made of oil, water and gelatine are popular recipes for preparing such phantoms, as they are cheap and easy to alter, but can suffer from dehydration.

We have constructed phantoms based on the materials and processes proposed in [32] for breast phantoms. Using different gelatine-oil concentrations compared to those for breast tissues, we have fabricated tissue-mimicking materials with specific dielectric properties that mimic average brain tissue, cerebrospinal fluid (CSF), blood and ischaemia. The process shown in Fig. 1.

Initially, a water-gelatine mixture was prepared and mixed at 70 °C, until the gelatine particles were fully dissolved into water and the mixture was transparent. Propanol was added to deal with the creation of air bubbles on the surface. Once heated to 70 °C, we added a 50% kerosene-safflower oil solution into the water-

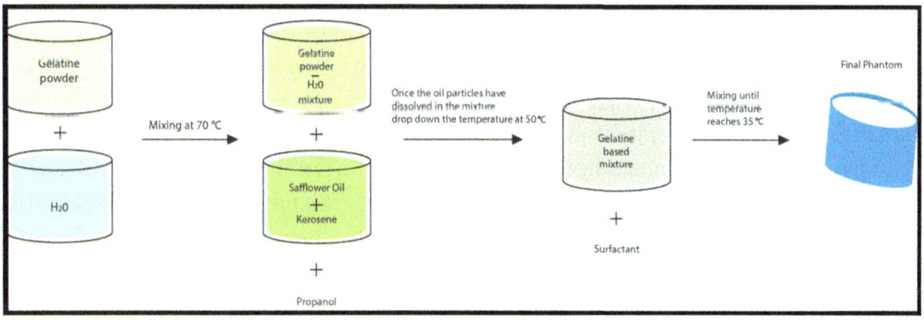

Fig. 1 Schematic representation of the phantoms' preparation process

Table 1 Concentrations of materials used for 100 ml of tissue mimicking phantoms

Phantom type	Water (ml)	Powder (g)	Kerosene (g)	Oil (g)	Propanol (ml)	Surfactant (ml)
Average brain	60	11	13	13	2.5	1.5
CSF/blood	80	16	–	–	4	–
Ischemia	50	8.5	20	20	1.5	1.5

gelatine mixture and stirred at the same temperature, until the emulsion had an opaque white colour and the oil particles were fully dissolved. We kept stirring the mixture until the temperature dropped to 50 °C, when we added the surfactant. Finally, when the prepared mixture reached 35 °C, we poured it into the mould, and let it set overnight before we conduct any measurements.

Table 1 shows the concentrations of the materials used for the phantoms imitating various brain tissues. We prepared the four tissue-mimicking phantoms in accordance with the properties reported in [33]. The in-vivo measurements in [35] revealed that properties of an ischaemic vessel in the brain can vary from − 10% up to − 25% relative to the dielectric properties of healthy brain tissue at 1.0 GHz. Hence, we chose to prepare ischaemic phantoms with the maximum reported difference of − 25% relative to the properties of the brain phantom. For the experimental testbeds it is assumed that, due to similar dielectric properties of CSF and blood, the same gelatine mixture can be used to mimic the two different tissues. Figure 2 shows the measured dielectric properties of the prepared phantoms for brain, CSF, blood and ischaemia, respectively. The measurements were performed using Keysight's dielectric spectroscopy kit, over a 0.5–2.5 GHz frequency range at different points of the phantoms. The plots in Fig. 2 reveal that the measured dielectric constant is in agreement with the reference data, but the measured conductivity has lower values in comparison to the reported reference conductivity. As MWI depends mostly on contrast in dielectric constant, the conductivity mismatch is not critical for this initial investigation.

After construction, phantoms were placed in elliptical plastic acrylonitrile butadiene styrene (ABS) moulds which mimic the brain's shape and multi-layer structure, as shown in Fig. 4. We initially poured the CSF phantom in the outer layer of 5 mm thickness. After the CSF phantom was solidified, we extracted the inner mould and filled the remaining cavity with the brain phantom. For the one-layer model without the CSF, we poured the phantom in the outer mould directly. We used an additional cylindrical mould to create a hole (diameter = 30 mm) in the phantom, which can be filled with the material mimicking either blood or ischaemia, reproducing the two cases of brain stroke named as h-stroke for haemorrhagic and i-stroke for ischaemic, respectively.

To investigate more the change in the dielectric properties over time, we fabricated two additional brain phantoms and measured their real and imaginary part of permittivity on day 1 and day 6 after their preparation. Table 2 presents that, over a

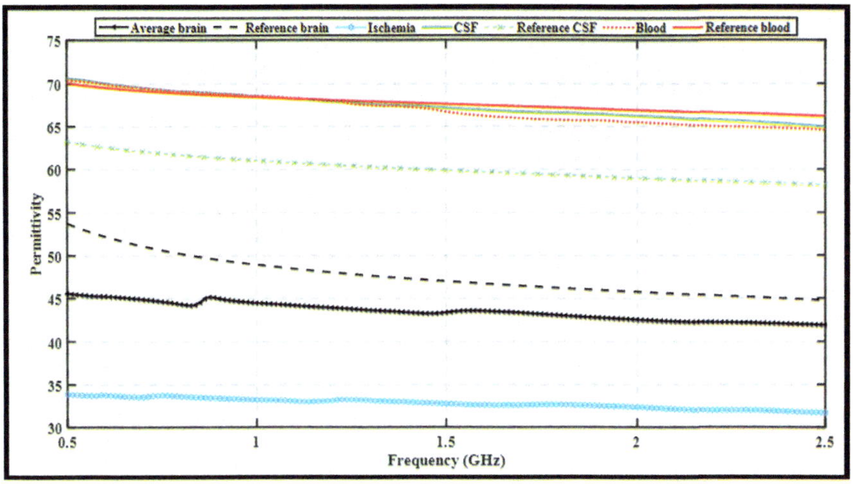

(a) Measured permittivity of the of the prepared tissue mimicking phantoms relative to their reference values.

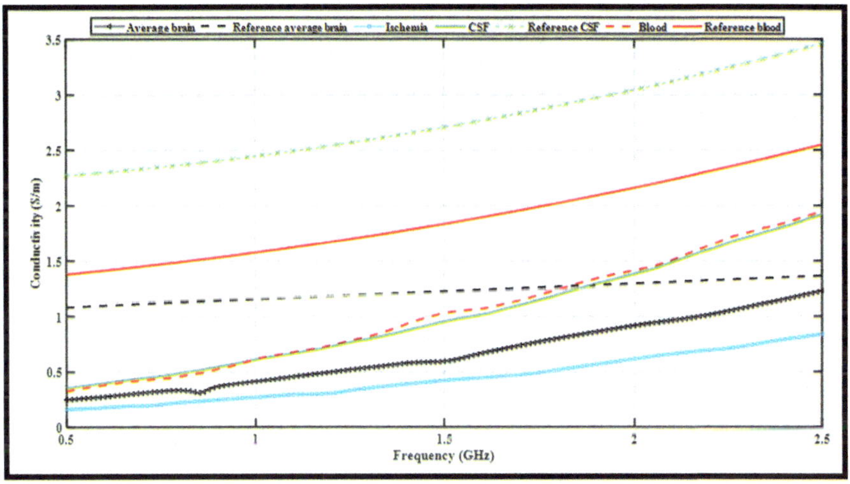

(b) Measured conductivity of the of the prepared tissue mimicking phantoms relative to their reference values.

Fig. 2 Dielectric properties of the fabricated tissue mimicking phantoms relative to reference values taken from [34] (to the authors' knowledge, there is no literature regarding the dielectric properties of ischaemia in a wide frequency range). Same type of tissues are presented with the same colour

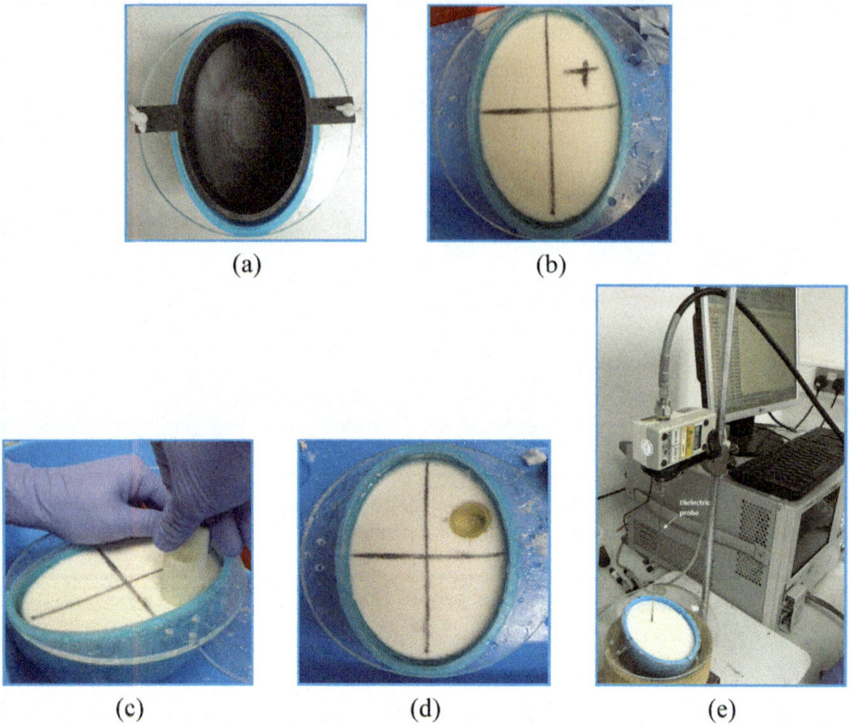

Fig. 3 Summary of the head model construction: **a** cylindrical 3-D printed mould to form the cerebrospinal fluid (CSF) and average brain layers; **b** the two-layer model after CSF and average brain phantoms are poured into the moulds (the CSF is a thin, transparent layer just inside the blue mould); **c** creating a hole in the phantom to insert the stroke-like target; **d** final two-layer phantom with a target of blood-mimicking phantom; **e** setup with Keysight's slim form probe to measure the dielectric properties of the phantoms

Table 2 Dielectric properties of brain phantoms over time at 1 GHz

Measured property	Day 1	Day 6
ϵ' sample 1	44.5	51.2
ϵ' sample 2	46.2	51.6
ϵ'' sample 1	0.43	0.51
ϵ'' sample 2	0.42	0.48

period of 6 day, the dielectric properties increased 10–15% compared to their initial values. This was expected as the water particles have scattered over time leading to higher dielectric properties.

3.2 Setup, Data Acquisition Process and Implementation of the Algorithm

Figure 4b presents the experimental setup, which was used for performing the measurements. It includes a 300 mm diameter cylindrical tank enclosed by an absorber (ECCOSORB MCS) and surrounded with a metallic shield to decrease surface waves propagating into the perimeter of the tank. The tank was filled with a 90% glycerol-water matching medium to enhance impedance matching with the phantom and widen the antenna operating frequency range [55]. An eight-antenna elliptical array (153 mm × 112 mm), formed by spear-shaped patch monopole antennas [56], was immersed inside the tank. Antennas were placed into vertical and horizontal rulers that allowed us to adjust their height and the array dimensions.

The phantoms were immersed in the bottom of the tank, and measurements of the S-parameters of the scattered signals were performed in a frequency range of 0.5–2.5 GHz using an eight-port vector network analyzer (VNA) from Keysight. The eight antennas acted both as transmitters and receivers, generating an 8×8 scattering matrix which was fed into DBIM-TwIST. The linear integral equation was discretized iteratively for the entire 8×8 transmit-receive pairs. Overall, two sets of measurements were conducted for each imaging scenario. The first set of measurements was performed for the phantom without a target (the "no target" scenario), and then the process was repeated for the phantom which included the 30 mm diameter h-stroke or i-stroke target ("target" scenario). The target was located in the upper-right section of the phantom between antennas 1 and 8. Following this process, the S-parameters for "no target" and "target" scenarios were processed by the DBIM-TwIST algorithm [39], which was analysed in the previous section.

The forward model for the algorithm is shown in Fig. 4a, and includes a simplified two-dimensional (2-D) depiction of the experimental scenario, which contains the

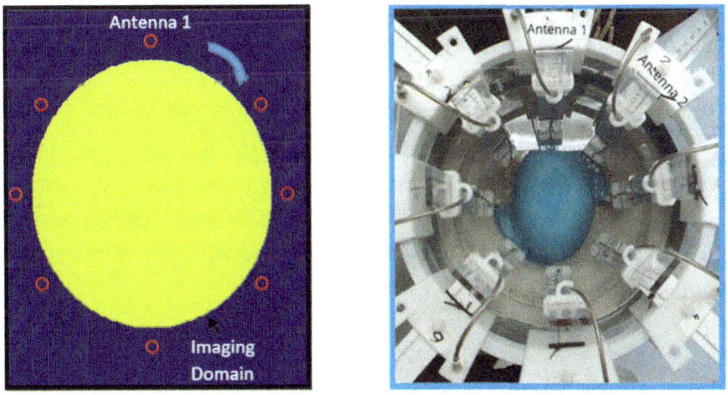

(a) (b)

Fig. 4 **a** Schematic of the simplified "initial guess" model for the inversion (the ellipsoid's axes are 153 and 112 mm long); **b** the hardware system prototype

Table 3 Debye parameters of materials after curve fitting to measured dielectric properties

Material type	$\Delta\epsilon$	ϵ_∞	σ_s
90% glycerol-water	6.56	16.86	0.3232
Average brain	30	10	0.147

tank immersed into the matching medium and an ellipsoid (153 × 112 mm) that imitates the average brain tissue. To calibrate the simulated initial model, we used a "no target" reference measurement, as in our previous work [57]. This calibration method relies on the signal difference between the experimental and the simulated "no target" scenarios, to calibrate the signals received from the "target" experiment. It is used to minimize the errors caused by the mismatch between the simulated initial model and the experiment.

We have used the Debye parameters of Table 3 in combination with a forward solver based on the FDTD method to compute the scattered and background fields, and then employed DBIM-TwIST to reconstruct the dielectric properties inside the area of interest with a grid voxel size of 2 mm. We have conducted 15 DBIM-TwIST iterations at each frequency for every reconstruction. The running time for each DBIM iteration was 8 s using MATLAB R2019b run on an Intel i7 processor with 16 GB RAM memory, resulting to an execution time of 2.5 min for 15 iterations of DBIM at each frequency.

3.3 *Results*

In our first set of experiments, our aim was to validate our method's potential to determine the type of stroke from estimating its dielectric properties. To this end, we have examined two phantoms of average brain tissue in which we inserted a target emulating h- and i-stroke, respectively. In the second set of experiments, our aim was to test detection performance in cases where the inverse experimental model was slightly different than our forward model. To this end, we have examined two additional phantoms with h-stroke targets. The first one was a two-layer phantom including a thin CSF-mimicking layer, and the second one was a brain phantom which we let it set for 6 d before conducting the "target" measurements, so that its properties were different in comparison with the first day when the "no-target" measurements were taken.

3.3.1 Detection and Classification of Stroke Targets

Figures 5 and 6 show the reconstructed real and imaginary parts of the complex permittivity for the phantom inside the prototype of Fig. 4b containing h-stroke and i-stroke targets as well as a target which has 50% lower dielectric constant than the average brain phantom as an extreme case (50% i-stroke). Images illustrate the whole tank, however, the reconstruction domain contains only the brain ellipsoid inside the antenna array. Representative single-frequency reconstructions in the range 0.7–1.5 GHz are presented in Fig. 5, while results from a frequency hopping approach in the same range are presented in Fig. 6. These images reveal that the dielectric constant (real part) of 25% i-stroke is detectable at low frequencies albeit with more artefacts in the images, whilst the h-stroke and the 50% i-stroke, is clearly visible in frequencies above 1.1 GHz for both real and imaginary parts. As the dielectric contrast between the constructed 25% i-stroke and brain phantoms is lower than that of h-stroke and 50% i-stroke, detecting the 25% i-stroke target is more challenging than the other cases, with unsuccessful results for the imaginary part of the complex permittivity. We note that the images are shown by properly adapting the color scale to a range of values which is in the order of the expected permittivity of interest for h-stroke and i-stroke, respectively. The most encouraging observation from these figures, however, is that there is a clear distinction in the reconstructed permittivity value in the area where the target is located for all the three targets: the frequency hopping results of Fig. 6 estimate the value of real permittivity ϵ' at 1.5 GHz as ϵ'=33.8, ϵ'=20.9 and ϵ'=56.25 for 25% i-stroke, 50% i-stroke and h-stroke, respectively.

3.3.2 Stroke Target Detection for Brain Phantoms with Unknown Properties

We have also evaluated experimentally the robustness of DBIM-TwIST when the structure and dielectric properties of the brain phantom are slightly different from the "initial guess" of Fig. 4a. To examine the effect of differences in the dielectric properties of the average brain phantom where the target is inserted, we prepared a phantom and performed "no target" measurements after the phantom had set, and then kept it for a week, in which we recorded the change of its dielectric properties over time. We then acquired "target" measurements for h-stroke on day 6, to ensure a change in the phantom's dielectric properties from the "no-target" measurements on day 1. In addition to this test, we also performed experiments with a two-layer phantom (CSF and average brain). To this end, we added a CSF layer of 5 mm thickness using the process described in Fig. 3, and with dielectric properties plotted in Fig. 2.

Figure 7 shows the reconstructed real and imaginary parts of complex permittivity value in the area where the target is located, using the data obtained by the measurements at days 1 and 6 after the preparation of the brain phantom, as well as the reconstructed images of the two-layer phantom including the CSF layer. These reconstructions were produced by frequency hopping in the range of 1.1–2.0 GHz. The reconstructed real and imaginary permittivity value in the area where the target is located for the h-stroke target at 2 GHz are $\epsilon' = 49.28$, $\epsilon'' = 12.19$ for the one-layer phantom, and $\epsilon' = 62.38$, $\epsilon'' = 8.48$ for the two-layer phantom, respectively. With the exception of the imaginary part for the two-layer phantom, these images provide

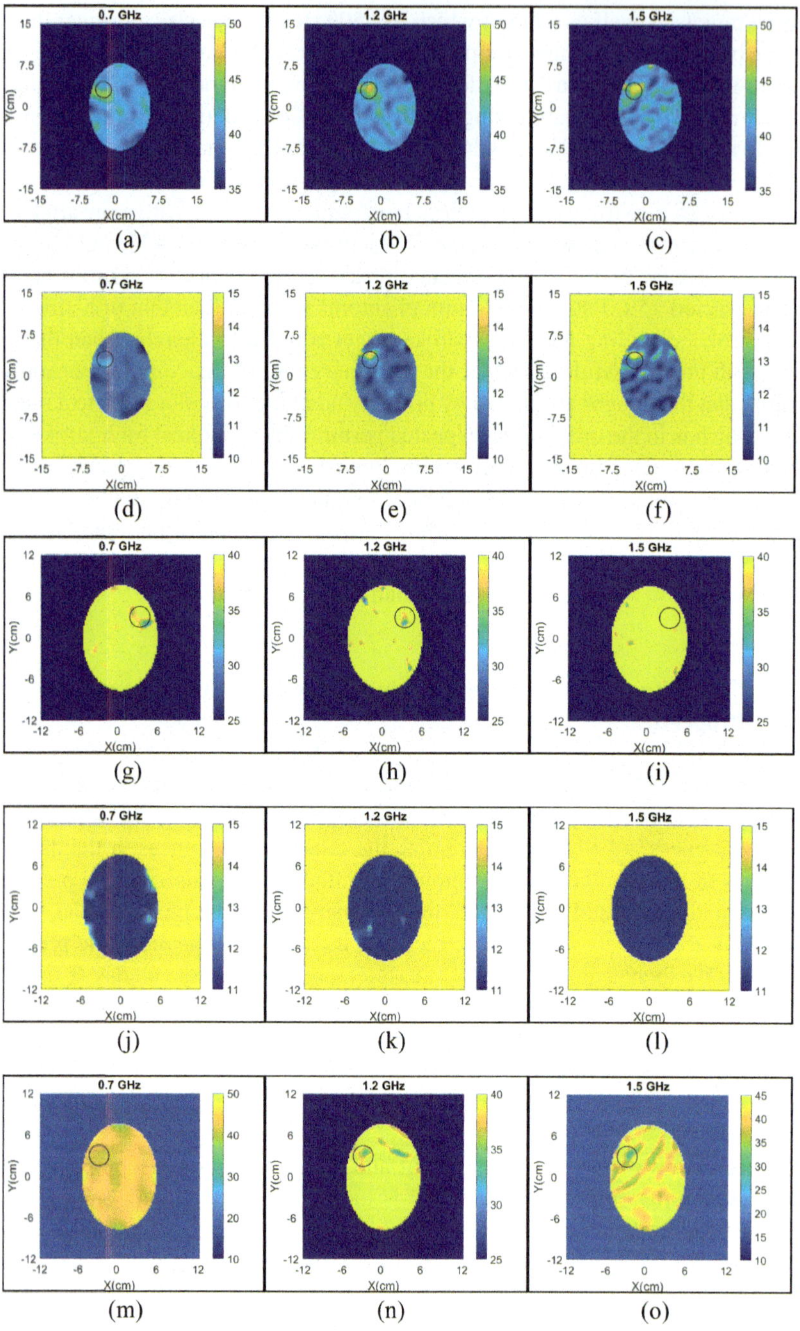

Fig. 5 Results from single frequency reconstructions of the real (first, third and fifth line) and imaginary (second, fourth and sixth line) part of the complex permittivity for: **a–f** h-stroke, **g–l** 25% i-stroke and **m–r** 50% i-stroke

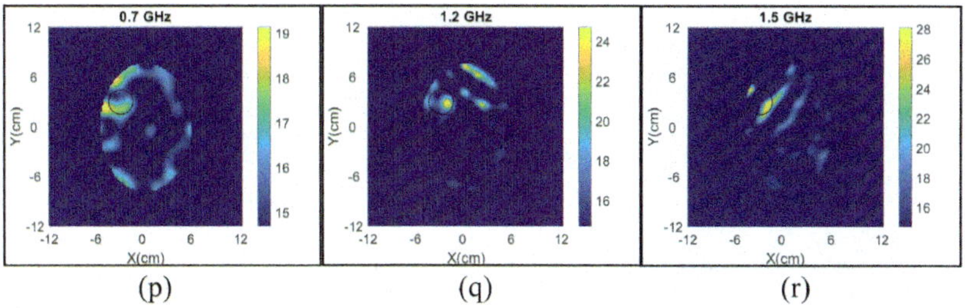

Fig. 5 (continued)

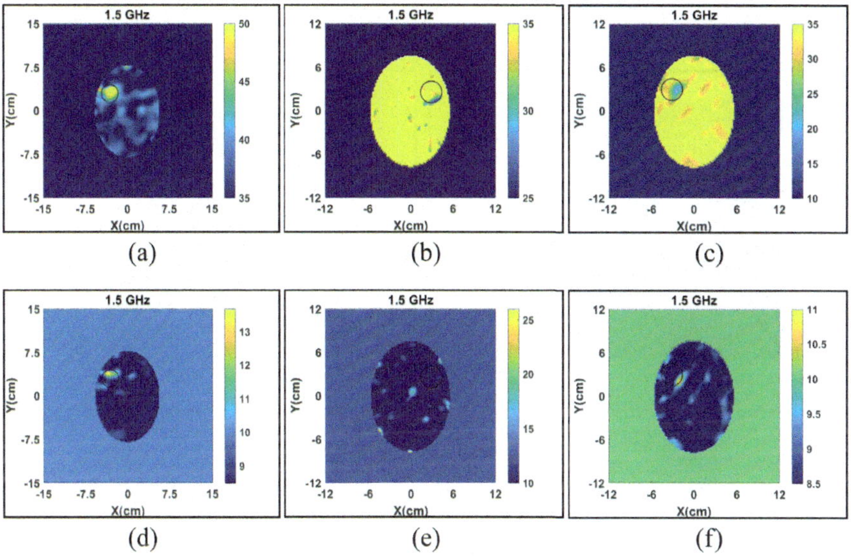

Fig. 6 Reconstructed: **a–c** real and, **d–f** imaginary part of the complex permittivity for h-stroke (left), 25% i-stroke (middle) and 50% i-stroke (right). The complex permittivity was calculated at 1.5 GHz using the frequency hopping approach in a frequency range of 0.7–1.5 GHz

a clear indication of where the h-stroke is located and estimate its dielectric constant with satisfactory accuracy. These reconstructions confirm that the algorithm is sufficiently robust to detect and localize the target successfully in cases with "mild" uncertainties in the true background medium in the experiment and its "initial guess" in the imaging algorithm.

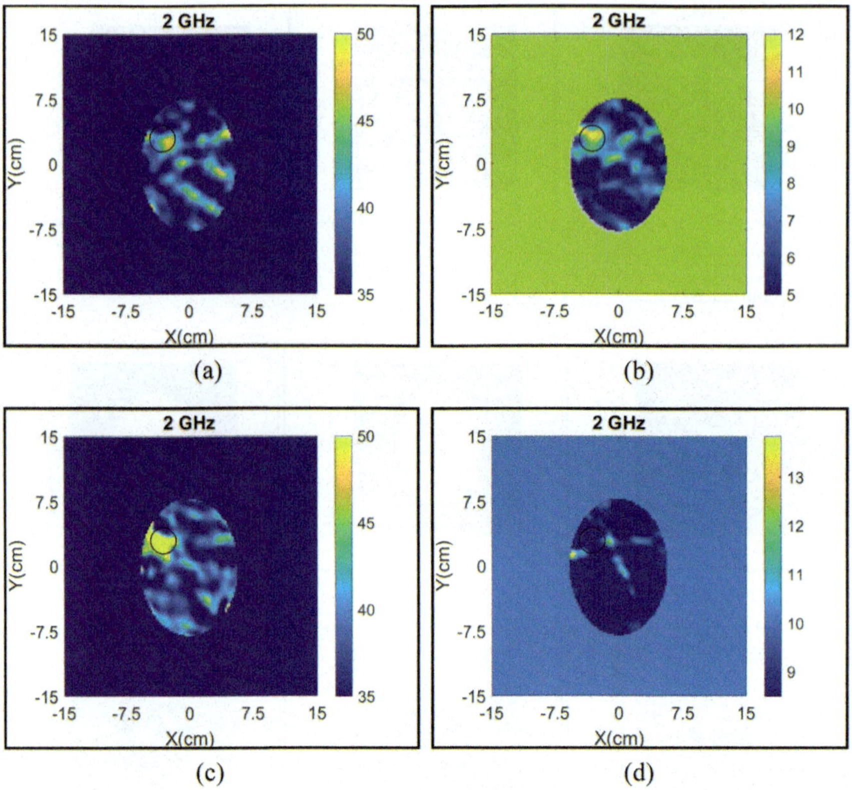

Fig. 7 **a** Reconstructed real and **b** imaginary part of the complex permittivity for the phantom inside the tank of Fig. 4b, when "no-target" and "h-stroke target" measurements were conducted at day 1 and day 6 after the preparation of phantom, respectively. **c** Reconstructed real and **d** maginary part of the complex permittivity for a two-layer phantom with an "h-stroke target" and a CSF layer which is unknown to the imaging algorithm. The complex permittivity was calculated at 2 GHz using a frequency hopping approach in 1.1–2 GHz

3.4 Discussion

We presented an initial experimental assessment of a microwave imaging prototype based on the DBIM-TwIST algorithm for brain stroke detection and classification. This MWI system was able to differentiate between targets that mimic haemorrhagic and ischaemic stroke, based on the difference in their estimated dielectric properties. However, we point out that the images are shown by properly adjusting the color scale to a range of values which is in the order of the expected permittivity of interest for h-stroke and i-stroke, respectively. Between the two stroke types, we observe more artifacts in the reconstructed i-stroke figures. This was expected, as the dielectric difference observed between the i-stroke and the average brain is less than the dielectric difference between the h-stroke and the average brain, resulting in a more complex brain structure. Moreover, the system was able to reconstruct the target inside a brain phantom even when its structure or dielectric properties are different from the "initial guess" used in the inversion. These results benefited from reconstructions in multiple frequencies based on the use of our in-house developed spear-shaped antennas which operate in a wide range from 0.5 to 2.5 GHz. In addition, the flexible tuning of the DBIM-TwIST parameters allow an easy adaptation of the algorithm to inverse problem at hand. We note that, as with most non-linear inverse methods, the DBIM-TwIST parameters must be tuned based on considerations that relate to both the imaging problem and the experimental prototype. Moreover, experimental measurement errors may result in certain frequencies providing more accurate reconstructions than others. In the past, we have proposed a method to discard data dominated by measurement error, based on a correlation technique [57]. Regardless of whether such a pre-processing step is applied to the measured data, frequency hopping is an excellent way to increase the accuracy and robustness of the algorithm [39].

However, knowing a priori information for brain imaging is very challenging. For this reason, we have assumed that the initial model is filled with homogeneous average brain tissue. Moreover, our previous work in [39] has confirmed that the DBIM-TwIST (as any other DBIM implementation) is very sensitive to the initial guess of the inverse model. While this has been partially addressed in this work with our investigation in Sect. 3.2, it will become a much more challenging problem in a realistic brain stroke detection application, where the structure of the brain is much more complex than a homogeneous average brain phantom. This issue is being examined more in depth in the next section.

Furthermore, our experimental validation required that the antenna array is placed at the same height as the target. Next section will be focused on a more realistic experimental validation process which involves acquiring data for more than one height and employing a three-dimensional (3-D) version of the DBIM-TwIST.

4 Validation of the 3-D DBIM-TwIST Algorithm for Brain Stroke Detection and Differentiation Using a Multi-layered Anatomically Complex Head Phantom

In this section we aim to validate the 3-D DBIM-TwIST algorithm for the problem of brain stroke detection and classification, using an anatomically accurate, multi-layer head phantom. To this end, we have developed an anatomically complex head phantom which mimics various brain tissues and also includes a target mimicking hemorrhagic or ischemic stroke. We simulated the model and setup using CST Microwave Studio and then used our experimental imaging setup to collect numerical and measured data, respectively. We then used our DBIM-TwIST algorithm to reconstruct the dielectric properties of the imaging domain for both simulated and measured data. Results from our CST simulations showed that we are able to locate and reconstruct the permittivity of different stroke targets using an approximate initial guess. Our experimental results demonstrated the potential and challenges for successful detection and differentiation of the stroke targets.

4.1 *Evaluation of the Initial Guess for Brain Stroke Detection Using Microwave Imaging*

In MWI, the arised EM problem is non-linear and ill-posed, therefore a robust algorithm is essential. As some a-priori information is needed, MWI algorithms should allow the tuning of the parameters depending on each examined scenario [36]. Specifically for brain imaging, the high dielectric contrast between the different head layers adds an additional challenge to the successful detection and differentiation of brain stroke, using MWI [30]. Therefore, before we validate DBIM-TwIST for more challenging scenarios, there is a need to explore what a-priori knowledge is required to achieve a successful differentiation of the types of the stroke for a complex scenario. In this section, we investigate whether an improved approximate model as the initial guess could result into stronger reconstructed results, using 3-D DBIM-TwIST and an anatomically accurate head phantom.

4.1.1 Zubal Model and Numerical Configuration

The proposed system and model were simulated using CST Microwave Studio. Figure 8 shows the simulated prototype that is composed by a tank with a diameter of 300 mm and filled with 90% glycerol-water mixture. The imaging system was presented in detail at [58] and in the previous section. Within the tank we placed a six-layer brain model with the shape of the axial slice of the original MRI-derived Zubal head phantom that emulated our experimental phantom [59]. In each layer we allocated the dielectric properties of the equivalent head tissues that were assigned from CST's library. The dispersive dielectric properties of the Zubal head phantom as well as the matching medium's are shown in Table 4. The phantom was surrounded by 16 monopole antennas [58] in two arrays and in an elliptical arrangement with major and minor axes equal to 205 mm and 148 mm respectively. A cylindrical target (diameter d = 20.3 mm and height h = 100 mm), which mimics either the hemorrhagic or the ischemic stroke, was placed at the left of the phantom, inside the white matter area. For each scenario we conducted three simulations to acquire the scattered field data: one without the target (NT), one with hemorrhagic (h-stroke) and one with ischemic (i-stroke) stroke, respectively. We increased the complexity of the Zubal phantom by adding one layer at each time and allocating the dielectric properties of average brain to the rest of the tissues. Our final six-layer phantom which includes skin, fat, bone, CSF, grey matter, white matter and the stroke-like target is presented at Fig. 8. Overall, we examined five scenarios of different complexities, resulting in a total of 15 CST simulations.

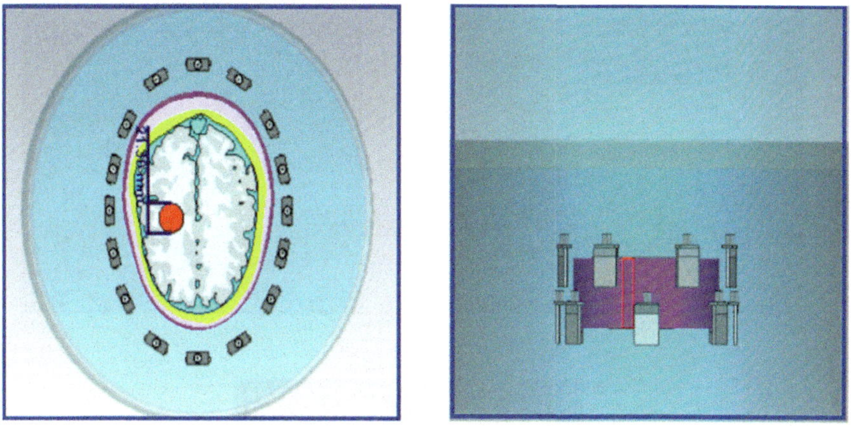

(a) (b)

Fig. 8 The simulation system prototype in CST, which uses a numerical head model based on the Zubal head phantom. The head model's dimensions are approximately $170 \times 130 \times 100$ mm

Table 4 Debye parameters of the forward model. The Debye parameters for the head tissues have been assigned by curve fitting from CST data

Material type	$\Delta\epsilon$	ϵ_∞	σ_s
90% glycerol-water	6.56	16.86	0.3232
Average brain	45.8	0.1244	0.7595
Skin	41.17	11.15	0.6672
Fat	5.331	0.6564	0.037
Bone	11.59	1.932	0.0788
CSF	67.71	5.46	0.21
Grey matter	49.63	11.66	0.7
White matter	36.67	8.6	0.39
H-stroke	59.23	6.142	1.374
I-stroke	25	10	0.01

Debye parameters of the 90% glycerol-water and of the i-stroke have been obtained from experimental dielectric measurements that have been presented in [58]

4.1.2 Implementation of the 3-D DBIM-TwIST Algorithm

The DBIM method which solves the non-linear EM inverse scattering problem by forming iteratively, a system of scattering equations under the Born approximation for every transmit-receive pair, was presented in the previous section. For the 3-D DBIM, the electric field and the dyadic Green's function have been calculated by taking into account an $x - y$ horizontal plane, considering that only electric fields at the z-polarized antennas are important [24, 45]. To capture the frequency-dependent

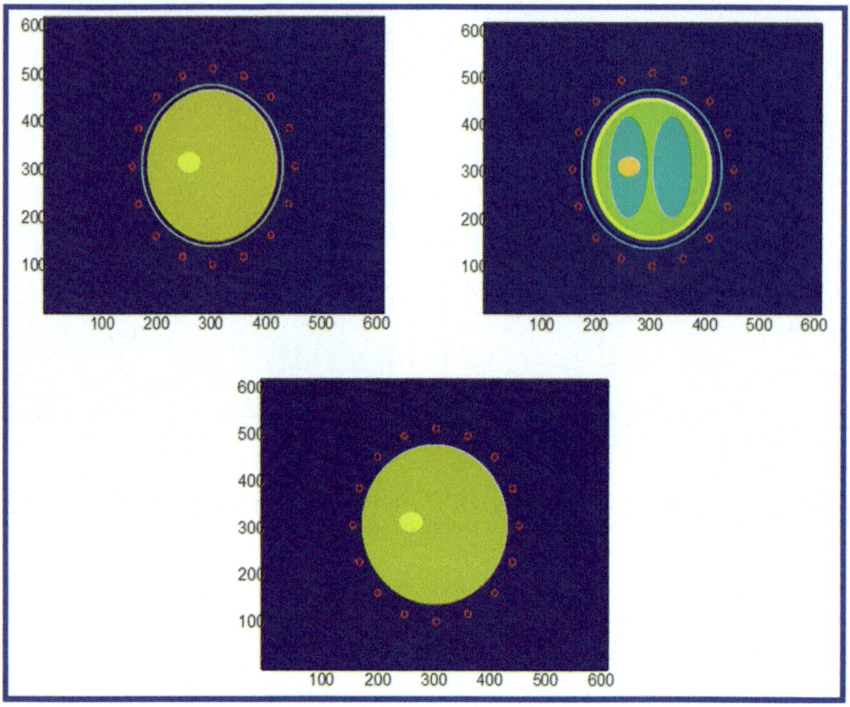

Fig. 9 Top left: top view of the forward model for the three-layer Zubal head phantom (skin, fat, bone); top right: top view of the forward model for the six-layer Zubal head phantom (skin, fat, bone, CSF, grey matter and white matter); bottom: top view of the forward model filled with average brain. The dimensions of the outer layer of the forward model are $170 \times 130 \times 100$ mm

behaviour of the human tissues, we implemented the single-pole Debye model for the complex permittivity which was also presented in Sect. 3.

For each case, we examined two forward models. The first one consisted of a simplified representation of the simulation setup, which includes the tank filled with the matching medium and an ellipsoid that mimics the average brain tissue. The latter one included a multi-layer ellipsoid that represents a simplified version of the multi-layer phantom. By means of an example, Fig. 9 presents the forward models for the three and the six-layer Zubal phantom, as well as the forward model filled only with the average brain dielectric properties. In each layer of the forward model we assigned the Debye parameters of the equivalent head tissue which are presented at Table 4. The algorithm considered the system's antennas as line sources, which were placed at the same locations as the simulation. To calibrate the simulated model, we used the "NT" reference measurement. This calibration method relies on the signal difference between the "with target" and the "no target" scenarios, to calibrate the signals received from the "NT" simulation.

For the examined scenarios, we measured the S-parameters from 0.5 to 1.5 GHz and we fed the scattered signals into the algorithm. The computation time for the forward solver for each antenna was approximately 17 s. The execution time for each antenna was 80 s resulting in a total execution time of 7.1 h for 20 iterations at each frequency. In this study, we employed the Nvidia GPU Tesla K20c to accelerate the 3-D FDTD implementation with CUDA which was developed by our our group [45], and Matlab MEX functions to integrate the GPU code with our DBIM code in Matlab. The implementation was based on Matlab R2020b, CUDA 8.0 and VS 2015.

4.1.3 Results

We present the reconstruction results using the 3-D version of DBIM-TwIST algorithm, for the region surrounded by the antenna array shown in Fig. 8. In addition, we define the relative error of ϵ_∞ in the target area, between the inverse and the forward model as:

$$\text{er} = \frac{||\epsilon_\infty \text{ original} - \epsilon_\infty \text{ reconstructed}||^2}{||\epsilon_\infty \text{original}||^2} \tag{17}$$

The error metrics are calculated at each DBIM iteration and can be used to compare the quality of the reconstructed images as well as choose the optimal number of iterations. For the examined cases, we chose 20 iterations as our previous work has shown that it is a good trade-off between the quality of the images and the total computational time.

Figures 10, 11, 12, 13 and 14 demonstrate the reconstructed results at the horizontal cross-section at the level Z = 60 achieved with 3-D DBIM-TwIST for 1 GHz, for the case of h- and i-stroke respectively, for both types of forward models. The size of the 3-D domain is 100 × 100 × 100 with a resolution of 2 mm. Figures 10, 11, 12 and 13 reveal that we achieve similar reconstructed images while using the multi-layer initial model for both stroke targets. We also note that the reconstructed values of the targets, are closer to their real values when using the homogeneous forward model. However, in Fig. 14 we observe that the quality of the reconstructed h-stroke is better when we apply a homogeneous initial model, while for the case of i-stroke, we achieve less artifacts when using a multi-layer forward model.

Figure 15 plots the e_r of ϵ_∞ in the target area, as a function of DBIM iterations. We observe that in most cases the values between the homogeneous and the multi-layer forward model are similar both for hemorrhagic and ischemic stroke. However, we also note that for the six-layer Zubal model, h-stroke's e_r for the multi-layer forward model is higher than the equivalent value for the homogeneous forward model, while we notice the opposite for the i-stroke. The plots of Fig. 15 further validate what we observed from the reconstructed images. More specifically, for the i-stroke cases, we observe an offset between the location and the shape of the real and the detected target, which translates into errors higher than 1.

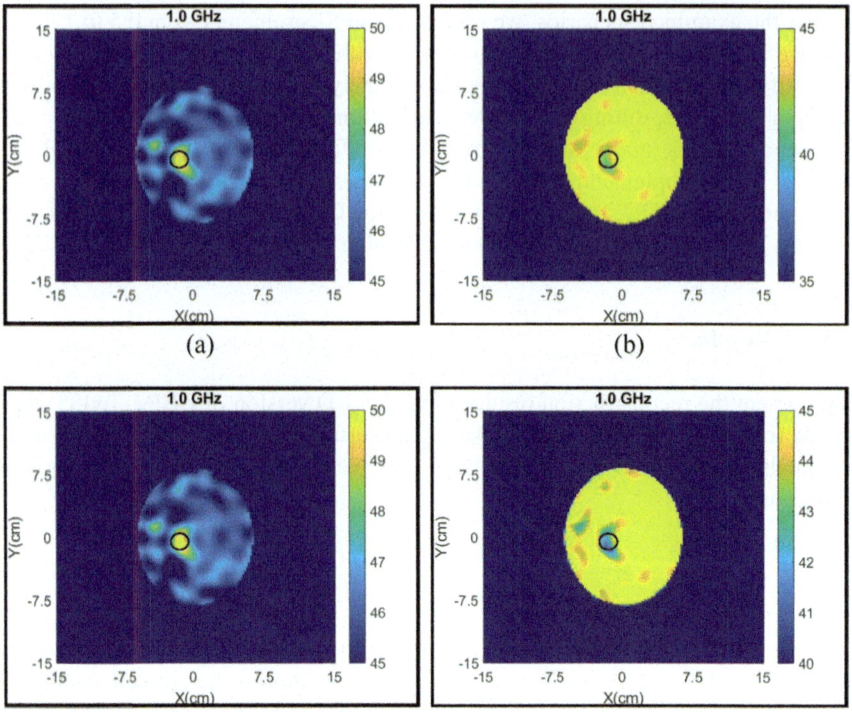

(a) (b)

(c) (d)

Fig. 10 Reconstructed results of the real part of the complex permittivity for the Zubal model including the skin and average brain layers, for the cases of hemorrhagic (left) and ischemic stroke (right). Top: homogeneous forward model; bottom: forward model that includes the skin and average brain layers

Overall, the results indicate that for most scenarios, the use of a homogeneous forward model produces slightly better reconstructed images than the multi-layer forward model. However, for the case of Zubal head phantom in the presence of all head tissues, we observe that we achieve better results while applying an improved forward model, for the case of ischemic stroke where the dielectric contrast between the target and the background tissue is less than the equivalent dielectric difference in the presence of hemorrhagic stroke. We note that for the DBIM-TwIST there is no "best parameter" that improves the quality of the reconstructed images. One of the advantages of this method lies on its adaptive parameters that can be adjusted according to each scenario.

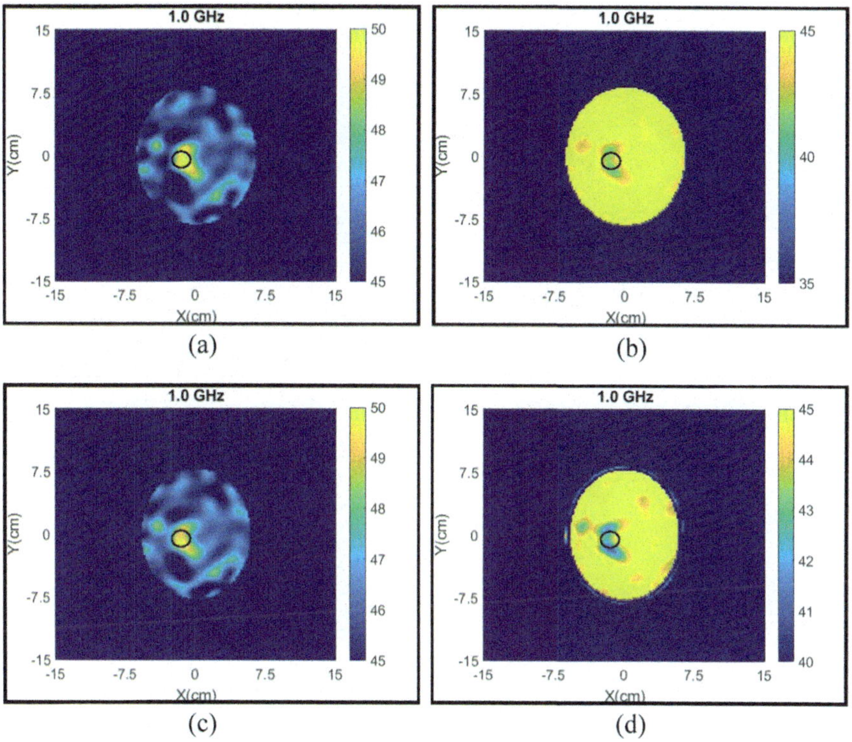

Fig. 11 Reconstructed results of the real part of the complex permittivity for the Zubal model including the skin, fat and average brain layers, for the cases of hemorrhagic (left) and ischemic stroke (right). Top: homogeneous forward model; bottom: forward model that includes the skin, fat and average brain layers

4.2 Comparison of 2-D and 3-D DBIM-TwIST for a 3-D Imaging Problem

In this section we present an initial validation and assessment of the 3-D algorithm for the case of a simplified homogeneous elliptical brain phantom in the presence of a small stroke-like target, surrounded by a 3-D antenna array arranged in two cylindrical rings. We also compare the capabilities of the 2-D and 3-D versions of the algorithm to confirm the hypothesis that 3-D DBIM-TwIST can result in more accurate reconstructions for 3-D inverse problems.

Figure 16 illustrates the brain phantom immersed in our imaging setup presented at [58]. The phantom is surrounded by 16 monopole antennas in two different arrays, which form two offset rings of 8 elements each. To mimic the h-stroke or the i-stroke, we have examined the same configuration in the presence of a 20 mm diameter and 20 mm height target, located between the two heights of the antennas. The Debye parameters of the target and of the average brain have been assigned by curve fitting

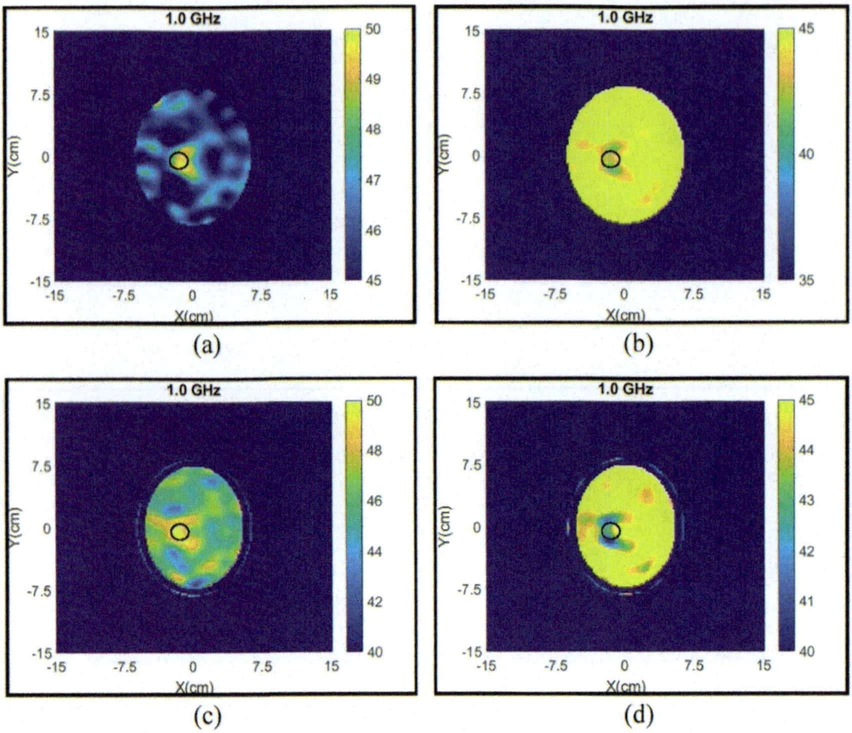

Fig. 12 Reconstructed results of the real part of the complex permittivity for the Zubal model including the skin, fat, bone and average brain layers, for the cases of hemorrhagic (left) and ischemic stroke (right). Top: homogeneous forward model; bottom: forward model that includes the skin, fat, bone and average brain layers

from CST data [58]. As explained before, the inverse model used in our imaging algorithm considers a 2-D (112×153 mm), or a 3-D ($112 \times 153 \times 100$ mm) ellipsoid, for the two versions of the algorithm respectively. The algorithm considers point sources which form an elliptical array and are located at the same height in 3-D case as the antennas in the CST simulation.

Figures 17, 18, 19 and 20 present the reconstructed permittivity and conductivity at 1 Ghz for the phantom of Fig. 16 for h- and i-stroke respectively, using both 2-D and 3-D versions of the DBIM-TwIST. More specifically, Fig. 17 shows the 2-D reconstructed permittivity and conductivity using either the top or the bottom array of antennas in our forward model. Figures 18, 19 and 20 demonstrate the sagittal, axial and coronal planes at Y-Z, X-Z and X-Y slices from the 3-D reconstructions, respectively.

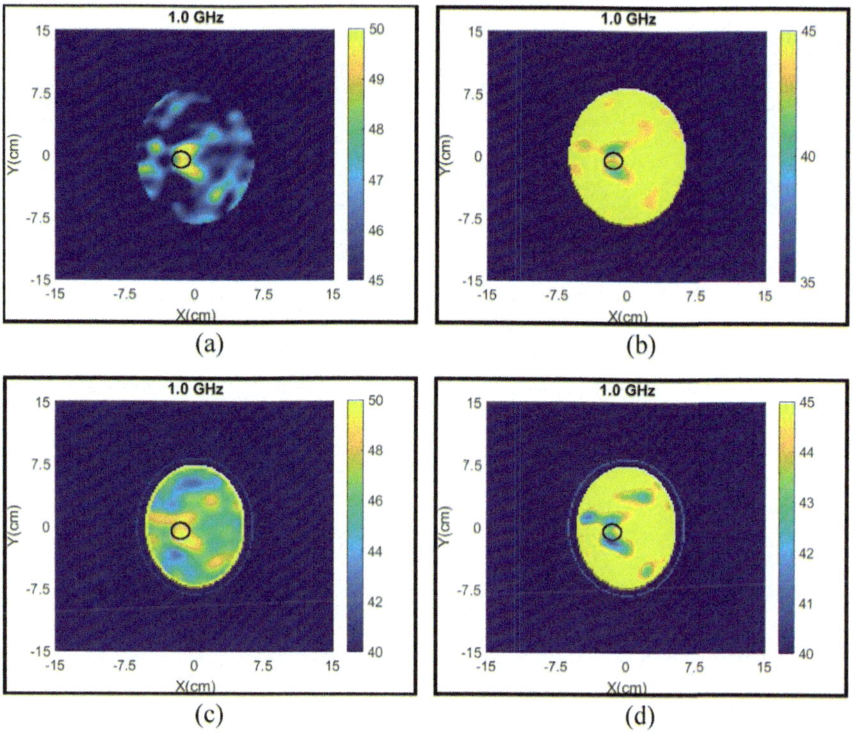

Fig. 13 Reconstructed results of the real part of the complex permittivity for the Zubal model including the skin, fat, bone, CSF and average brain layers, for the cases of hemorrhagic (left) and ischemic stroke (right). Top: homogeneous forward model; bottom: forward model that includes the skin, fat, bone, CSF and average brain layers

We observe that both versions of the algorithm can successfully locate and detect the dielectric permittivity of the target. However, the 3-D algorithm achieves more accurate results for both cases of stroke, as it combines information from both rings in an efficient way. Importantly, the 3-D DBIM-TwIST leads to a clear improvement in estimating the target's conductivity relative to 2-D reconstruction based on either ring. Having established the benefits of 3-D imaging, our subsequent results focus on validating the 3-D DBIM-TwIST experimentally, as well as compare its performance with the 2-D version of the algorithm. The experimental setup and respective CST model, however, does not include two antenna rings due to the smaller size of the experimental phantom. In this respect, the setup in these results resembles a 2-D configuration, but the 3-D results are still relevant as they are derived from the 3-D algorithm which reconstructs a fully 3-D region.

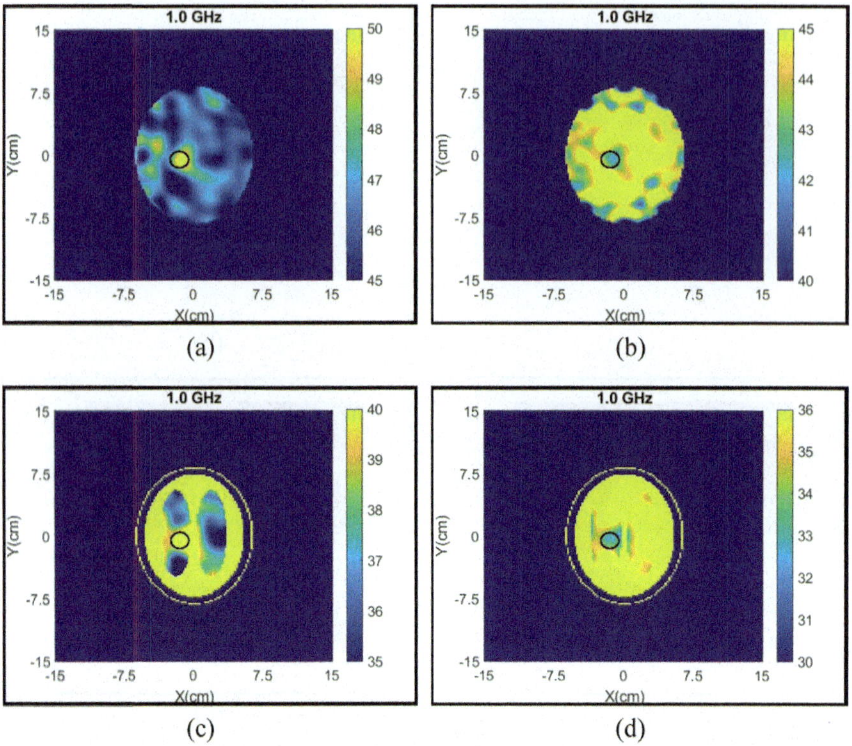

Fig. 14 Reconstructed results of the real part of the complex permittivity for the Zubal model including all the head layers, for the cases of hemorrhagic (left) and ischemic stroke (right). Top: homogeneous forward model; bottom: forward model that includes all the head layers

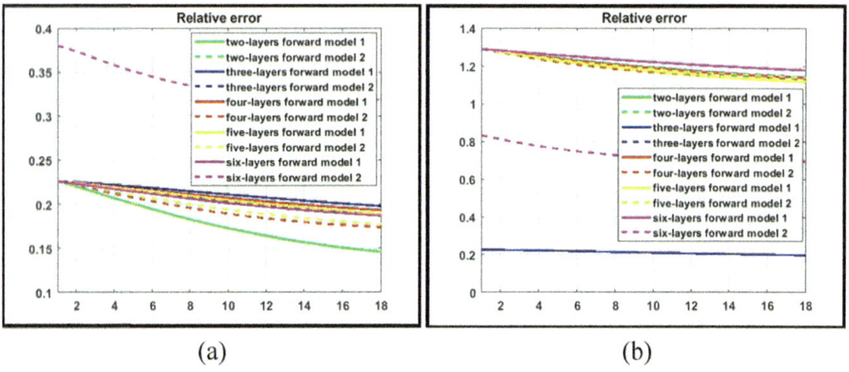

Fig. 15 Top: relative error of the ϵ_∞ in the target area for h-stroke at 1 GHz; bottom: relative error of the ϵ_∞ in the target area for i-stroke at 1 GHz

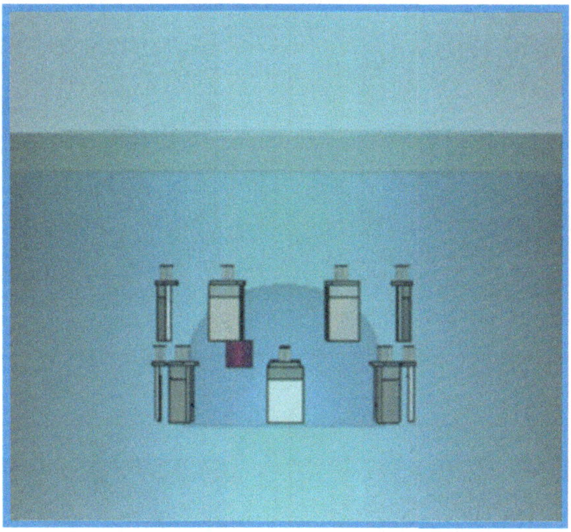

Fig. 16 CST model of a simplified brain phantom surrounded by 16 antennas in 2 elliptical arrays

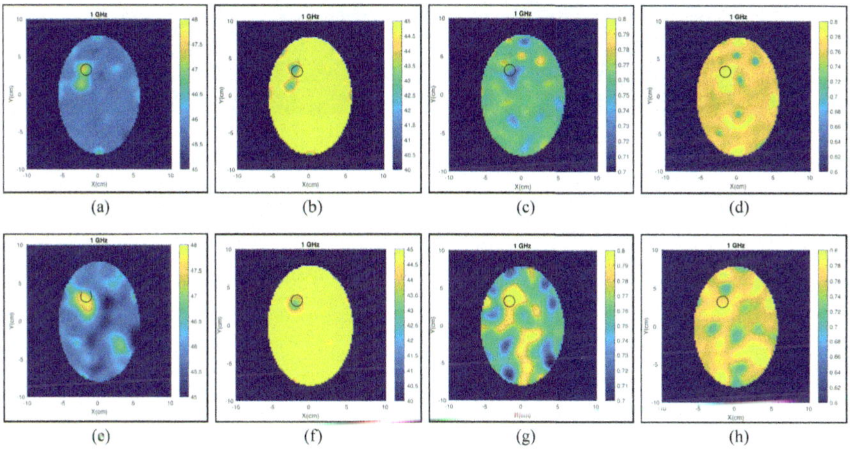

Fig. 17 First row **a**, **b** reconstructed permittivity for the phantom of Fig. 16, for h- and i-stroke using 2-D DBIM-TwIST and the bottom array of antennas; first row **c**, **d** reconstructed conductivity for the phantom of Fig. 16, for h- and i-stroke using 2-D DBIM-TwIST and the bottom array of antennas; second row **e**, **f** reconstructed permittivity for the phantom of Fig. 16, for h- and i-stroke using 2-D DBIM-TwIST and the top array of antennas; second row **g**, **h** reconstructed conductivity for the phantom of Fig. 16 for h- and i-stroke using 2-D DBIM-TwIS and the top array of antennas

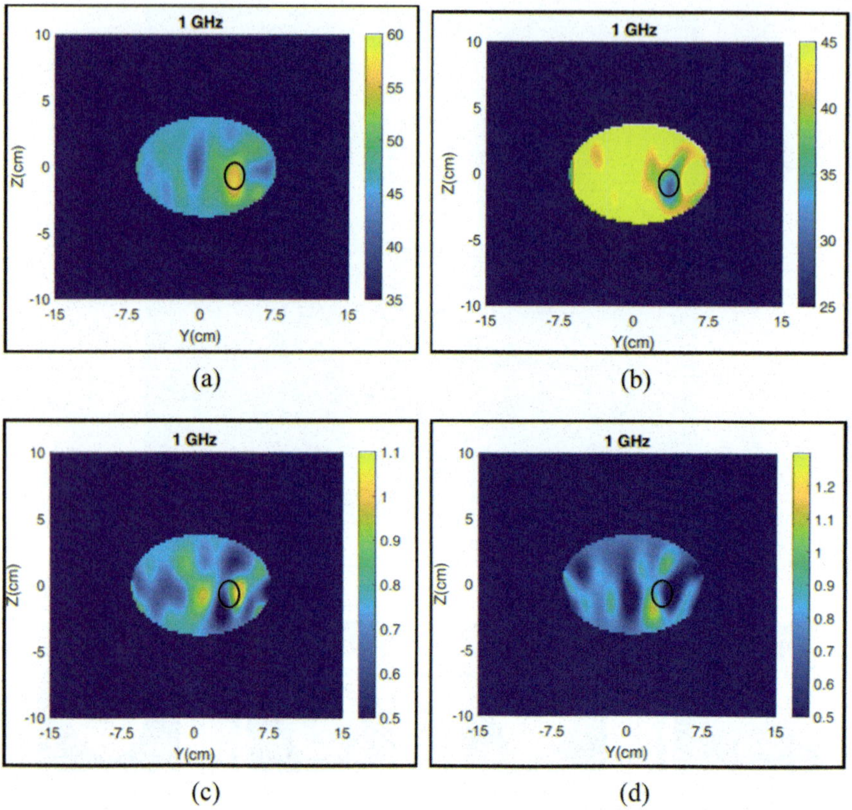

Fig. 18 First row: Y-Z slice of 3-D reconstructed permittivity for the phantom of Fig. 16 for h- and i-stroke; second row: Y-Z slice of 3-D reconstructed conductivity for the phantom of Fig. 16 for h- and i-stroke

4.3 Experimental Methodology

4.3.1 Head Phantom Construction

We derived our Zubal phantom from a 2-D axial $x - y$ slice taken from the original MRI-derived Zubal head phantom [59], which was then extended by a few cm along the z axis. We imported this 3-D model in CAD software to simplify its structure so that it is easy to fabricate and work with, without sacrificing much of its complexity (Fig. 21a, b). This process was conducted by Ioannis Sotiriou, a member of our research group in King's College London. For our simulation, we imported the CAD model on CST which from now on we will refer to as the "Zubal model". The simulated phantom and setup are shown in Fig. 21c, d). We filled each layer of the phantom with the dielectric properties of different head tissues having created a anatomically accurate 5-material head model. We placed a cylindrical target at the

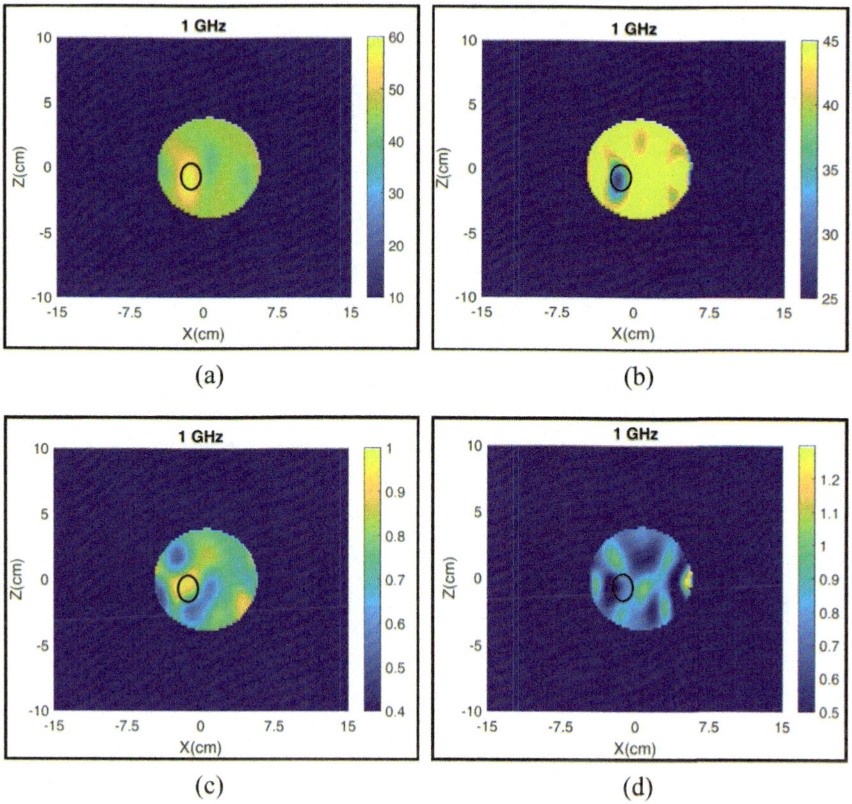

Fig. 19 First row: X-Z slice of 3-D reconstructed permittivity for the phantom of Fig. 16 for h- and i-stroke; second row: X-Z slice of 3-D reconstructed conductivity for the phantom of Fig. 16 for h- and i-stroke

top right, inside the white matter area, with radius ρ = 12.5 mm and height h = 100 mm. The phantom was immersed in our imaging setup presented in [58], which includes eight antennas placed in an elliptical array configuration with major and minor axes equal to 205 mm and 148 mm, respectively.

To validate the algorithm experimentally, we prepared a novel gelatine-based head phantom in accordance with the anatomy and the dimensions of the Zubal model. The mould form was designed and 3-D printed by Ioannis Sotiriou and consists of six compartments which create the sequence of the tissue mimicking materials needed for our assembly. The moulds were fabricated using a Zortrax M200 FDM 3-D printer in a complementary manner using two moulds shown in Fig. 22a, b. This methodology for the creation of the phantom allows a more accurate validation as it avoids the presence of 3-D printed infill patterns that irregularly encapsulate fluids. The adverse effect of the presence of plastic in phantoms which use it as support was demonstrated in [60].

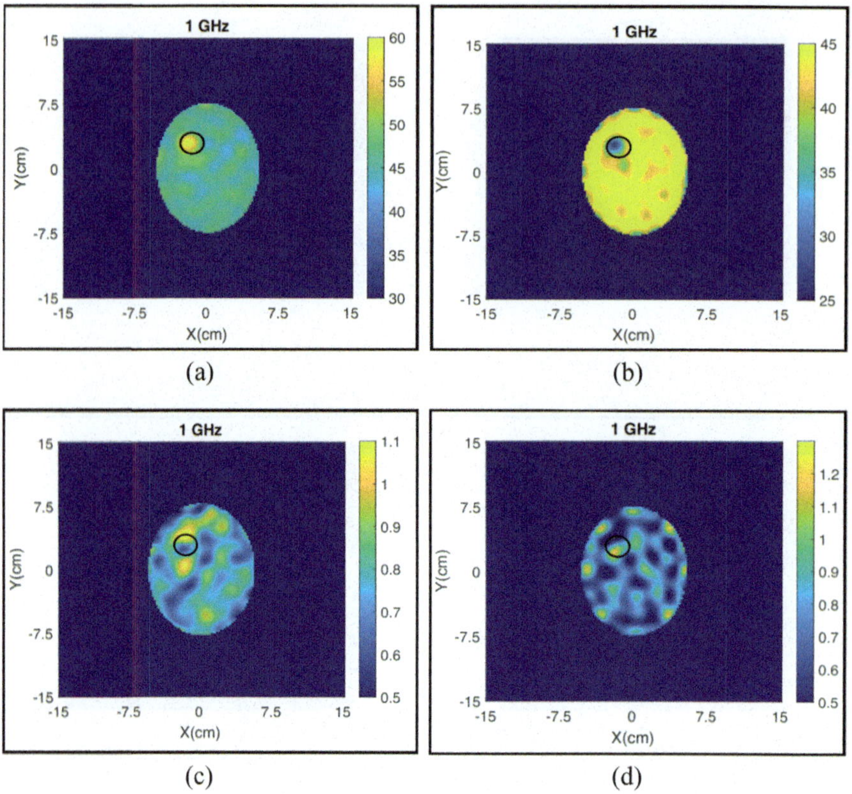

Fig. 20 First row: X-Y slice of 3-D reconstructed permittivity for the phantom of Fig. 16 for h- and i-stroke; second row: X-Y slice of 3-D reconstructed conductivity for the phantom of Fig. 16 for h- and i-stroke

Having prepared the tissue-mimicking gels, we poured them in the desired compartments in Fig. 22a, b, and we allowed them to dry for at least 8 h depending on their concentration. To extract the gels from the moulds we initially submerged the moulds in warm water so the outer layers could slide out easier, whilst we were applying small amount of air pressure at the lower part of the mould (Fig. 23). Once the gels were extracted, we stacked the materials within each other and wrapped the assembly using a thin plastic film to prevent it from drying. The final Zubal model is presented in Fig. 22c.

We prepared the tissue-mimicking materials for skin, bone, CSF, grey matter, white matter, blood and ischemia by following the process presented in [58]. Table 5 shows the concentrations of the materials used for preparing each layer of the phantom.

Figure 24 displays the measured dielectric properties of the prepared phantoms for skin, bone, grey matter, white matter and ischemia, respectively. We note that the figure does not report the measured properties of CSF/h-stroke, as the material

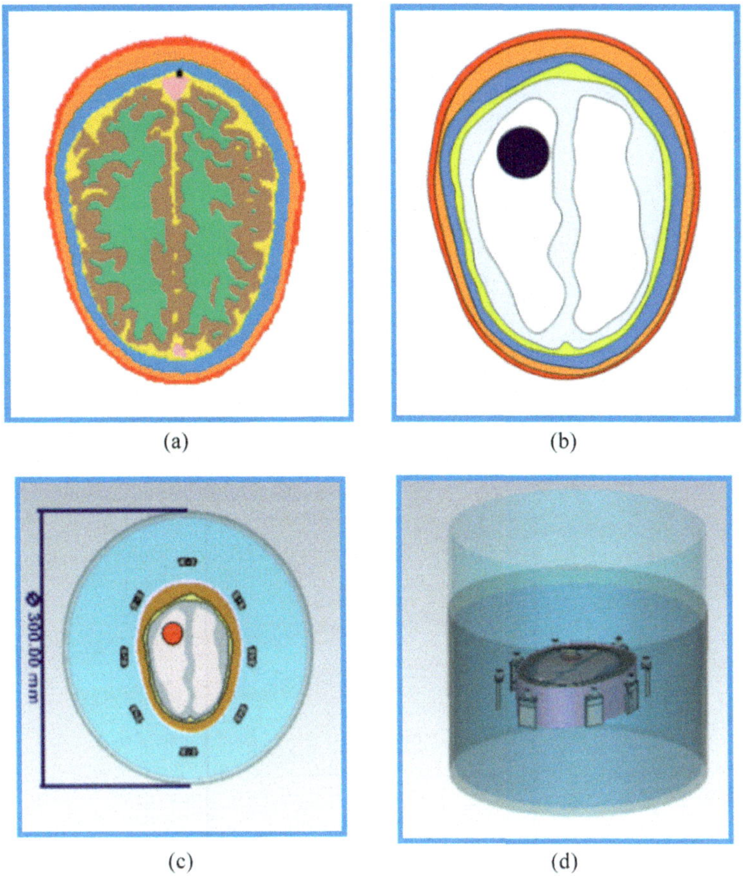

Fig. 21 **a** The Zubal head model section; **b** simplified CAD model for fabrication; **c** top view of the Zubal model on CST; **d** side view of the Zubal model on CST. The phantom includes the following head layers: skin, bone, CSF, grey matter, white matter and the target

Table 5 Concentrations of materials used for 100 ml of tissue mimicking phantoms

	Water (ml)	Gelatine powder (g)	Kerosene + safflower oil (ml)
Skin	73	13	15
Bone	45	5	50
CSF/h-stroke	80	17	–
Grey matter	75	12	13
White matter	70	10	20
i-stroke	55	10	35

For each tissue mimicking phantom we also use 1 ml of propanol and 1 ml of surfactant

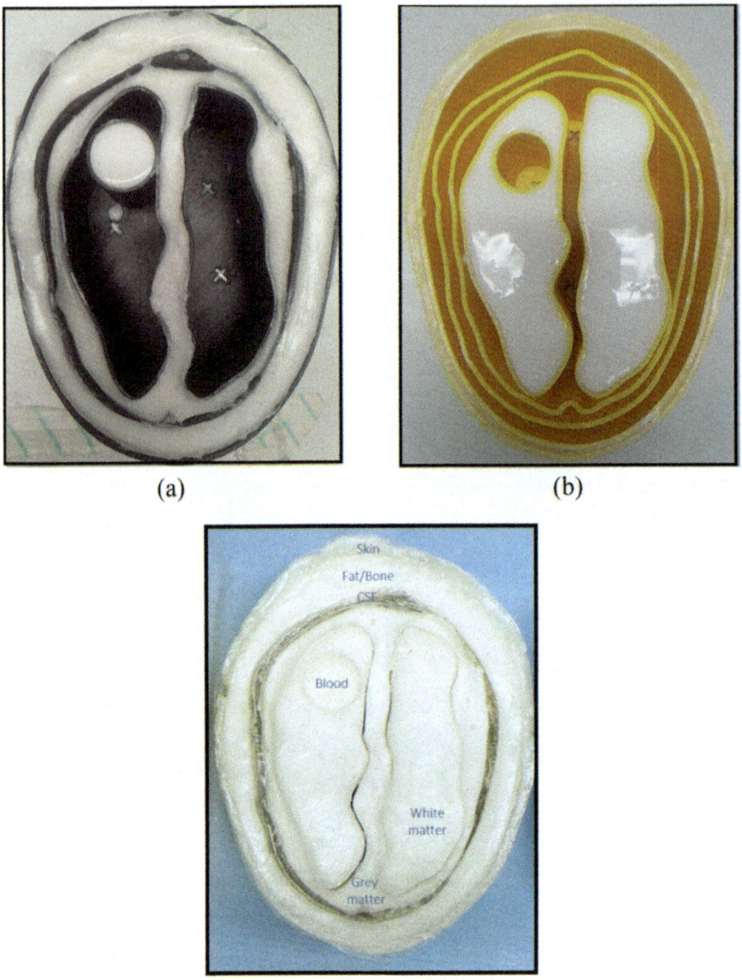

Fig. 22 Top view of the moulds constructing the Zubal head phantom from gelatine-based tissue mimicking materials: **a** first mould (bone, grey matter, and target); **b** second mould (skin, CSF and white matter); **c** final phantom with five tissues and target

was not set at the time of the dielectric measurements. The preparation of CSF/h-stroke does not require the addition of oil, which is the most challenging step in the preparation process and can lead to variations in the final dielectric properties of the phantoms. Preparing phantoms that mimic the dielectric properties of CSF or h-stroke requires only dissolving the gelatine within the water. Therefore, we can safely assume that we can achieve the properties reported at [58]. For the CSF/h-stroke fabricated phantom of [58] the permittivity at 1 GHz is 68, while the conductivity is 0.3 S/m and 1.2 S/m for h-stroke and CSF respectively. The measurements were

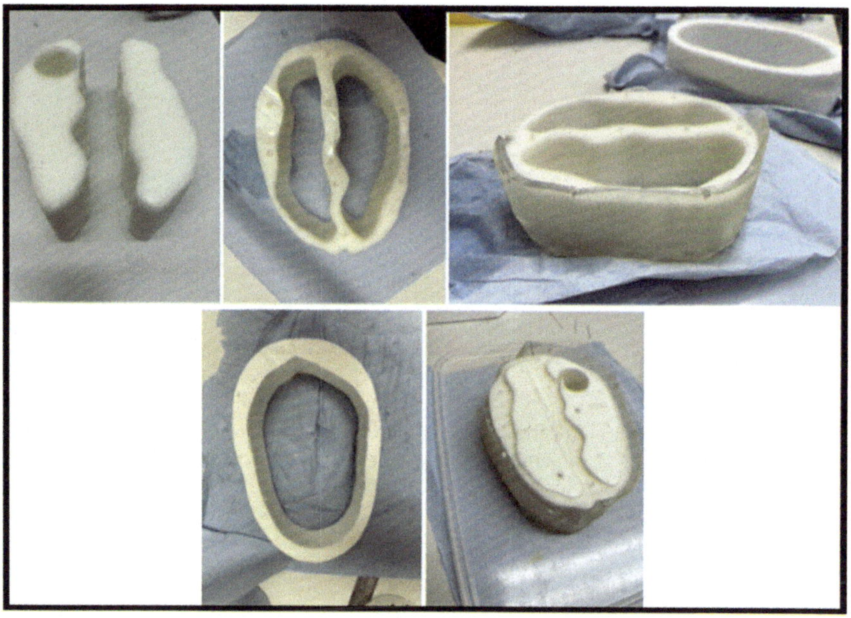

Fig. 23 Demoulding process

performed using Keysight's dielectric spectroscopy kit, over a 0.5–1.5 GHz frequency range at different points. The plots reveal a very good agreement with the reference permittivity taken from CST (Fig. 24a), but lower conductivity values for skin and grey matter phantoms (Fig. 24b). As there is no literature regarding the dielectric properties of ischemia in a wide frequency range, we made the assumption that the ischemic properties are 25% less than the dielectric properties of healthy brain tissue [35].

4.3.2 Imaging Methodology

The Zubal model was positioned in the middle of the 300 mm diameter cylinder tank, as shown in Fig. 25. It was surrounded by eight spear-shaped antennas [55] in an elliptical array, and the imaging tank was filled with 90% glycerol-water. The experimental system, the choice of number of antennas, as well as the choice of the matching medium have been extensively reported in our previous chapter. The setup was surrounded by vacuum and the measurements were performed in the frequency range of 0.5–1.5 GHz. For every studied case, two scenarios were examined. First, we considered a scenario without a target ("NT"), where the target area was filled with the dielectric properties of the white matter. For the "WT" scenario we filled the equivalent target territory (radius $\rho = 12.5$ mm and height h = 100 mm) with the dielectric properties either of blood or of ischemia. For each scenario and for M

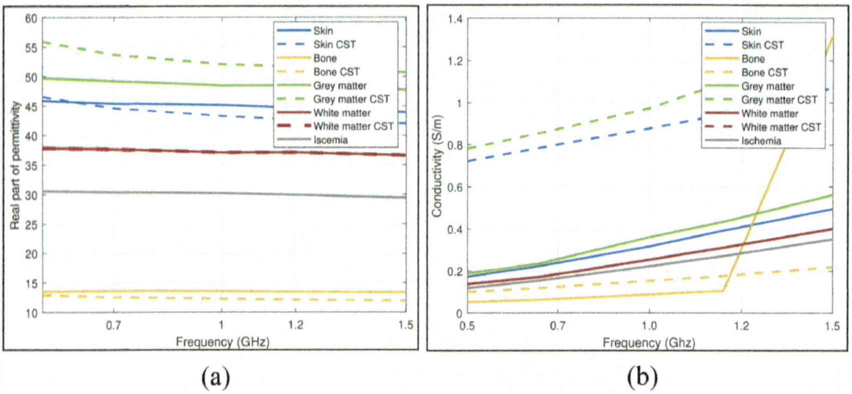

Fig. 24 Measured dielectric properties of the produced tissue mimicking materials against reference values taken from CST: **a** permittivity, and **b** conductivity. Note that there appears to be a measurement error for the bone phantom above 1.2 GHz

Fig. 25 The imaging tank with the immersed Zubal model

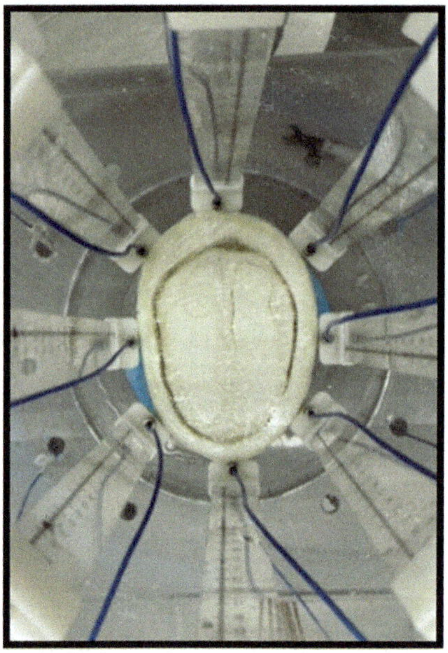

frequency points, the eight antenna-array captured the MWI data, resulting in a $8 \times 8 \times M$ scattering matrix, which was then fed and processed by our algorithm.

Our forward model consisted of both a 2-D and 3-D depiction of the set-up for the two versions of the DBIM-TwIST respectively. It included the tank filled with the dielectric properties of 90% glycerol-water and an ellipsoid that represented the Zubal model filled with the dielectric properties of the average brain. The axes of the 2-D ellipsoid were 153 and 112 mm long and they represented the actual size of the

Table 6 Debye parameters of the forward model

Material type	$\Delta\epsilon$	ϵ_∞	σ_s
90% glycerol-water	6.56	16.86	0.3232
Average brain	45.8	0.1244	0.7595

phantom's axial slice at the height where the antennas were located. The dimensions of the 3-D ellipsoid were $170 \times 130 \times 100$ mm long.

As the forward models are simplified versions of the experiment, they are not identical and hence there are differences between the real and the FDTD simulated models. The algorithm simulated the setup's antennas with line sources placed at the same location as their feeding point. To minimize the mismatch between the simulation or the experiment and the forward model, we calibrated our forward model by employing a "no target" scenario, as in our previous work (e.g., [58]). We calculated the signal difference between the inverse and the forward "no target" cases and we used this difference to calibrate the scattered data taken from the "with target" scenario.

To model the frequency-dependent behaviour of the human tissues, we applied the single-pole Debye model for the complex permittivity ϵ_b [58]. Table 6 presents the Debye parameters of the "initial guess" model that were used with the FDTD forward solver to calculate the 2-D (transverse magnetic wave) and 3-D (full wave) fields. The calculation of full wave fields is the main difference between the 2-D and the 3-D version of the algorithm. The Debye parameters are calculated and updated at each iteration until we reach a fixed number of iterations or the difference between the "measured" and calculated electric fields is minimised. We employ DBIM-TwIST to reconstruct the Debye parameters with a 2 mm resolution grid. We have observed that the reconstructed results do not improve significantly when we apply more than 20 iterations, and hence we chose to perform 20 DBIM iteration for both 2-D and 3-D versions of DBIM-TwIST. This number is a trade-off between the accuracy and the total execution time.

For the 3-D code, we employed the Nvidia GPU Tesla K20c to accelerate the 3-D FDTD execution with CUDA, and Matlab MEX functions to incorporate the GPU code with the DBIM code in Matlab. The environment was based on Matlab R2020b, CUDA 8.0 and VS 2015. The implementation of DBIM-TwIST with a GPU-based FDTD forward solver was extensively described in [45]. By the means of an example, using the 3-D DBIM-TwIST, for the case of the Zubal model with h-stroke, the elapsed time for each antenna was 81 s, meaning that the running time for each iteration was roughly 11 min when using 8 monopole antennas. For the 2-D code we do not need GPU acceleration for the forward solver. The running time for each iteration, when using 8 antennas, was 12 s using Matlab R2020b on an Intel i7 processor with 16 GB RAM memory.

4.4 Results with the Zubal Phantom

This section presents reconstruction results from our Zubal model in the presence of a hemorrhagic and an ischemic stroke. Our aim was to assess our imaging system's ability to determine the type of stroke from estimating its dielectric properties using the 2-D and 3-D DBIM-TwIST algorithm. We have examined both CST-calculated and experimental data, and we have compared our 2-D and 3-D results. As previous work in MWI has shown that thermal noise does not affect reconstruction quality as much as the mismatch between the true experiment and the inverse model, we did not include noise in the CST data, to focus on the effect of the model mismatch errors [23, 39]. To compare the quality of the reconstructed images we calculated the relative errors of the permittivity and conductivity in the target area at each iteration, as in Eq. 17.

4.4.1 Reconstructions with Simulation Data

Two CST simulations were conducted to acquire the scattered field data, with and without the target. The dielectric properties of each tissue were imported from the CST library. Specifically for the ischemic target, as there are no frequency-dependent reported values in the literature, we used our measured dielectric properties reported in [58].

Figures 26 and 27 present the reconstructed permittivity and conductivity at four frequencies for the Zubal model of Fig. 21 inside our prototype, in the presence of h-stroke and i-stroke targets. More specifically, Fig. 26 shows the reconstructed permittivity and conductivity for the 2-D version of DBIM-TwIST, while Fig. 27 shows the equivalent reconstructed dielectric properties for the 3-D version. Overall, more accurate results are produced for the lower frequencies of 0.9 and 1 GHz, and artifacts increase significantly for frequencies above 1 GHz. This can be attributed to the increased non-linearity of the EM inverse scattering problem in higher frequencies, where multiple scattering effects become stronger.

We also observe that there is a clear distinction in the estimation of the reconstructed permittivity value in the area where the target is located for the two targets for both versions of the code. For 2-D DBIM-TwIST, the results of Fig. 26 estimate the value of permittivity ϵ' at 1 GHz as $\epsilon' = 50.31$ and $\epsilon' = 43.15$ for hemorrhagic and ischemic stroke, respectively. For the 3-D DBIM-TwIST, Fig. 27 shows that the equivalent values of permittivity are $\epsilon' = 61.58$ and $\epsilon' = 28.52$, respectively. We also observe that 3-D DBIM-TwIST estimates more accurately the reconstructed values for the target area. The equivalent reconstructions of Fig. 26 reveal that we are not able to successfully detect the target from the conductivity reconstructions. The quality of conductivity reconstructions improves using the 3-D DBIM-TwIST algorithm, for which the reconstructed conductivity values are significantly different for the two stroke-type targets.

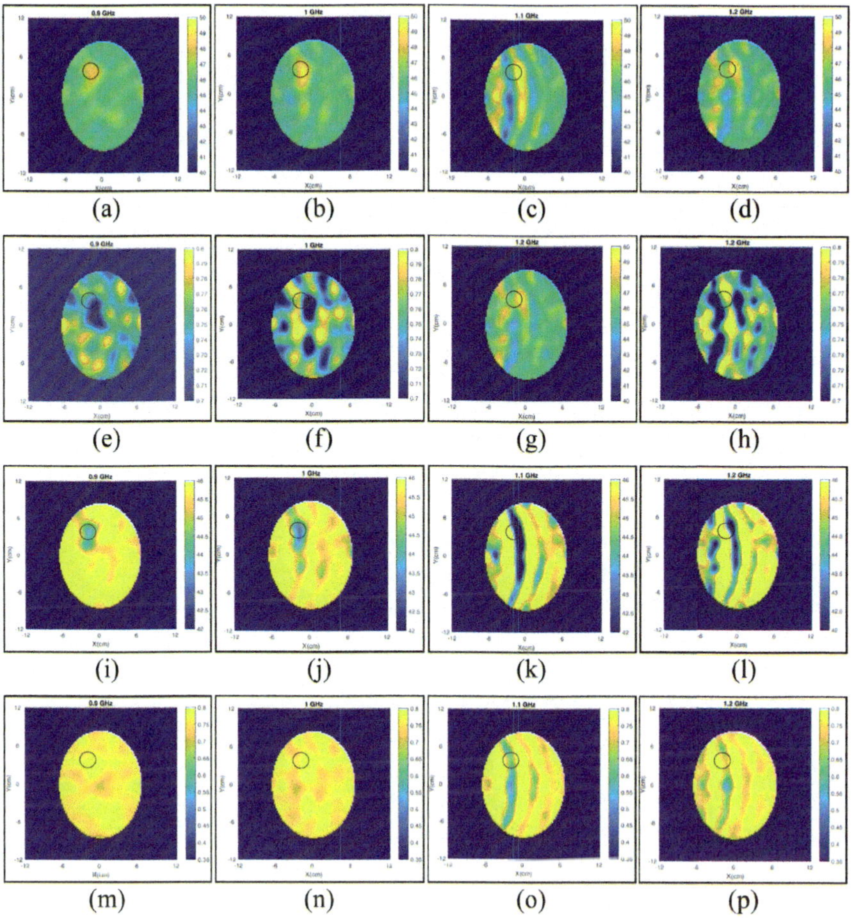

Fig. 26 First row: CST reconstructed results of the permittivity for h-stroke using 2-D DBIM-TwIST; second row: CST reconstructed results of the conductivity for h-stroke using 2-D DBIM-TwIST; third row: CST reconstructed results of the permittivity for i-stroke using 2-D DBIM-TwIST; fourth row: CST reconstructed results of the conductivity for i-stroke using 2-D DBIM-TwIST

4.4.2 Reconstructions with Experimental Data

Figures 28 and 29 present complex permittivity reconstructions in the presence of the hemorrhagic or ischemic stroke target, using measured data and the 2-D and 3-D DBIM-TwIST, respectively. These images suggest that we are able to detect and estimate the dielectric properties of h-stroke for both versions of the algorithm. The 3-D DBIM-TwIST reconstructions of the permittivity have more artifacts that could be misleading regarding the target location in the absence of a-priori information. Interestingly, the 3-D version of the algorithm leads to conductivity reconstructions with less artifacts than the equivalent permittivity results for the h-stroke case at lower

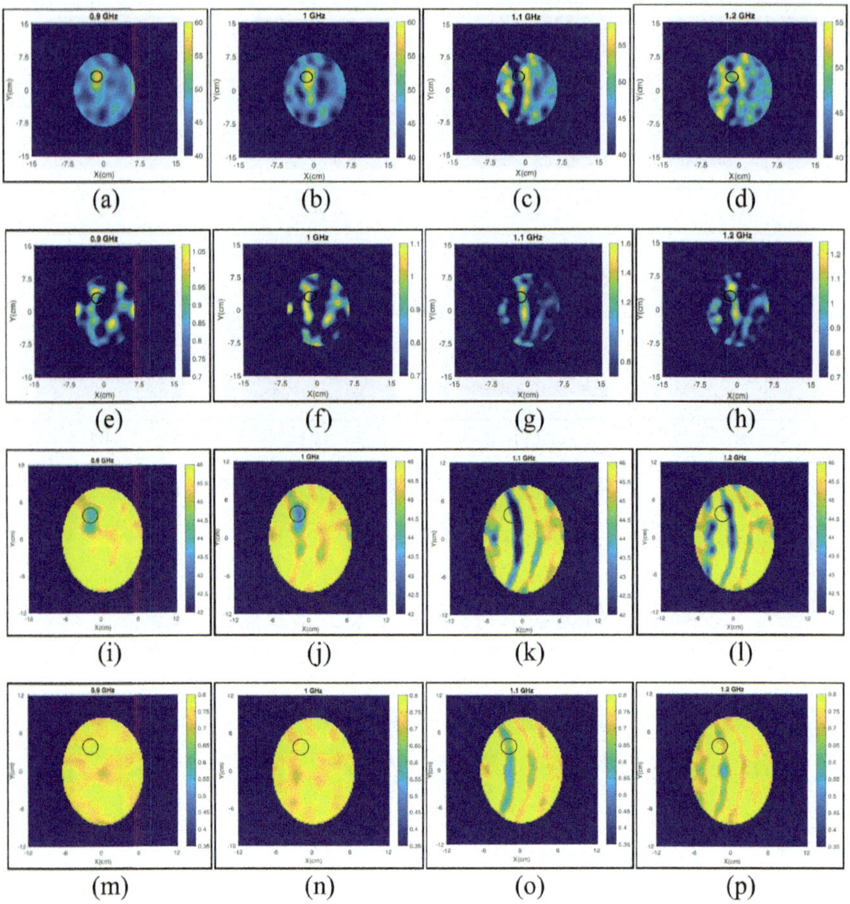

Fig. 27 First row: CST reconstructed results of the permittivity for h-stroke using 3-D DBIM-TwIST; second row: CST reconstructed results of the conductivity for h-stroke using 3-D DBIM-TwIST; third row: CST reconstructed results of the permittivity for i-stroke using 3-D DBIM-TwIST; fourth row: CST reconstructed results of the conductivity for i-stroke using 3-D DBIM-TwIST

frequencies. Reconstructed permittivity values in the area where the target is located at 1 GHz are $\epsilon' = 51.6$ and $\epsilon' = 61.57$, for the 2-D and 3-D DBIM-TwIST respectively. However, we observe artifacts in the reconstruction images for reconstructions with the ischemic stroke. More specifically and as shown in Figs. 28i–l–29i, we are able to detect an area with values lower than the average brain, which is close to the target area. There is an offset, however, between the location of the real and the detected target. The results in Figs. 28m–p–29m for the conductivity reconstructions reveal similar tendencies.

The dielectric difference observed between the i-stroke and the white matter is less than the dielectric difference between the h-stroke and the white matter, resulting

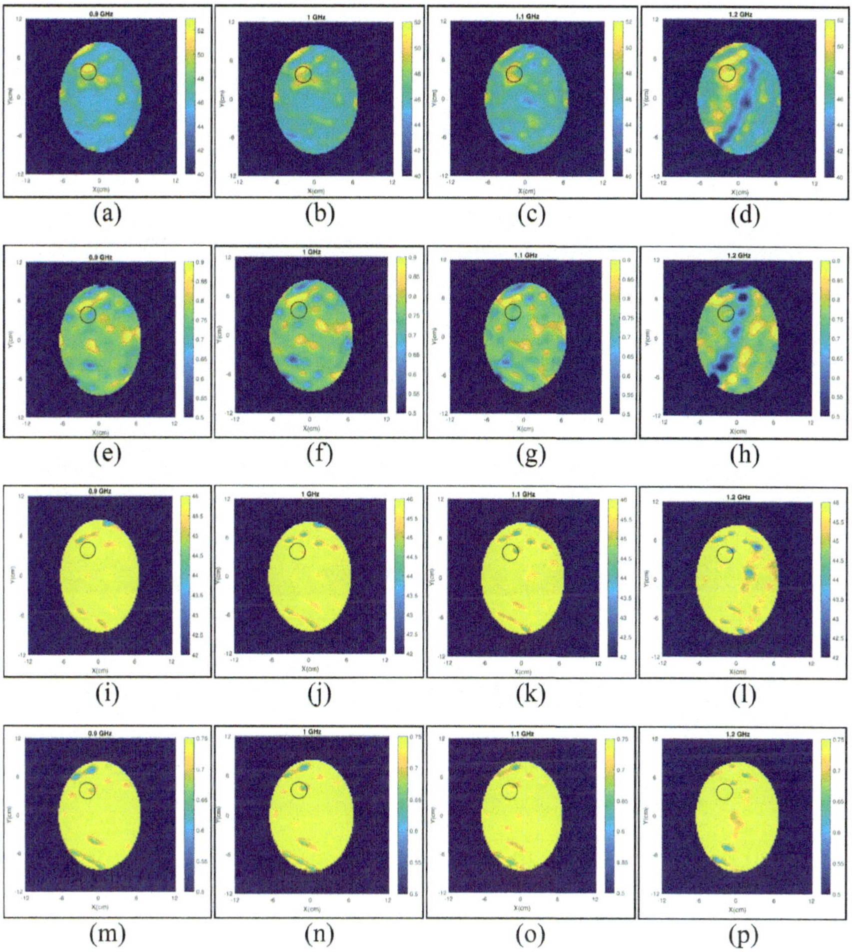

Fig. 28 First row: experimental reconstructed results of the permittivity for h-stroke using 2-D DBIM-TwIST; second row: experimental reconstructed results of the conductivity for h-stroke using 2-D DBIM-TwIST; third row: experimental reconstructed results of the permittivity for i-stroke using 2-D DBIM-TwIST; fourth row: experimental reconstructed results of the conductivity for i-stroke using 2-D DBIM-TwIST

in a more complex brain structure. In addition, differences between our experimental model and the FDTD forward model, in combination with different software environments, result in significant model mismatch. Overall, both 2-D and 3-D algorithms have detected accurately the h-stroke, although we observe more artifacts for the 3-D version. Overall, experimental measurement errors lead to less accurate reconstructions than their CST equivalent.

The error metrics calculated by Eq. 17 can be used to evaluate the accuracy of the reconstruction. These errors are summarized in Tables 7, 8, 9 and 10. We observe

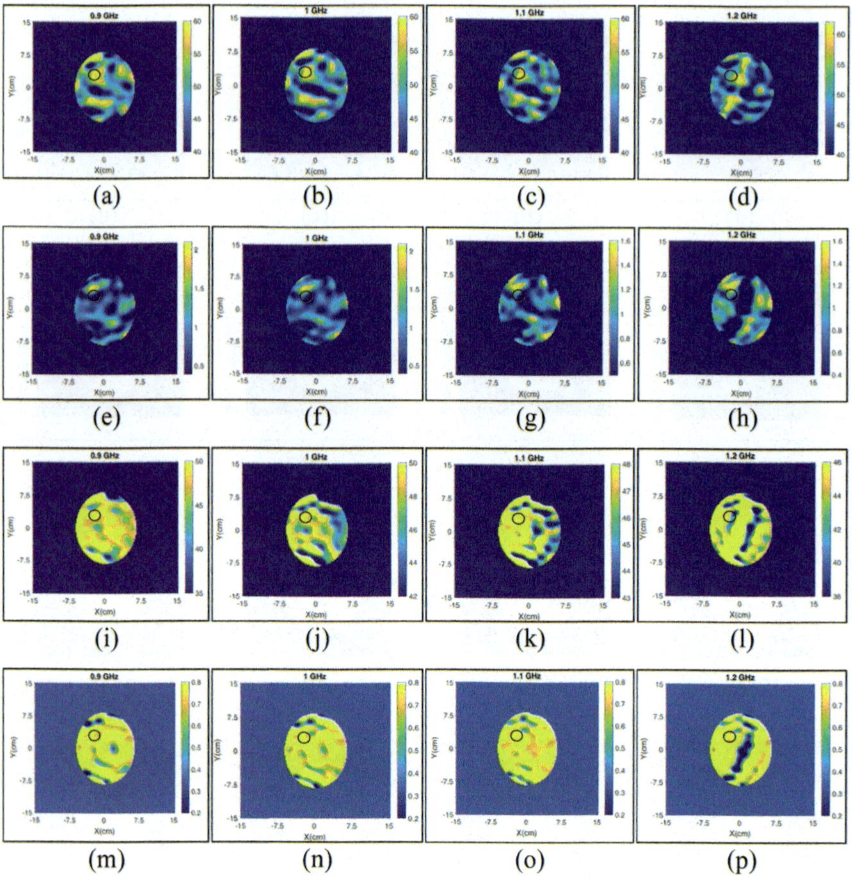

Fig. 29 First row: experimental reconstructed results of the permittivity for h-stroke using 3-D DBIM-TwIST; second row: experimental reconstructed results of the conductivity for h-stroke using 3-D DBIM-TwIST; third row: experimental reconstructed results of the permittivity for i-stroke using 3-D DBIM-TwIST; fourth row: experimental reconstructed results of the conductivity for i-stroke using 3-D DBIM-TwIST

that relative errors fall below 1 for all cases except from the conductivity errors of the 2-D i-stroke cases, as well as the 3-D i-stroke scenarios, where image artifacts are dominant. The high conductivity errors for the 2-D experimental i-stroke suggest that the estimation of the target is not accurate. This is due to an offset between the real and the reconstructed target area for the experimental i-stroke reconstructions. The lowest values are observed for the cases of 2-D and 3-D h-stroke with CST data, at 1 and 0.9 GHz, respectively. Comparing all the images, these are the scenarios with the best estimated reconstructed target and the weakest artifacts.

Table 7 Relative errors of the permittivity in the target area for the 2-D examined scenarios

	0.9 GHz	1 GHz	1.1 GHz	1.2 GHz
CST h-stroke	0.2051	0.1877	0.2259	0.2121
CST i-stroke	0.7853	0.7595	0.7843	0.8303
Exp h-stroke	0.1951	0.1929	0.1944	0.1460
Exp i-stroke	0.8613	0.8524	0.8465	0.8414

Table 8 Relative errors of the conductivity in the target area for the 2-D examined scenarios

	0.9 GHz	1 GHz	1.1 GHz	1.2 GHz
CST h-stroke	0.4717	0.4586	0.4195	0.477
CST i-stroke	1.68	1.66	1.33	1.601
Exp h-stroke	0.4644	0.4687	0.4750	0.4595
Exp i-stroke	1.4939	1.4937	1.5228	1.624

Table 9 Relative errors of the permittivity in the target area for the 3-D examined scenarios

	0.9 GHz	1 GHz	1.1 GHz	1.2 GHz
CST h-stroke	0.1236	0.1271	0.244	0.1906
CST i-stroke	0.9714	1.0099	1.3393	1.2201
Exp h-stroke	0.096	0.1081	0.1241	0.1277
Exp i-stroke	1.4114	1.3885	1.3826	1.29

Table 10 Relative errors of the conductivity in the target area for the 3-D examined scenarios

	0.9 GHz	1 GHz	1.1 GHz	1.2 GHz
CST h-stroke	0.4524	0.438	0.373	0.4482
CST i-stroke	2.2492	1.7673	1.7737	3.09
Exp h-stroke	0.6892	0.6587	0.6166	0.4834
Exp i-stroke	2.7054	2.7975	3.1453	3.1617

5 Conclusions

In this chapter, we have examined the performance of an MWI setup for the problem of brain stroke detection and differentiation, using the 2-D and 3-D DBIM-TwIST algorithm and a novel, anatomically accurate, multi-layer head phantom. We called this a Zubal model, as it extends a slice from the Zubal numerical phantom to create 3-D moulds that can be combined to construct the final phantom. We note that the phantom is an extended 2-D cut of the head surrounded by a 2-D configuration of antennas, which was confined to a ring by our experimental setup. This phantom and antenna configuration allow a 2-D inversion to produce results of comparable accuracy. However, it is still critical to validate the 3-D algorithm for such scenarios by comparing with 2-D images, and this has been the main objective of this study. Preliminary simplified simulations in which the antenna array contains more than one ring, in Sect. 4.2, have shown the superior performance of the 3-D algorithm.

The phantom mimics the high dielectric difference in the brain tissues by comprising gelatine-based phantoms that mimic the dielectric permittivity of various head tissues. Importantly, we used a construction process that eliminates the presence of plastic in the phantom, which would distort the signals processed by our algorithm.

We placed the phantom in our experimental prototype as well as in a CST model that simulates the experiment. CST simulation results indicated that we are able to successfully localize and differentiate stroke targets by estimating their dielectric properties. Our measured data reconstructions confirmed that stroke detection and differentiation are more challenging with experimental data. Results for the hemorrhagic-like target were more accurate than for the ischemic-like target as the latter has a lower dielectric contrast with the surrounding brain matter. In accordance with previous work [58], we observed that permittivity estimation is more accurate than conductivity estimation. Importantly, we also observed that the differences in the dielectric estimation of i-stoke and h-stroke targets is higher if reconstructions are performed with the 3-D DBIM-TwIST algorithm. We specify that the images are presented by modifying the color scale to a range of values which is in the order of the anticipated permittivity of interest for h-stroke and i-stroke, respectively.

We also note that our calibration method relies on the signal difference between the experimental and the simulated NT scenarios. Section 4.1 has examined what prior knowledge is needed to successfully differentiate the two types of the stroke when using a complex model, and whether an improved forward model could produce better image quality. Our preliminary results indicated that the addition of thin head layers in the forward model is not critical for the successful reconstruction of the target's dielectric properties.

The purpose of this study was to experimentally validate our 2-D and 3-D DBIM-TwIST algorithms for stroke detection and differentiation, but also to further assess experimentally the performance of the 3-D algorithm for the examined application. For some of the studied scenarios, the 2-D DBIM-TwIST can lead to more accurate reconstructions. This is because the 2-D inverse problem approximates well the 3-D problem in these cases, and it is less ill-posed than the 3-D inverse problem as it contains a much smaller number of unknowns. However, and as showed by our numerical results with 2 rings of antennas and a smaller target placed between the two rings, the 3-D inverse algorithm can provide more accurate results for problems with significant variation along all three dimensions, as can be the case of brain imaging.

Acknowledgements This research was supported by the EMERALD project funded from the European Union's Horizon 2020 under the Marie Skłodowska-Curie grant agreement No. 764479.

References

1. K.K. Shung, M.B. Smith, B.M. Tsui, *Principles of Medical Imaging* (2015)
2. R. Chandra, H. Zhou, I. Balasingham, R.M. Narayanan, On the opportunities and challenges in microwave medical sensing and imaging. IEEE Trans. Biomed. Eng. **62**(7), 1667–1682 (2015)
3. M. Hopfer, R. Planas, A. Hamidipour, T. Henriksson, S. Semenov, Electromagnetic tomography for detection, differentiation, and monitoring of brain stroke: a virtual data and human head phantom study. IEEE Antennas Propag. Mag. **59**(5), 86–97 (2017)
4. Y.-H. Shao, K. Tsai, S. Kim, Y.-J. Wu, K. Demissie, Exposure to tomographic scans and cancer risks. JNCI Cancer Spectrum **4**(1), pkz072 (2020)
5. R. Benny, T.A. Anjit, P. Mythili, An overview of microwave imaging for breast tumor detection. Progr. Electromagn. Res. B **87**, 61–91 (2020)
6. R.C. Heros, Stroke: early pathophysiology and treatment. Summary of the fifth annual decade of the brain symposium. Stroke **25**(9), 1877–1881 (1994)
7. L. Alon, S. Dehkharghani, A Stroke Detection and Discrimination Framework Using Broadband Microwave Scattering on Stochastic Models with Deep Learning (2021)
8. N.K. Nikolova, *Introduction to Microwave Imaging* (Cambridge University Press, 2017)
9. S.C. Hagness, E.C. Fear, A. Massa, Guest editorial: special cluster on microwave medical imaging. IEEE Antennas Wirel. Propag. Lett. **11**, 1592–1597 (2012)
10. K.R. Foster, H.P. Schwan, et al., Dielectric properties of tissues, in *CRC Handbook of Biological Effects of Electromagnetic Fields*, pp. 27–96 (1986)
11. J.L. Schepps, K.R. Foster, The uhf and microwave dielectric properties of normal and tumour tissues: variation in dielectric properties with tissue water content. Phys. Med. Biol. **25**(6), 1149 (1980)
12. N.H. Jamali, K.H. Ping, S. Sahrani, T. Takenaka, Image reconstruction based on combination of inverse scattering technique and total variation regularization method. Indonesian J. Electr. Eng. Comput. Sci. **5**(3), 569–576 (2017)
13. M. Pastorino, *Microwave Imaging*, vol. 208 (John Wiley & Sons, 2010)
14. P.M. Meaney, M.W. Fanning, D. Li, S.P. Poplack, K.D. Paulsen, A clinical prototype for active microwave imaging of the breast. IEEE Trans. Microw. Theor. Techn. **48**(11), 1841–1853 (2000)
15. J.A. Tobon Vasquez, R. Scapaticci, G. Turvani, G. Bellizzi, D.O. Rodriguez-Duarte, N. Joachimowicz, B. Duchêne, E. Tedeschi, M.R. Casu, L. Crocco, et al., A prototype microwave system for 3D brain stroke imaging. Sensors **20**(9), 2607 (2020)
16. S. Candefjord, J. Winges, A.A. Malik, Y. Yu, T. Rylander, T. McKelvey, A. Fhager, M. Elam, M. Persson, Microwave technology for detecting traumatic intracranial bleedings: tests on phantom of subdural hematoma and numerical simulations. Med. Biol. Eng. Comput. **55**(8), 1177–1188 (2017)
17. M. Persson, A. Fhager, H.D. Trefná, Y. Yu, T. McKelvey, G. Pegenius, J.-E. Karlsson, M. Elam, Microwave-based stroke diagnosis making global prehospital thrombolytic treatment possible. IEEE Trans. Biomed. Eng. **61**(11), 2806–2817 (2014)
18. R. Scapaticci, L. Di Donato, I. Catapano, L. Crocco, A feasibility study on microwave imaging for brain stroke monitoring. Progr. Electromagn. Res. B **40**, 305–324 (2012)
19. R. Scapaticci, J. Tobon, G. Bellizzi, F. Vipiana, L. Crocco, Design and numerical characterization of a low-complexity microwave device for brain stroke monitoring. IEEE Trans. Antennas Propag. **66**(12), 7328–7338 (2018)
20. L. Crocco, I. Karanasiou, M.L. James, R.C. Conceição, *Emerging Electromagnetic Technologies for Brain Diseases Diagnostics* (Springer, Monitoring and Therapy, 2018)
21. O. Karadima, N. Ghavami, I. Sotiriou, P. Kosmas, Performance assessment of microwave tomography and radar imaging using an anthropomorphic brain phantom, in *2020 XXXIIIrd General Assembly and Scientific Symposium of the International Union of Radio Science* (IEEE, 2020), pp. 1–4
22. N. Ghavami, G. Tiberi, D.J. Edwards, A. Monorchio, UWB microwave imaging of objects with canonical shape. IEEE Trans. Antennas Propag. **60**(1), 231–239 (2011)

23. M. Azghani, P. Kosmas, F. Marvasti, Microwave medical imaging based on sparsity and an iterative method with adaptive thresholding. IEEE Trans. Med. Imaging **34**(2), 357–365 (2014)
24. J.D. Shea, P. Kosmas, S.C. Hagness, B.D. Van Veen, Three-dimensional microwave imaging of realistic numerical breast phantoms via a multiple-frequency inverse scattering technique. Med. Phys. **37**(8), 4210–4226 (2010)
25. A. Fedeli, C. Estatico, M. Pastorino, A. Randazzo, Microwave detection of brain injuries by means of a hybrid imaging method. IEEE Open J. Antennas Propag. **1**, 513–523 (2020)
26. M. Salucci, A. Polo, J. Vrba, Multi-step learning-by-examples strategy for real-time brain stroke microwave scattering data inversion. Electronics **10**(1), 95 (2021)
27. F. Wang, H. Zhang, J. Bao, H. Li, W. Peng, J. Xu, J. Yang, W. Zhuang, X. Ning, L. Xu et al., Experimental study on differential diagnosis of cerebral hemorrhagic and ischemic stroke based on microwave measurement. Technol. Health Care **28**(S1), 289–301 (2020)
28. G. Zhu, A. Bialkowski, L. Guo, B. Mohammed, A. Abbosh, Stroke classification in simulated electromagnetic imaging using graph approaches. IEEE J. Electromagn. RF Microw. Med. Biol. **5**(1), 46–53 (2020)
29. X. Chen, Z. Wei, M. Li, P. Rocca, A review of deep learning approaches for inverse scattering problems (invited review). Progr. Electromagn. Res. **167**, 67–81 (2020)
30. V. Zhurbenko, Challenges in the design of microwave imaging systems for breast cancer detection. Adv. Electr. Comput. Eng. **11**(1), 91–96 (2011)
31. H. Iida, Y. Hori, K. Ishida, E. Imabayashi, H. Matsuda, M. Takahashi, H. Maruno, A. Yamamoto, K. Koshino, J. Enmi et al., Three-dimensional brain phantom containing bone and grey matter structures with a realistic head contour. Ann. Nuclear Med. **27**(1), 25–36 (2013)
32. M. Lazebnik, E.L. Madsen, G.R. Frank, S.C. Hagness, Tissue-mimicking phantom materials for narrowband and ultrawideband microwave applications. Phys. Med. Biol. **50**(18), 4245 (2005)
33. C. Gabriel, *Compilation of the Dielectric Properties of Body Tissues at RF and Microwave Frequencies* (1996)
34. S. Gabriel, R. Lau, C. Gabriel, The dielectric properties of biological tissues: III. Parametric models for the dielectric spectrum of tissues. Phys. Med. Biol. **41**(11), 2271 (1996)
35. S. Semenov, T. Huynh, T. Williams, B. Nicholson, A. Vasilenko, Dielectric properties of brain tissue at 1 GHz in acute ischemic stroke: experimental study on swine. Bioelectromagnetics **38**(2), 158–163 (2016)
36. V. Mariano, J.A.T. Vasquez, R. Scapaticci, L. Crocco, P. Kosmas, F. Vipiana, Comparison of reconstruction algorithms for brain stroke microwave imaging, in *2020 IEEE MTT-S International Microwave Biomedical Conference (IMBioC)* (IEEE, 2020), pp. 1–3
37. P.M. Van Den Berg, R.E. Kleinman, A contrast source inversion method. Inverse Probl. **13**(6), 1607 (1997)
38. M. Bertero, P. Boccacci, *Introduction to Inverse Problems in Imaging* (CRC Press, 2020)
39. Z. Miao, P. Kosmas, Multiple-frequency DBIM-twist algorithm for microwave breast imaging. IEEE Trans. Antennas Propag. **65**(5), 2507–2516 (2017)
40. T.M. Grzegorczyk, P.M. Meaney, P.A. Kaufman, K.D. Paulsen et al., Fast 3-D tomographic microwave imaging for breast cancer detection. IEEE Trans. Med. Imaging **31**(8), 1584–1592 (2012)
41. Q. Fang, P.M. Meaney, K.D. Paulsen, Viable three-dimensional medical microwave tomography: theory and numerical experiments. IEEE Trans. Antennas Propag. **58**(2), 449–458 (2009)
42. T.J. Colgan, S.C. Hagness, B.D. Van Veen, A 3-D level set method for microwave breast imaging. IEEE Trans. Biomed. Eng. **62**(10), 2526–2534 (2015)
43. D. Ireland, A. Abbosh, Modeling human head at microwave frequencies using optimized Debye models and FDTD method. IEEE Trans. Antennas Propag. **61**(4), 2352–2355 (2013)
44. A.M. Qureshi, Z. Mustansar, A. Maqsood, Analysis of microwave scattering from a realistic human head model for brain stroke detection using electromagnetic impedance tomography. Progr. Electromagn. Res. M **52**, 45–56 (2016)
45. P. Lu, P. Kosmas, Three-dimensional microwave head imaging with GPU-based FDTD and the DBIM method. Sensors **22**(7), 2691 (2022)

46. K. Chadan, D. Colton, L. Päivärinta, W. Rundell, *An Introduction to Inverse Scattering and Inverse Spectral Problems* (SIAM, 1997)
47. W.C. Chew, *Waves and Fields in Inhomogenous Media*, vol. 16 (Wiley-IEEE Press, 1995)
48. M. Born, E. Wolf, *Principles of Optics: Electromagnetic Theory of Propagation, Interference and Diffraction of Light* (Elsevier, 2013)
49. P. Kosmas, J.D. Shea, B.D. Van Veen, S.C. Hagness, Three-dimensional microwave imaging of realistic breast phantoms via an inexact gauss-newton algorithm, in *2008 IEEE Antennas and Propagation Society International Symposium* (IEEE, 2008), pp. 1–4
50. P. Mojabi, J. LoVetri, Microwave biomedical imaging using the multiplicative regularized Gauss-Newton inversion. IEEE Antennas Wirel. Propag. Lett. **8**, 645–648 (2009)
51. M.R. Hestenes, E. Stiefel et al., *Methods of Conjugate Gradients for Solving Linear Systems*, vol. 49 (NBS Washington, DC, 1952)
52. W.C. Chew, Y.-M. Wang, Reconstruction of two-dimensional permittivity distribution using the distorted born iterative method. IEEE Trans. Med. Imaging **9**(2), 218–225 (1990)
53. D.W. Winters, J.D. Shea, P. Kosmas, B.D. Van Veen, S.C. Hagness, Three-dimensional microwave breast imaging: dispersive dielectric properties estimation using patient-specific basis functions. IEEE Trans. Med. Imaging **28**(7), 969–981 (2009)
54. J.M. Bioucas-Dias, M.A. Figueiredo, Two-step algorithms for linear inverse problems with non-quadratic regularization, in *2007 IEEE International Conference on Image Processing*, vol. 1 (IEEE, 2007), pp. 1–105
55. S. Ahsan, Z. Guo, Z. Miao, I. Sotiriou, M. Koutsoupidou, E. Kallos, G. Palikaras, P. Kosmas, Design and experimental validation of a multiple-frequency microwave tomography system employing the DBIM-twist algorithm. Sensors **18**(10), 3491 (2018)
56. W. Guo, S. Ahsan, M. He, M. Koutsoupidou, P. Kosmas, Printed monopole antenna designs for a microwave head scanner, in *2018 18th Mediterranean Microwave Symposium (MMS)* (IEEE, 2018), pp. 384–386
57. Z. Miao, P. Kosmas, S. Ahsan, Impact of information loss on reconstruction quality in microwave tomography for medical imaging. Diagnostics **8**(3), 52 (2018)
58. O. Karadima, M. Rahman, I. Sotiriou, N. Ghavami, P. Lu, S. Ahsan, P. Kosmas, Experimental validation of microwave tomography with the DBIM-twist algorithm for brain stroke detection and classification. Sensors **20**(3), 840 (2020)
59. I.G. Zubal, C.R. Harrell, E.O. Smith, Z. Rattner, G. Gindi, P.B. Hoffer, Computerized three-dimensional segmented human anatomy. Med. Phys. **21**(2), 299–302 (1994)
60. T. Rydholm, A. Fhager, M. Persson, S.D. Geimer, P.M. Meaney, Effects of the plastic of the realistic GeePS-L2S-Breast phantom. Diagnostics **8**(3) (2018)

Deep Learning Enhanced Medical Microwave Imaging

Álvaro Yago Ruiz, Alexandra Prokhorova, Darko Ninković, Marko Helbig, Marija N. Stevanovic, Marta Cavagnaro, and Lorenzo Crocco

Abstract This chapter is devoted to the thriving field of deep learning enhanced medical microwave imaging. Microwave imaging is a technology with applications in different fields involving the inspection of hidden or nested targets. Thanks to its non-ionizing nature and low-cost, it represents an emerging modality for non invasive medical diagnostics, with potential to compensate the limitations of well-established medical imaging modalities. However, microwave imaging faces several challenges that are slowing down its adoption. In particular, the underlying problem at the core of microwave imaging is a non-linear and ill-posed inverse scattering problem. Recently, employing deep learning techniques to address the difficulties in solving inverse scattering problems has received a significant attention. In this regard, this chapter reviews the main aspects concerned with the adoption of deep learning to enhance microwave imaging and then provides two examples to outline

Á. Yago Ruiz · M. Cavagnaro
DIET, Department of Information Engineering, Electronics and Telecommunications, University of Rome "La Sapienza", Rome, Italy
e-mail: marta.cavagnaro@uniroma1.it

Á. Yago Ruiz (✉) · M. Cavagnaro · L. Crocco
IREA-CNR, Institute for Electromagnetic Sensing of the Environment, National Research Council of Italy, Naples, Italy
e-mail: yago.a@irea.cnr.it

L. Crocco
e-mail: crocco.l@irea.cnr.it

A. Prokhorova · M. Helbig
Technische Universität Ilmenau, Ilmenau, Germany
e-mail: alexandra.prokhorova@tu-ilmenau.de

M. Helbig
e-mail: marko.helbig@tu-ilmenau.de

D. Ninković · M. N. Stevanovic
University of Belgrade, Belgrade, Serbia
e-mail: Darko@etf.rs

M. N. Stevanovic
e-mail: Mnikolic@etf.bg.ac.rs

F. Vipiana and L. Crocco (eds.), *Electromagnetic Imaging for a Novel Generation of Medical Devices*, Lecture Notes in Bioengineering,
https://doi.org/10.1007/978-3-031-28666-7_6

its potential in medical imaging applications. The first application is concerned with monitoring of thermal treatments (hyperthermia) whereas the second is dedicated to imaging the brain.

Keywords Microwave imaging · Deep learning · Inverse scattering · Medical imaging

1 Introduction

Microwave imaging (MWI) is a technology that provides images of otherwise inaccessible objects that may be buried, like in archaeology or geology [1], or whose content is to be inspected, without requiring the opening of the object to look inside. This is, for instance, the case of food monitoring [2].

As a medical imaging modality, MWI is an emerging technology that aims to compensate the drawbacks of other established modalities [3, 4]. At present, the most widely used medical imaging modalities are computed tomography (CT) and positron emission tomography (PET) on the ionizing technologies group, and magnetic resonance imaging (MRI) and ultrasounds (US) on the non-ionizing one.

The greatest feature of CT is an excellent spatial resolution, but its application in soft tissues limited due to the weak contrast. Conversely, MRI provides accurate images of soft tissues with a high contrast resolution. Both MRI and PET can be employed in functional imaging, though have expensive operating costs. All three techniques present longer examination times and are bulkier than US, which is an inexpensive alternative. However, US cannot achieve the MRI or CT spatial resolution and cannot penetrate beyond bones. Moreover, US is heavily dependent on the expertise of the physician. In view of the aforementioned limitations, MWI may represent a valid complementary or alternative imaging technology, being safe, portable and inexpensive.

The use of MWI for medical purposes is enabled by the different dielectric properties (i.e., permittivity and conductivity) that biological tissues exhibit in the microwave spectrum, due to the different water content. Furthermore, the dielectric properties of pathological tissue differ from those of healthy tissue. Therefore, diagnostic images can be, in principle, tailored via the images of the dielectric properties provided by an MWI device. To date, most of the efforts in the pursuit of medical MWI applications have been directed towards two diagnostic applications: breast cancer imaging [4] and brain stroke imaging [3].

The technical difficulty of MWI technologies can be found in the nature of electromagnetic (EM) scattering [5, 6]. Actually, MWI is formulated as an inverse scattering problem (ISP) [7], a category of very challenging problems found in applications like seismology, quantum theory and, of course, medical imaging [7]. The complexity of ISPs lies in their non-linear behaviour and, most notably, ill-posedness. The non-linearity affecting the ISP is not convex, meaning a solution finding method may be trapped in false solutions. Concerning ill-posedness, a solution for the ISP cannot be provided in a stable way, i.e., little changes in the data lead to large changes in the solution. Therefore in the presence of unavoidable noise, like measurement errors or background noise, the solution may be devoid of any meaning, unless proper regularization methods are employed. Of course, these issues represent a significant drawback towards the adoption of MWI for medical diagnostics.

Deep learning (DL) is a developing field in many applied sciences. Its tremendous success in solving highly non-linear problems, especially in image processing [8], has led to a flourishing research for biomedical applications, where published papers have been on an exponential rise in the recent years [9]. Different DL architectures may be chosen depending on the data types. A list of DL-based biomedical applications jointly with common data types, in bold, employed in each category is given below.

- *Omics*: the study of protein and DNA/RNA. In this group, the main data used as inputs are DNA/RNA **sequences** [10] and **graph-like** protein structures [11].
- *Bioimaging*: DL works mainly on microscope **images** in the study of different healthy and pathological tissues [12].
- *Medical imaging*: applications related to processing **images** of any imaging modality like MRI [13] or x-rays [14]. This category comprises the closest group of applications to the works on microwave imaging discussed in this chapter (see Sect. 4).
- *Body-machine interface*: the analysis of biological **signals** such as EEG [15] or ECG [16].
- *Public health-care management*: the study of the logistics and decision-making within health systems relying on medical **big data** [17].

Regarding MWI, the excellent capabilities of DL to solve non-linear problems can be applied to the ISP to face its difficulties in an efficient and reliable way. Additionally, once trained, DL architectures operate in real-time thus allowing real-time imaging results. In contrast, iterative solvers traditionally used to solve the ISP in MWI tend to be computationally expensive [18]. Current applications of DL-MWI appear to be limited to laboratory prototypes with phantoms. Particularly, the detection of brain anomalies is studied in [19, 20] and a multi-modal US-MWI framework for breast imaging in [21, 22]. As an exception, the classification of brain stroke in [23] reached clinical trials, though strictly speaking, it is not an imaging device, but a sensing one. Considering a broader scope, works on ISPs via DL can be seen in electric impedance tomography [24–26] and regular US [27].

In this chapter, the main aspects concerned with the adoption of deep learning to enhance microwave imaging are reviewed and discussed. Then, two examples are presented to outline, in simplified conditions, how DL-MWI can be exploited in medical imaging applications. The first application is concerned with hyperthermia treatment monitoring, whereas the second is dedicated to brain imaging.

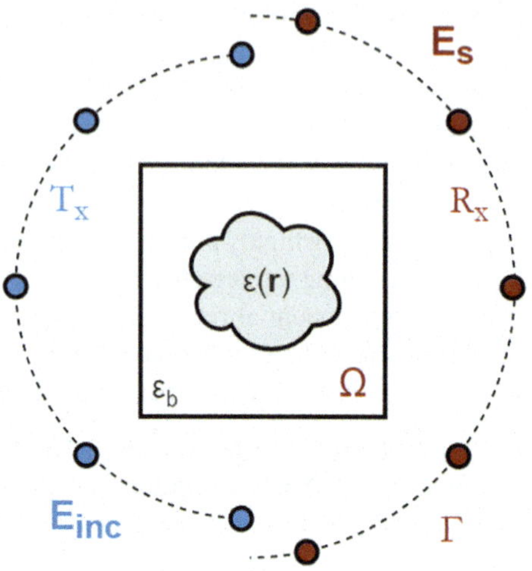

Fig. 1 Typical 2D scalar MWI problem. Under transverse magnetic polarization the electric fields E_{inc}, E_s have only one component

2 Microwave Imaging

The aim of MWI is the retrieval of the properties of a target from the measure of the electromagnetic (EM) field it scatters under known incidence conditions. The target's properties which are of interest are the morphological ones, like shape, size and position, and the electromagnetic ones, permittivity and conductivity. These properties are typically encoded in the *contrast* function, $\tau(\mathbf{r}) = \varepsilon_{eq}(\mathbf{r})/\varepsilon_b - 1$, a complex function which relates the target's electromagnetic properties $\varepsilon_{eq}(\mathbf{r}) = \varepsilon(\mathbf{r}) - j\sigma(\mathbf{r})/\omega\varepsilon_0$, to those of the background medium, ε_b, where the target is hosted. Here ε denotes the real part of the relative permittivity, σ the conductivity, measured in $[S/m]$, ω the angular frequency, and ε_0 the permittivity of vacuum.

A typical 2D MWI problem is shown in Fig. 1, where several "antennas" T_x, R_x are located in different $\mathbf{r}_t$, $\mathbf{r}_m$ positions, respectively, around the imaging domain Ω where the target is enclosed. The imaging procedure begins with T_x illuminating Ω with time-harmonic incident fields, E_{inc}. From the interaction of E_{inc} with the inspected target, a scattering phenomenon happens, which gives raise to a scattered field E_s. The scattered field encodes the perturbation of the incident field due to the target, which can be measured at R_x in the form of the total field $E = E_s + E_{inc}$. Also, the scattered field can be seen as the field radiated by the current density $W = \tau E$ induced in the target by its interaction with the incident field E_{inc} radiated by the MWI antennas. The ISP at the core of MWI problems is an inverse problem in the sense that observing the effect, i.e., the measured scattered field E_s, the goal is to retrieve its cause, i.e., the target that, interacting with E_{inc} has generated the induced current density W.

Casting the MWI problem in the 2D scalar form is a broadly adopted formulation to cope with the ISP in a simple yet useful manner [28]. For the case of EM waves, the scalar fields are achieved under two conditions. The first one is having targets that are infinitely long, thus ensuring no property changes of the target along the longitudinal axis (2D problem). Additionally, the transverse magnetic mode (TM), in which the electric field is polarized along the invariance axis, is employed. In doing so, the electric field has only the longitudinal component (scalar problem) whereas the magnetic field has the transverse components.

The contrast function τ can be determined by solving two paired integral equations:

$$E_s(\mathbf{r}_m, \mathbf{r}_t) = \int_{\Omega} G(\mathbf{r}_m, \mathbf{r}')W\left(\mathbf{r}', \mathbf{r}_t\right) d\mathbf{r}' = \mathcal{A}_e(W), \tag{1}$$

$$E\left(\mathbf{r}, \mathbf{r}_t\right) = E_{inc}\left(\mathbf{r}, \mathbf{r}_t\right) + \int_{\Omega} G\left(\mathbf{r}, \mathbf{r}'\right) W\left(\mathbf{r}', \mathbf{r}_t\right) d\mathbf{r}' = \mathcal{A}_i(W) \tag{2}$$

wherein G denotes the Green's function i.e., the impulse response of the assumed homogeneous background medium and $\mathcal{A}_e$ and $\mathcal{A}_i$ are short notations for the *external* and *internal* radiation operators, respectively.

Even though the 2D scalar problem has advantages over its 3D vectorial counterparts, solving it is not straightforward since it is intrinsically ill-posed and non-linear. In practical applications, the issue of ill-posedness is the effect of the unavoidable measurement errors. As a matter of fact, since the solution does not depend continuously on the data, little variations in the measurements may provide solutions corresponding to very different scenarios. In this regard, regularization techniques are adopted to filter out the noise affecting the solution [29].

Typically, ISPs are solved through an optimization procedure, i.e., casting the problem as a minimization one [28]. These approaches explore the solution space locally and iteratively converge towards the optimal solution. However, due to the fact that the non-linearity of the ISP is not convex, local iterative solvers may be trapped in false solutions, what in optimization terminology is known as local minima. Alternatively, global optimization techniques are capable of escaping local minima and reach the actual solution (global minimum), but their implementation is limited by their computational burden [30].

2.1 *Quantitative Versus Qualitative Microwave Imaging*

In MWI literature, techniques are frequently clustered in two groups depending on the final objective of the reconstruction [28]. On one hand, qualitative imaging techniques refer to those algorithms that provide a partial characterization of the targets being inspected. On the other hand, quantitative imaging aims to deal with the more ambitious task of giving a full characterization of the objects. However, this categorization is rather stringent. In fact, with the appropriate a-priori information even qualitative imaging techniques could provide a full characterization of the targets.

Instead of using a strict definition of qualitative and quantitative imaging, it is more suitable to understand the goal of a given application and follow with a decision on the properties that should be retrieved. Although ideally a full characterization is the ultimate goal, knowing the electromagnetic properties of an inspected object might not be needed [1]. Two examples of this situation are listed here regarding the characterization of morphological properties:

1. Localization of buried objects: based on radar technology and the estimation of the direction of arrival [31].
2. Non-destructive quality control: shaping a defect to decide if a product must be discarded [32].

In the first example, only the position of the target, thus a morphological property, is required. Similarly, in the second example, the requirement for a successful implementation is the retrieval of the shape of the object, yet again a morphological property. In short, depending on the application, qualitative imaging techniques may be valid options.

3 Deep Learning Basics

Applied sciences have been experiencing an expansion of artificial intelligence (AI) research. Although AI is used interchangeably with machine learning (ML) and DL, there are important differences among them. Actually, AI is an umbrella term for any technique that bestows a computer with the ability to "think" like a human. For instance, the successful implementation of algorithms capable of defeating humans at chess [33] is an example of AI but not of ML nor DL. These latter two indeed belong to AI, as depicted in Fig. 2: they represent evolved forms with additional characteristics. Most notably, they attempt to incorporate human-like "learning".

The reasoning behind ML algorithms is tackling problems that are not easily described by mathematical rules. For instance, face recognition is one of such problems. Humans are excellent at recognizing familiar faces, but most of us would struggle to effectively describe the procedure we internally run to solve the problem. Humans are innately capable of learning a recognition pattern by performing the task many times, i.e., by analysing the information (data) we receive through our eyes.

ML can be defined as a problem-solving paradigm in which the algorithm learns how to solve a problem by inspecting the data. In ML, the pieces of relevant information present in the data that are required to solve a problem are called *features*. Thus, features provide a representation of the data in the context of a problem, e.g., in the problem of assessing the risk of a patient with diabetes, blood glucose concentration is a feature.

To this point, it is clear that having access to features is important if an ML algorithm is to be implemented. However, in some circumstances features are unknown, because the problem at hand is new or very complex. In particular, this is common in image processing, because features are not obvious in the image (like in the face recognition problem). To this end, prior to the explosion of DL, an expert with a proper understanding of a problem would tailor *engineered features* for ML algorithms.

Employing engineered features in a problem is not ideal, since it is limited to the expertise of the person that tailored them. Oppositely, DL algorithms learn the best features available in the data (*learned features*). Such property of DL algorithms is known as *representation learning* and it often leads to better performances than their ML counterparts based on engineered features [34].

DL algorithms (and ML ones) have two separate stages: training and deployment. *Training* is the part of the implementation where the algorithm learns how to solve the problem. In general, it is an off-line process that is run once. During this stage, the internal parameters (weights) of the DL algorithm are tuned for the specific task at hand via the optimization of a suitable loss function. Usually, local optimization approaches based on the gradient of a loss function are employed [35].

In the *deployment*, the DL algorithm has become competent at solving the problem and therefore can automatically process new instances. For example, considering again the problem of assessing the risk of a diabetic patient, the training stage would comprise the part where the algorithm learns to assign the risk based on historical data of different patients. Then, during the deployment new patients could be assessed.

Regarding the training stage, depending on the availability of *labelled* data, two learning modes are defined: supervised or unsupervised learning [36]. If each instance of the historical data are accompanied by their desired results (human supervision), the problem is suitable for supervised learning (SL). Conversely, if only *unlabelled* data are available, unsupervised learning (UL) must be used.

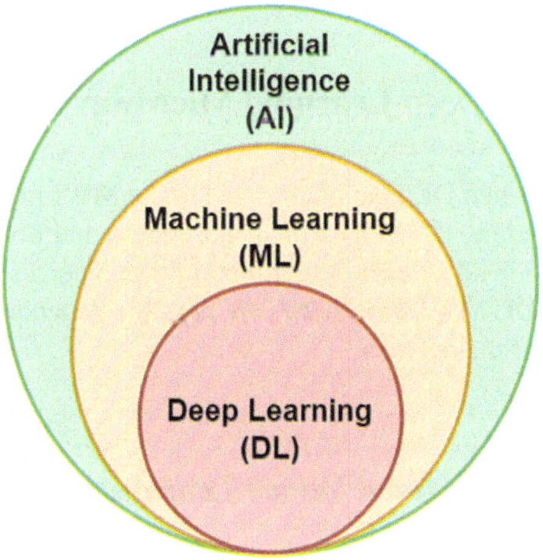

Fig. 2 Deep learning with respect to machine learning and artificial intelligence

In SL, there are two well-defined tasks:

- *Classification*: Categorical labels are provided. For instance, determining if an X-ray image has a tumor in it or not is a classification problem with two possible answers (yes/no), whose correct solution can be provided by a human, the physician. Image segmentation tasks are considered classification problems as well. The two examples of DL enhanced MWI implementations shown in this chapter are classification tasks.
- *Regression*: The labels are continuous values, more precisely referred to as target values. For instance, estimating the concentration of a substance in a tissue using MRI scans. The desired concentration could be estimated by other means but has to be measurable from the images.

UL exploits patterns that form in the data, i.e., instances with similar attributes may show "closer" values. Most prominent tasks within UL are:

- *Clustering*: Gathering instances with similar characteristics into groups with the number of groups as an unknown. E.g., find the best way of separating patients into age ranges.
- *Visualization*: Providing 2D or 3D representations of the data. For instance, the synthetic image generation of pathological tissues.
- *Dimensionality reduction*: Slightly different from visualization tasks. Find a compact representation of the data. E.g., a compression algorithm for MRI scans.

One inconvenient about DL is the need for large amounts of data for the training stage. This has led to creative implementations of the algorithms with less internal parameters [34]. For instance, in the case of ISPs of interest in this chapter, efforts on synthetic data generation and speed improvements in EM forward simulators via DL are under research to address the data availability limitation [37, 38].

4 Deep Learning Microwave Imaging Approaches

When DL is applied to solve the MWI problem, the goal is to alleviate, if not overcome, the impact of ill-posedness and non-linearity. If the DL architecture is conceived as a transfer function that receives an input and provides an output, several DL-MWI approaches are possible depending on the data types supplied at both ends of the network.

4.1 Model Versus Learning Microwave Imaging

From a general perspective, the conventional way of solving an ISP can be described as the minimization of a loss function Ψ which encodes the mismatch between the measured data and the one obtained through the mathematical model describing the scattering phenomenon (1)–(2):

$$\arg\min_{\tau} \ \Psi\left\{\mathcal{A}_e(E\tau), E_s\right\} + R(\tau) \tag{3}$$

In (3), $R(\tau)$ is the regularization term introduced to deal with ill-posedness. Typically, such a minimization is performed by means of local optimization iterative algorithms. Alternatively, a DL architecture can be used to solve the problem, moving the slow iterative process ahead of the deployment of the algorithm. If a learning procedure is followed, the actual solutions of an arbitrary number of samples are provided in the training stage of the algorithm. In this case the minimization problem is formulated like:

$$\arg\min_{\theta} \ \Psi\left\{\tau, F_\theta\left(E_s\right)\right\} + R(\theta) \tag{4}$$

where $F_\theta\left(E_s\right)$ is the output of a DL architecture whose parameters θ must be optimized. The regularization term $R(\theta)$ acts on the parameters but it is not mandatory. This is an important difference with (3), where the regularization term is necessary to guarantee a stable reconstruction. Here, the purpose of the regularization is adding more variability to the data in order to increase the performance of the algorithm (i.e., improve the generalization capabilities of the algorithm). Regularization is an effective way of boosting the performance when a reduced number of training samples are available [34].

Learning approaches to ISP are grouped in 3 families [18]. The first and more straightforward approach belongs to the class of *direct learning methods*, which directly employ the measurements E_s as inputs to the architecture. It is worth noting that τ can be represented in pixel basis, therefore images could be used as outputs.

When the purpose of the architecture is assisting an ISP iterative solver which relies on a objective function, i.e., finding the best parameters ξ of the iterative solver of choice, the problem cast in eq. (4) becomes:

$$\arg\min_{\theta} \ \Psi\left\{\xi, F_\theta\left(E_s\right)\right\} + R(\theta) \tag{5}$$

If E_s is preprocessed to facilitate its manipulation by the DL architecture, the approach is said to be a *physics-assisted method*. These methods typically imply using an imaging algorithm $\mathcal{I}$ to transform the data from the measurement domain to the contrast domain. Therefore the problem in eq. (4) is cast like:

$$\arg\min_{\theta} \ \Psi\left\{\tau, F_\theta\left(\mathcal{I}\left(E_s\right)\right)\right\} + R(\theta) \tag{6}$$

As can be foreseen, the output of the network is modifiable depending on the needs of a given application. For instance, instead of a complete characterization via the retrieval of τ, just the nominal shape of the targets could be the goal [39]. In

Sect. 5, examples of these potential changes to the output are discussed in two of the most popular DL implementations, classification and segmentation problems.

5 Examples of DL-Enhanced Medical MWI

In this Section, two examples of exploitation of physics-assisted DL MWI for medical imaging applications are discussed. Both approaches are cast in terms of SL classification problems, with one application in therapy and one in diagnosis. The first application is concerned with the monitoring of hyperthermia anti-cancer treatments, whereas the second application is related to imaging of brain tissue, which is relevant to the diagnosis of ischemic and hemorrhagic strokes.

It is worth noting that the simulated scenarios considered to build the examples are remarkably simplified with respect to the actual situations. This is only a partial shortcoming, since the focus here is to address the potential of DL-MWI by describing how these techniques can be implemented to tackle the specific challenges relevant to applications.

5.1 Hyperthermia Treatment Monitoring

Hyperthermia treatment (HT) is an adjuvant cancer therapy in which cancerous tissues are heated up to 40–43 °C to help damaging malignant cells. This is achieved via a sensitization to radiation in radiotherapy, or an increased permeability of the cell membrane to drugs in chemotherapy [40]. Local, regional or whole-body HT may be considered depending on the size of the treatment area. In Fig. 3, the concept of HT is illustrated.

During HT, tracking the temperature is needed to verify that the therapeutic temperature is reached in the target tissue and that the surrounding healthy tissue is not heated (and thus not damaged), due to the so-called hot-spots. In the clinical practice, this task is performed by means of fiber optics thermometers, once inserted in the area under treatment, as well as relying on patients feedback. However, this approach has the two-fold disadvantage of being invasive (due to the need of inserting the probes), limited to local measurements (since probes are sensitive to temperature in the insertion location) and subject to the individual perception of the patient. For this reason, there is a huge interest towards the development of fully non-invasive temperature monitoring approaches capable of overcoming the above issues. To this end, attempts to employ magnetic resonance (MR) thermometry have been reported [41, 42], but its application is only feasible in 15 % of the treatments, mostly when the treatment area is in the limbs or the HT device fits inside the bore of the MR system.

In this framework, MWI may represent a potential candidate technology owing to the known relationship between temperature and tissue electromagnetic properties

[43]. Notably, unlike MRI, MWI-based thermometry would rely on portable, cost-effective devices, which could be easily integrated with HT systems.

However, appraising the temperature from MWI not only implies solving the ISP but also to relate the retrieved contrast values to the temperature. On the other hand, rather than the exact temperature value, the indication of the heating status of the tissue of interest can be pursued via MWI, as this is the information which is ultimately needed. From the perspective of DL, this latter goal can be naturally tackled turning the problem into a classification task in which the regions of interest (i.e., the tumor region and the surrounding healthy tissue) are *labelled* according to their temperature. In particular, in the tumor region, the expected outcome is reaching the therapeutic temperature, whereas in the healthy tissue the goal is to avoid occurrence of hot-spots.

In the following, a DL-MWI framework based on this concept is described and developed for a scenario related to the case of neck tumors. In such an application, the objective is heating the tumor while avoiding heating the spinal cord, given the known

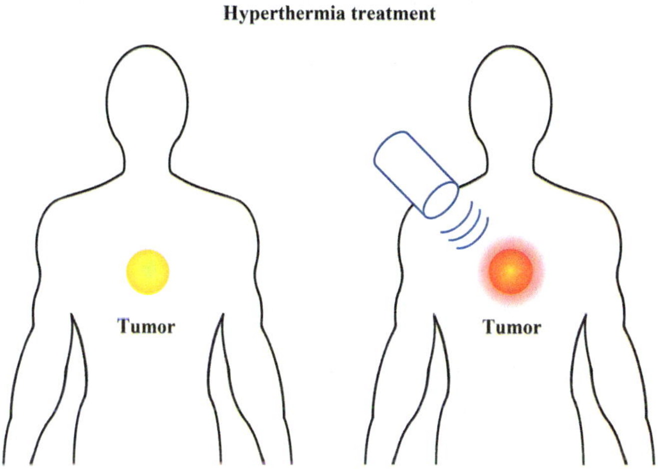

Fig. 3 Local hyperthermia treatment diagram showing the application of heat to a tumor

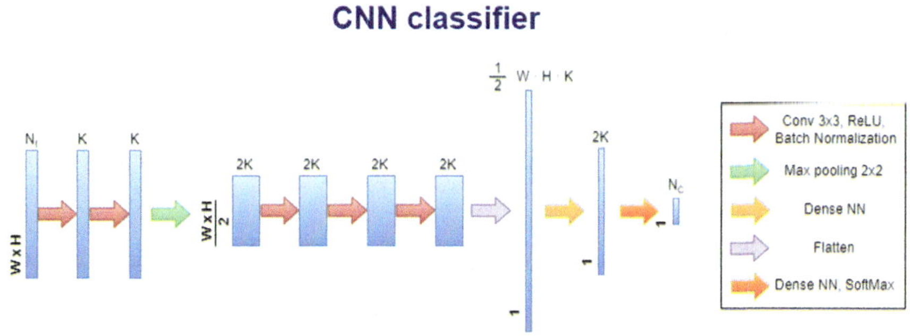

Fig. 4 CNN classifier

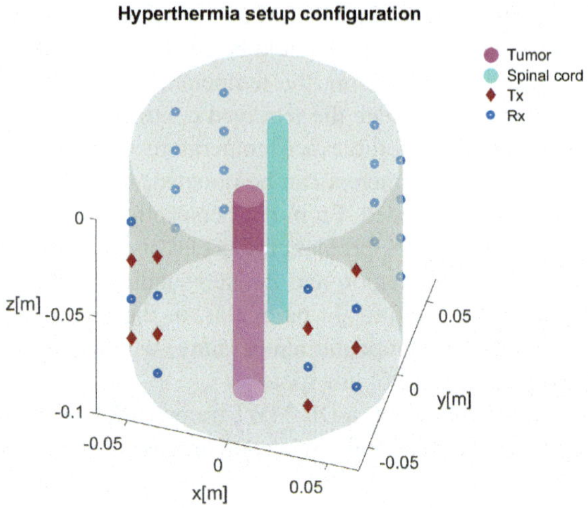

Fig. 5 Setup configuration for hyperthermia monitoring. Here the only targets present are a tumor (pink) and a spinal cord (cyan) which have been modelled for simplicity as elongated cylinders

sensitivity of nervous tissue to heat [40]. In this regard, the proposed framework consists of two convolutional neural networks (CNNs), like the one shown in Fig. 4; one in charge of monitoring the tumor and one in charge of monitoring the spinal cord. The choice of the CNN as DL architecture is due to its powerful image processing capabilities [34] that are convenient for the purpose of enhancing the images coming from an MWI algorithm.

The scenario considered to model the problem at hand in a simple way is shown in Fig. 5. The neck is modeled as a homogeneous cylinder with two smaller cylinders inside, one representing the tumor, the other representing the spinal cord. The antennas used to generate the MWI data are modelled as point sources located on the surface of the cylinder. The considered geometry allows casting the ISP in terms of the 2D model presented in Sect. 2.

The reconstructed image consists of a 2D slice of the neck, from which the heating status of the tumor and the spinal cord is to be determined. In this regard, it is worth noting that the temperature increase ΔT during HT corresponds to a small variation of electromagnetic properties $\delta\tau$. Such a circumstance allows modeling the EM scattering behavior exploiting the Born-approximation [5]. According to this approximation, the total field in the imaging domain can be well approximated with the incident field $E \approx E_{inc}$. Under this assumption, the system of Eqs. (1–2) becomes linear and reduce to the single equation:

$$\Delta E_s(\mathbf{r}_m, \mathbf{r}_t) = \int_{\Omega} G(\mathbf{r}_m, \mathbf{r}')\, E_{inc}(\mathbf{r}', \mathbf{r}_t)\, \delta\tau(\mathbf{r}')d\mathbf{r}' = L_{\Omega}\delta\tau(\mathbf{r}') \tag{7}$$

In (7), ΔE_s is the field scattered with respect to the unheated condition, Ω denotes the half-height cross-section of the cylinder (at $z = -0.05$ m) and L_{Ω} denotes the scattering operator linearized according to the Born approximation. Note that given the arrangement of the antennas (see Fig. 5), the considered model is not exactly 2D.

In particular, the L_Ω operator is built considering the z-component of the incident field and of the Green's function for the cross-section Ω. Moreover, the incident field and the Green's function are computed with respect to the unheated scenario.

By means of (7), an ill-posed linear problem has been obtained. The straightforward way of dealing with this type ISP is using the pseudo-inverse L_Ω^+:

$$L_\Omega^+ \Delta E_s = \delta\tau^+ \tag{8}$$

The pseudo-inverse is generally calculated using the singular value decomposition (SVD). However, this does not provide a stable reconstruction per se due to the exponentially decaying behavior of singular values, which results in an ill-conditioned pseduo-inverse. To address this issue, truncating the singular values can be exploited to enforce a regularization effect on the inversion [29]. In the following, reconstructions via this type of regularization jointly with SVD are referred to as TSVD.

Considering pixel basis, a TSVD image I is generated. The general imaging approach would consist in having the whole neck as the imaging domain Ω. However, HT treatments are usually preceded by a treatment planning stage that involves using MRI followed by an accurate segmentation of the tissues in the treatment area to determine the excitation of the HT applicator suitable for the treatment at hand [40]. Accordingly, the dimensions and positions of the tumor and the spinal cord as well as the overall distribution of tissues in the scenario are known a-priori. This piece of information allows to reduce the ISP to the sole identification of the changes in EM properties at those locations. In addition to this, given the fact the scattering phenomenon can be cast through the Born Approximation, linearization allows to directly consider two separate sub-domains enclosing only the tumor and the spinal cord, respectively. Hence, two TSVD images are achieved restricting the scattering operator L_Ω to each sub-domain (see Fig. 6).

The two images encoding the MWI processing of the tumor and the spinal cord represent the input of the two CNN used to implement the physics-assisted DL framework herein developed to perform HT monitoring. The CNNs are trained to tell the user the heating status of the tumor and the spinal cord by analysing the sub-domains. The categorical nature of the outputs the CNNs should retrieve turns the problem into a classification. However, there is a difference between the expected outcome for the two tissues. The tumor is expected to stay in the therapeutic regime between two thresholds, therefore 3 possible classification outcomes can be defined: $\{unheated, therapeutic, hot\}$. Whereas, the spinal cord must remain unheated during the whole treatment. Therefore only two classes are required: $\{unheated, hot\}$.

For each network, the optimization problem in (6) is properly adapted like:

$$\arg\min_\theta \ \Psi\{C, F_\theta(I)\} + R(\theta) \tag{9}$$

where I is the TSVD image, with $I = I_T$ or I_S depending on whether the image refers to the tumor or the spine. C is the ground truth classes supplied during the

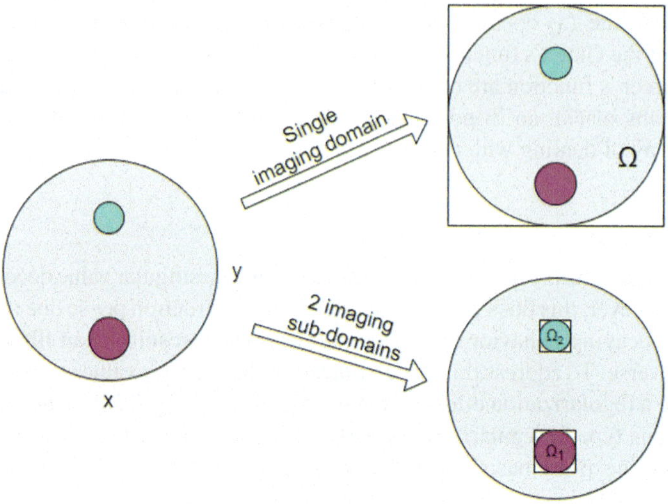

Fig. 6 Using sub-domains of the neck instead of the whole imaging domain Ω. The only targets present are a tumor (pink) and a spinal cord (cyan) which have been modelled for simplicity as elongated cylinders

training of the CNNs, with $C = C_T$ or C_S depending on whether the tumor or the spine is considered. In classification problems, the mismatch between ground truths and predictions is carried out by means of the cross-entropy [44]:

$$\arg\min_{\theta} \left[-\sum_{c=1}^{N_c} C_c^{[n]} \cdot log\ \hat{C}_c^{[n]} \right] \tag{10}$$

being N_c the number of categories the CNN has to learn to distinguish for each sample $n = 1, ..., N$, with N being the number of samples used in the training. While ground truth labels $C_c^{[n]}$ are either 0 or 1, the values of the predicted labels $\hat{C}_c^{[n]}$ fall within the range [0, 1]. A diagram summarizing the training stage of the spinal cord classifier is shown in Fig. 7.

To provide an example of the proposed approach, data samples were generated via a proprietary software based on the method of moments (MoM) [45, 46] and the conjugate gradient fast Fourier transform (CG-FFT) [47]. The EM properties of the tumor were modelled as $\varepsilon_T = 64.4$, $\sigma_T = 1.6$ S/m whereas the spinal cord was modelled as $\varepsilon_S = 49.7$, $\sigma_S = 1.5$ S/m with background medium $\varepsilon_b = 33.0$, $\sigma_b = 1.0$ S/m. Temperature changes were modelled as little variations in the permittivities values, e.g., $\Delta\varepsilon_T = \Delta\varepsilon_S = [0.1, 0.2, ..., 1.0]$. Lastly, the dataset was corrupted with different noise levels {20, 30, 40, 50} dB to increase the variability. In total, $N_1 = 1118$ samples for the tumor CNN and $N_2 = 1980$ samples for the spinal cord CNN were simulated. The setup configuration of antennas for both datasets was the same and consisted of point sources. As depicted in Fig. 5, $T_x = 8$, $R_x = 24$ and their displacement in an aspect limited configuration is motivated by the presence of the HT heating waveguides that do not allow any antennas in those areas. For this reason, aspect limitation was compensated by using $N_f = 11$ frequencies, $f = [1, 1.1, ..., 2]$ GHz for the calculation of the TSVD imaging results.

Thresholds considered for the creation of the ground truth classes C_T, C_S were based on spectroscopy measurements aimed at assessing the relationship between temperature and EM properties [43], see also Chap. "Microwave Ultra-Wideband Imaging for Non-invasive Temperature Monitoring During Hyperthermia Treatment" of this book. Thus, the employed tumor thresholds were:

$$C_T(\Delta\varepsilon_T) = \begin{cases} 0 & \text{if } \Delta\varepsilon_T < 0.4 \\ 1 & \text{if } 0.4 \leq \Delta\varepsilon_T < 0.8 \\ 2 & \text{if } 0.8 \leq \Delta\varepsilon_T \end{cases} \tag{11}$$

corresponding to $\{0, 1, 2\} = \{unheated, therapeutic, hot\}$. Similarly for the spinal cord:

$$C_S(\Delta\varepsilon_S) = \begin{cases} 0 & \text{if } \Delta\varepsilon_S < 0.2 \\ 1 & \text{if } 0.2 \leq \Delta\varepsilon_S \end{cases} \tag{12}$$

with $\{0, 1\} = \{unheated, hot\}$.

Before training, 15% test splits were taken from the tumor and the spinal cord sets. The performance assessment of the CNNs was based on these test splits. The results of the assessment are provided by means of a categorical metric called *Matthew's correlation coefficient* (MCC) [48], which is defined like:

$$MCC = \frac{TP \cdot TN - FP \cdot FN}{\sqrt{DM}} \tag{13}$$

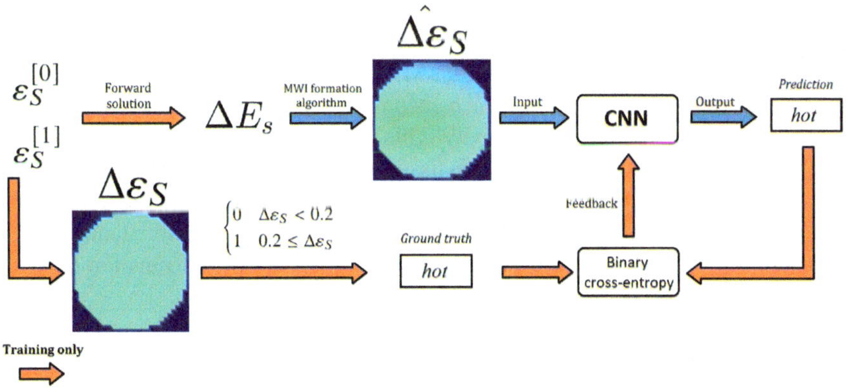

Fig. 7 Spinal cord CNN classifier training diagram

where $DM = (TP + FP)(TP + FN)(TN + FP)(TN + FN)$. Here the TP, TN correspond to the statistical count of positives and negatives and their corresponding false FP, FN. A true positive would be, for instance, a variation of $\Delta\varepsilon_T = 0.6$ correctly labelled as *therapeutic*.

The results of the performance assessment reported an average $MCC = 0.945$ over the test split with the tumor classification CNN and $MCC = 0.906$ with the spinal cord classification CNN. Difference in performances should be expected since the number of classes in each CNN are different. Moreover, the spinal cord classifier might have lagged behind the tumor classifier due to the forced unbalance of the ground truth classes in (12). Note that $MCC = 1.000$ is the maximum achievable score, therefore both CNNs show compelling capabilities.

5.2 Brain Imaging

The use of MWI to image brain tissue is of interest in various applications [3]. For instance, in brain stroke diagnosis, MWI can be employed to detect brain tissues affected by the stroke, but also determine if the stroke is hemorrhagic or ischemic. Moreover, in early assessment of Alzheimer disease, images of the amyloid plaques formed on the gray matter can be obtained [49].

However, unless a prior image of the brain is available as reference (like for instance in the HT case discussed in the previous section), these applications correspond to a significantly difficult task in terms of ISP solution. As a matter of fact, since the whole head reacts to the probing incident fields, the signal associated to the pathology (e.g., the stroke or the amyloid plaques) is embedded into the signal scattered by the whole head, which in turn is affected by the different tissues. As such, extrapolation of the useful signal is not straightforward and a complex ISP involving both the estimation of the healthy and pathological tissue has to be faced.

To cope with this difficulty, the capabilities of DL to handle classification tasks can be exploited to build a segmented map of the brain tissue. Such a map can then provide an approximated but faithful model to be used by ISP algorithms focused on the diagnostic task. In particular, an MWI-DL framework can be developed to provide an image of the head in which each tissue is associated to a single value of dielectric permittivity and conductivity. This map represents a simplification of the actual scenario, since EM properties in tissue are not exactly homogeneous and vary locally within the same tissue. Nevertheless, associating each tissue with an effective property value is useful to provide a sufficiently accurate model for standard MWI procedures focused on the retrieval of the clinically relevant information (e.g., the stroke or the plaques).

As detailed in the following, the developed MWI-DL framework is composed of two parts. In the first part, the segmentation task is carried via DL to label each voxel of the brain with a tissue. In the second part, an ISP is solved to associate each tissue with an effective complex permittivity.

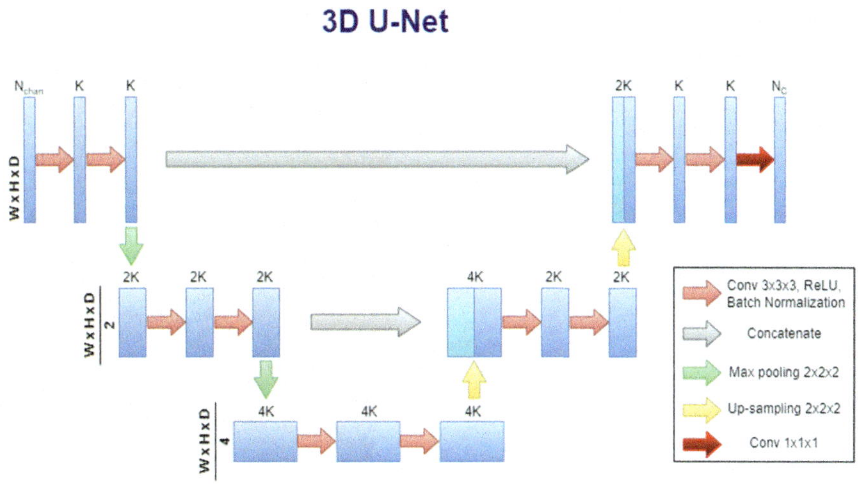

Fig. 8 3D U-Net diagram

To implement the segmentation task a physic-assisted framework is exploited. The domain knowledge is supplied to the neural network in terms of a low-resolution qualitative image of the head. This latter is obtained by processing the scattered field data using the Born approximation and the TSVD algorithm. It is worth noting that in this case the ISP cannot be linearized, so that the TSVD results cannot encode the EM properties and it is therefore only qualitative. The qualitative image is then processed by a peculiar CNN known as U-Net [12], depicted in Fig. 8, that is capable of automatically retrieving the solution as an image and is particularly suited to perform processing tasks involving image segmentation. In particular, U-Net is herein exploited to turn the incoming TSVD images into a segmented image of the head with several domains.

In the second step, the effective complex permittivity estimation is implemented applying the distorted Born iterative method (DBIM) [50]. This iterative inverse scattering approach is suitable for this purpose as it very easily embeds the spatial prior supplied by the segmented image and takes advantage of the fact the actual value of the effective permittivity in each tissue is expected to be close to the average one given in the literature. As such, this latter provides a convenient starting guess for a local optimization procedure based on a linear approximation like the one exploited in DBIM.

To give an example of the described MWI-DL framework, a simplified case in which the head is made only of two 2 domains was considered. Such domains were $\{head, brain\}$, as illustrated in Fig. 9. Other tissues like skin, fat, bone, cerebrospinal fluid were not considered or included in these two domains.

The corresponding pixel-wise classification problem amounts to minimize the cross-entropy:

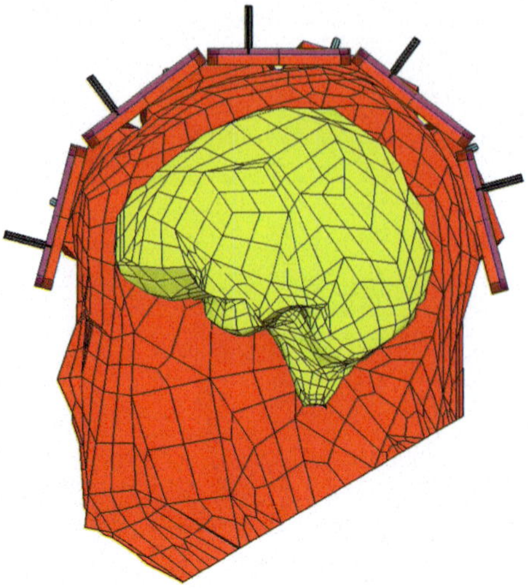

Fig. 9 Head model with two domains: head (red) and brain (yellow). Around it several micro-strip trapezoidal patch antennas are positioned (left side removed for visualization) in a helmet-like manner

$$\arg\min_{\theta}\left[-\sum_{i=1}^{N_p}\sum_{c=1}^{N_c} C_{i,c}^{[n]}\cdot\log\ \hat{C}_{i,c}^{[n]}\right] \tag{14}$$

which is indeed optimized in the training of U-Net to determine its parameters.

The training is carried out generating a dataset of $N = 500$ samples. Models included brains with simple geometries and different permittivity values of the head and brain. The head permittivities were from the set of values within $\varepsilon_H = [33 - j11,\ 36 - j14]$ while the brain permittivities were from the set of values in the range $\varepsilon_B = [41 - j13,\ 50 - j17.5]$.

Regarding the setup configuration, $T_x = 21$, $R_x = 21$ micro-strip trapezoidal patch antennas were used. Those narrow band antennas were matched to work at $f = 1\,GHz$. Forward simulations were carried out using WIPL-D Pro software (WIPL-D d.o.o., Belgrade, Serbia) [51].

The simulated data were then processed via TSVD to provide the input to the U-Net. In doing so, a reference model was made of a homogeneous head of permittivity $\varepsilon_{ref} = 33.5 - j11.5$ (the exact value of the head permittivity was considered unknown).

A test split of 15% was selected for the a-posteriori performance assessment which resulted in a $MCC = 0.931$. Such a compelling accuracy can be seen in one of the test samples shown in Fig. 10.

After achieving a reasonable accuracy, the output segmentation volume jointly with ε_{ref} as initial guess can be supplied to the DBIM algorithm. For instance, a random simulated sample with $\varepsilon_H = 35 - j12$ and $\varepsilon_B = 48.88 - j16.46$, whose corresponding initial guess (shown in Fig. 11) was obtained via the trained U-net, converged to $\hat{\varepsilon}_H = 34.9136 - j9.4128$ for the head and $\hat{\varepsilon}_B = 48.5259 - j15.995$

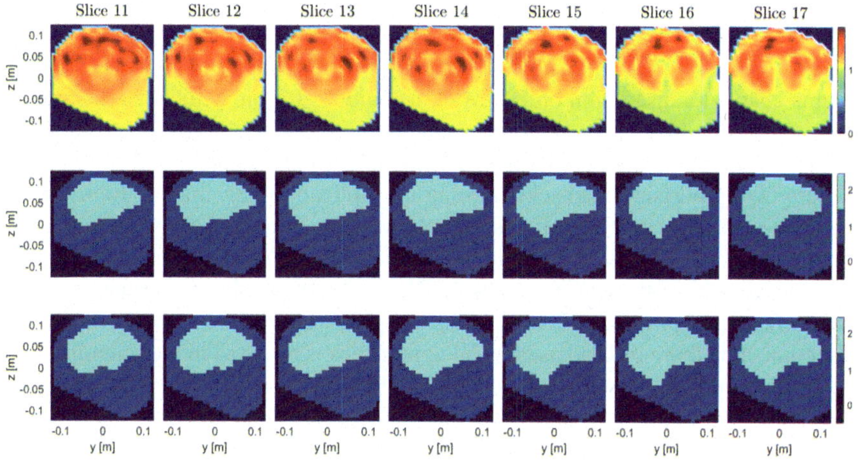

Fig. 10 Slices of TSVD volumes of a head with an inner brain. Second row shows the predictions provided by U-Net (head in dark blue and brain in cyan) while last row depicts the ground truths values

Fig. 11 Segmentation output of U-Net superimposed on a head model: head (red) and brain (yellow)

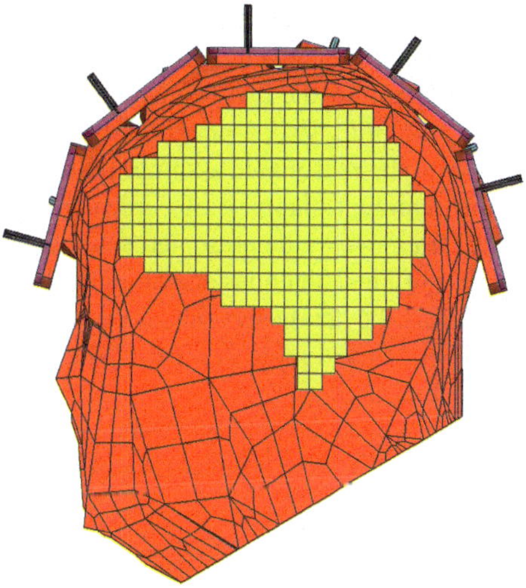

for the brain. A *mean absolute percentage error* (MAPE) [52] can be defined to analyse the performance of the DBIM algorithm:

$$MAPE = 100\% \left| \frac{x_i - \hat{x}_i}{x_i} \right| \tag{15}$$

Computing MAPEs on the selected random sample gave errors of 0.25% and 0.72% for the real parts of the head and the brain, respectively. The imaginary values, as usual, were more cumbersome to discern, with errors of 21.56 and 2.82%.

6 Conclusions

In this chapter, a discussion on DL-MWI techniques for biomedical applications is reported. While MWI is an emerging imaging modality, it still has some challenges ahead before mass adoption. Particularly, the underlying problem of MWI, the ISP, suffers two important drawbacks limiting its applicability as imaging tool. On one hand, ill-posedness severely affects the solution of the imaging problem which may become unstable. On the other hand, non-linearity may lead to false solutions.

Motivated by a thriving research in DL for biomedical applications, some of the implementations on medical imaging are permeating MWI as well [18]. In this direction, a plethora of DL-based MWI approaches is reported in the scientific literature. Most notably, physics-assisted approaches, where images are generated from scattering measurements prior to undergoing DL processing, dominate the contributions.

In the chapter, the main aspects related to the exploitation of DL in MWI have been discussed and then two physics assisted DL-MWI applications related to medical MWI are presented. In both cases, measurements were transformed into images via TSVD to provide the domain knowledge input to the DL model.

The first case exploited 2D CNNs to provide a classification of hyperthermia measurements into labels reporting the temperature status in different areas of interest. In this example, the EM behaviour in the temperature range of HT allowed a linear approximation. Also, a-priori information on the shapes of the targets was exploited based on the reasonable assumption that such information would be obtained in the prior treatment planning of clinical HT. However, in the development of the DL-MWI framework, the presented case considered a number of simplifications such as the use of ideal antennas and elongated targets with respect to the realistic scenario. Therefore the obtained results need to be verified with anthropomorphic permittivity distributions and more realistic EM conditions to ensure an eventual application in clinical scenarios.

The second case employs a special 3D CNN called U-Net to provide segmentation maps of the brain to be used as initial guesses of the ISP iterative solver called DBIM. As opposed to the HT case, here the problem was formulated in more realistic conditions, with a 3D approach and micro-strip antennas instead of point sources. However, the validation of the concept has also been presented in a simplified case, in which the head model only consists of two domains. As such, clinical applicability of the approach is far from being demonstrated. Nevertheless, it can be expected that the increase in the complexity to be faced in actual scenarios (e.g., a brain model considering all involved tissue) could be technically addressed by an increase in the size of the training sets. In this regard, obtaining large datasets is a known limitation associated to DL techniques. In ISP forward simulations, the more realistic the scenarios the higher the computational burden, especially in 3D. To this end, research is moving towards exploiting DL in the pursuit of efficient EM forward simulators and synthetic data generation so as to lift the computational burden with the objective of having datasets with arbitrary large sizes.

Acknowledgements This research was supported by the EMERALD project funded from the European Union's Horizon 2020 under the Marie Sklodowska-Curie grant agreement No. 764479.

References

1. A.S. Turk, K.A. Hocaoglu, A.A. Vertiy, *Subsurface Sensing*, vol. 197 (John Wiley & Sons, 2011)
2. F. Vipiana, L. Crocco, J. LoVetri, Electromagnetic imaging and sensing for food quality and safety assessment [guest editorial]. IEEE Antennas Propag. Mag. **62**(5), 16–17 (2020)
3. L. Crocco, I. Karanasiou, M.L. James, R.C. Conceição, *Emerging Electromagnetic Technologies for Brain Diseases Diagnostics* (Springer, Monitoring and Therapy, 2018)
4. R.C. Conceição, J.J. Mohr, M. O'Halloran, et al., *An Introduction to Microwave Imaging for Breast Cancer Detection* (Springer, 2016)
5. M. Pastorino, A. Randazzo, *Microwave Imaging Methods and Applications* (Artech House, 2018)
6. C.A. Balanis, *Advanced Engineering Electromagnetics* (John Wiley & Sons, 2012)
7. D.L. Colton, R. Kress, R. Kress, *Inverse Acoustic and Electromagnetic Scattering Theory*, vol. 93 (Springer, 1998)
8. Y. LeCun, Y. Bengio, G. Hinton, Deep learning. Nature **521**, 436–444 (2015)
9. R. Zemouri, N. Zerhouni, D. Racoceanu, Deep learning in the biomedical applications: recent and future status. Appl. Sci. **9**(8), 1526 (2019)
10. G. Lo Bosco, M.A. Di Gangi, Deep learning architectures for DNA sequence classification, in *International Workshop on Fuzzy Logic and Applications* (Springer, 2016), pp. 162–171
11. A.W. Senior, R. Evans, J. Jumper, J. Kirkpatrick, L. Sifre, T. Green, C. Qin, A. Žídek, A.W. Nelson, A. Bridgland et al., Improved protein structure prediction using potentials from deep learning. Nature **577**(7792), 706–710 (2020)
12. O. Ronneberger, P. Fischer, T. Brox, U-net: convolutional networks for biomedical image segmentation, in *Medical Image Computing and Computer-Assisted Intervention—MICCAI 2015*, eds. by N. Navab, J. Hornegger, W.M. Wells, A.F. Frangi (Springer International Publishing, Cham, 2015), pp. 234–241
13. S. Wang, Z. Su, L. Ying, X. Peng, S. Zhu, F. Liang, D. Feng, D. Liang, Accelerating magnetic resonance imaging via deep learning, in *2016 IEEE 13th International Symposium on Biomedical Imaging (ISBI)* (IEEE, 2016), pp. 514–517
14. K. Yasaka, H. Akai, A. Kunimatsu, S. Kiryu, O. Abe, Deep learning with convolutional neural network in radiology. Jpn. J. Radiol. **36**(4), 257–272 (2018)
15. A. Craik, Y. He, J.L. Contreras-Vidal, Deep learning for electroencephalogram (EEG) classification tasks: a review. J. Neural Eng. **16**(3), 031001 (2019)
16. S.M. Mathews, C. Kambhamettu, K.E. Barner, A novel application of deep learning for single-lead ECG classification. Comput. Biol. Med. **99**, 53–62 (2018)
17. D. Ravì, C. Wong, F. Deligianni, M. Berthelot, J. Andreu-Perez, B. Lo, G.-Z. Yang, Deep learning for health informatics. IEEE J. Biomed. Health Inform. **21**(1), 4–21 (2016)
18. X. Chen, Z. Wei, M. Li, P. Rocca, A review of deep learning approaches for inverse scattering problems (invited review). Progr. Electromagn. Res. **167**, 67–81 (2020)
19. A. Brankovic, A. Zamani, A. Trakic, K. Bialkowski, B. Mohammed, D. Cook, J. Walsham, A.M. Abbosh, Unsupervised algorithm for brain anomalies localization in electromagnetic imaging. IEEE Trans. Comput. Imaging **6**, 1595–1606 (2020)
20. A. Al-Saffar, A. Bialkowski, M. Baktashmotlagh, A. Trakic, L. Guo, A. Abbosh, Closing the gap of simulation to reality in electromagnetic imaging of brain strokes via deep neural networks. IEEE Trans. Comput. Imaging **7**, 13–21 (2021)
21. P. Mojabi, V. Khoshdel, J. Lovetri, Tissue-type classification with uncertainty quantification of microwave and ultrasound breast imaging: a deep learning approach. IEEE Access **8**, 182092–182104 (2020)

22. P. Mojabi, M. Hughson, V. Khoshdel, I. Jeffrey, J. LoVetri, CNN for compressibility to permittivity mapping for combined ultrasound-microwave breast imaging. IEEE J. Multiscale Multiphys. Comput. Techn. **6**, 62–72 (2021)
23. S. Candefjord, J. Winges, A.A. Malik, Y. Yu, T. Rylander, T. McKelvey, A. Fhager, M. Elam, M. Persson, Microwave technology for detecting traumatic intracranial bleedings: tests on phantom of subdural hematoma and numerical simulations. Med. Biol. Eng. Comput. **55**, 1177–1188 (2017)
24. Z. Wei, D. Liu, X. Chen, Dominant-current deep learning scheme for electrical impedance tomography. IEEE Trans. Biomed. Eng. **66**(9), 2546–2555 (2019)
25. S.J. Hamilton, A. Hauptmann, Deep D-bar: real-time electrical impedance tomography imaging with deep neural networks. IEEE Trans. Med. Imaging **37**(10), 2367–2377 (2018)
26. M. Capps, J.L. Mueller, Reconstruction of organ boundaries with deep learning in the d-bar method for electrical impedance tomography. IEEE Trans. Biomed. Eng. **68**(3), 826–833 (2021)
27. A.C. Luchies, B.C. Byram, Deep neural networks for ultrasound beamforming. IEEE Trans. Med. Imaging **37**(9), 2010–2021 (2018)
28. M. Pastorino, *Microwave Imaging*, vol. 208 (John Wiley & Sons, 2010)
29. M. Bertero, Linear inverse and iii-posed problems, vol. 75, in *Advances in Electronics and Electron Physics* (Academic Press, 1989), pp. 1 – 120
30. G.C. Onwubolu, B. Babu, *New Optimization Techniques in Engineering*, vol. 141 (Springer, 2013)
31. M. Pastorino, A. Randazzo, A smart antenna system for direction of arrival estimation based on a support vector regression. IEEE Trans. Antennas Propag. **53**(7), 2161–2168 (2005)
32. M. Pastorino, A. Salvadè, R. Monleone, T. Bartesaghi, G. Bozza, A. Randazzo, Detection of defects in wood slabs by using a microwave imaging technique, in *2007 IEEE Instrumentation & Measurement Technology Conference IMTC 2007* (IEEE, 2007), pp. 1–6
33. M. Campbell, A.J. Hoane Jr., F.-H. Hsu, Deep blue. Artif. Intell. **134**(1–2), 57–83 (2002)
34. I. Goodfellow, Y. Bengio, A. Courville, *Deep Learning* (MIT Press, 2016). http://www.deeplearningbook.org
35. R. Rojas, The backpropagation algorithm, in *Neural Networks* (Springer, 1996), pp. 149–182
36. A. Géron, Hands-on machine learning with scikit-learn and tensorflow: concepts, in *Tools, and Techniques to build intelligent systems* (2017)
37. Z. Ma, K. Xu, R. Song, C.-F. Wang, X. Chen, Learning-based fast electromagnetic scattering solver through generative adversarial network. IEEE Trans. Antennas Propag. **69**(4), 2194–2208 (2021)
38. M. Wang, H. Li, Y. Shuang, L. Li, High-resolution three-dimensional microwave imaging using a generative adversarial network, in *2019 International Applied Computational Electromagnetics Society Symposium—China (ACES)*, vol. 1, pp. 1–3 (2019)
39. A. Y. Ruiz, M. Cavagnaro, L. Crocco, A physics-assisted deep learning microwave imaging framework for real-time shape reconstruction of unknown targets. IEEE Trans. Antennas Propag. **70**(8,) pp. 6184–6194 (2022)
40. G.M. Hahn, *Hyperthermia and Cancer* (Springer Science & Business Media, 2012)
41. L. Winter, E. Oberacker, K. Paul, Y. Ji, C. Oezerdem, P. Ghadjar, A. Thieme, V. Budach, P. Wust, T. Niendorf, Magnetic resonance thermometry: methodology, pitfalls and practical solutions. Int. J. Hyperthermia **32**(1), 63–75 (2016)
42. H.P. Kok, E.N. Cressman, W. Ceelen, C.L. Brace, R. Ivkov, H. Grüll, G. Ter Haar, P. Wust, J. Crezee, Heating technology for malignant tumors: a review. Int. J. Hyperthermia **37**(1), 711–741 (2020)
43. S. Ley, S. Schilling, O. Fiser, J. Vrba, J. Sachs, M. Helbig, Ultra-wideband temperature dependent dielectric spectroscopy of porcine tissue and blood in the microwave frequency range. Sensors **19**(7), 1707 (2019)
44. K.P. Murphy, *Machine Learning: A Probabilistic Perspective* (MIT Press, 2012)
45. R. Harrington, Origin and development of the method of moments for field computation. IEEE Antennas Propag. Mag. **32**(3), 31–35 (1990)

46. J. Richmond, Scattering by a dielectric cylinder of arbitrary cross section shape. IEEE Trans. Antennas Propag. **13**(3), 334–341 (1965)
47. M.F. Catedra, E. Gago, L. Nuno, A numerical scheme to obtain the RCS of three-dimensional bodies of resonant size using the conjugate gradient method and the fast fourier transform. IEEE Trans. Antennas Propag. **37**(5), 528–537 (1989)
48. B.W. Matthews, Comparison of the predicted and observed secondary structure of t4 phage lysozyme. Biochimica et Biophysica Acta (BBA)-Protein Struct. **405**(2), 442–451 (1975)
49. H. Braak, E. Braak, Neuropathological stageing of Alzheimer-related changes. Acta Neuropathologica **82**(4), 239–259 (1991)
50. W. Chew, Y. Wang, Reconstruction of two-dimensional permittivity distribution using the distorted born iterative method. IEEE Trans. Med. Imaging **9**(2), 218–225 (1990)
51. B. Kolundzija, J. Ognjanovic, T. Sarkar, R. Harrington, WIPL: a program for electromagnetic modeling of composite-wire and plate structures. IEEE Antennas Propag. Mag. **38**(1), 75–79 (1996)
52. A. De Myttenaere, B. Golden, B. Le Grand, F. Rossi, Mean absolute percentage error for regression models. Neurocomputing **192**, 38–48 (2016)

Towards a Microwave Imaging Device for Cerebrovascular Diseases Monitoring: from Numerical Modeling to Experimental Testing

David O. Rodriguez-Duarte, Jorge A. Tobón Vasquez, Cristina Origlia, Rosa Scapaticci, Giovanna Turvani, Mario R. Casu, Lorenzo Crocco, and Francesca Vipiana

Abstract This chapter presents the design, realization, and validation of a full-fledged device for imaging-based detection and monitoring of cerebrovascular diseases via microwave technology. The designed system is a compact and low-complexity prototype with a wearable twenty-two-element antenna array able to perform continuous real-time follow-up of a brain stroke evolution. The imaging algorithm exploits a differential scheme that receives the scattering matrices collected at two different instants and reconstructs tridimensional images relying on the distorted Born approximation. Furthermore, it employs the singular value decomposition of the discretized scattering operator computed via full-wave simulations of high-realistic models and an EM solver based on an in-house finite element method.

D. O. Rodriguez-Duarte (✉) · J. A. Tobón Vasquez · C. Origlia · G. Turvani · M. R. Casu · F. Vipiana
Department of Electronics and Telecommunications, Politecnico di Torino, Turin, Italy
e-mail: david.rodriguez@polito.it

J. A. Tobón Vasquez
e-mail: jorge.tobon@polito.it

C. Origlia
e-mail: cristina.origlia@polito.it

G. Turvani
e-mail: giovanna.turvani@polito.it

M. R. Casu
e-mail: mario.casu@polito.it

F. Vipiana
e-mail: francesca.vipiana@polito.it

R. Scapaticci · L. Crocco
IREA-CNR, Institute for Electromagnetic Sensing of the Environment, National Research Council of Italy, Naples, Italy
e-mail: scapaticci.r@irea.cnr.it

L. Crocco
e-mail: crocco.l@irea.cnr.it

F. Vipiana and L. Crocco (eds.), *Electromagnetic Imaging for a Novel Generation of Medical Devices*, Lecture Notes in Bioengineering,
https://doi.org/10.1007/978-3-031-28666-7_7

The system is numerically and experimentally tested on post-onset scenarios, showing the capability to localize and monitor the progression of hemorrhage and ischemia zones with centimetric spatial resolution. Moreover, it details the manufacturing of the custom radiating elements and the human-like head and non-static stroke testing phantoms used during the experimental testing.

Keywords Biomedical electromagnetic imaging · Microwave imaging · Brain stroke monitoring · Distorted Born approximation · Flexible antennas · Hemorrhagic stroke · Ischemic stroke · Inverse scattering · Microwave antenna array · Propagation

1 Introduction

Cerebrovascular diseases are disorders in which an area of the brain is temporarily or permanently affected by ischemia or bleeding, involving one or more cerebral blood vessels in the pathological process. The flow alterations may occur from the stenosis (vessel narrowing), thrombosis (clot formation), or embolism (blockage) that provokes a lack of sufficient blood flow, affecting brain tissue and causing an ischemic (IS) stroke. This latter is the most frequent stroke category, counting for 85 % of all cases [1]. Commonly, during the IS stroke, the arteries are not entirely obstructed, and a meager bloodstream trickles to the brain, leading to the appearance of two distinct areas, the core, and the penumbra. The blood is almost completely cut off in the core, starving the cells within minutes. The penumbra is a much more extensive area surrounding the dead cells in the core, consisting of injured and malfunctioning cells (idling cells), which can remain in this state for about three hours before they die. Depending on the time window, physicians treat the IS stroke by removing the blocking and restoring blood flow to the brain using thrombolytic drugs or mechanical devices. However, the medication increases the risk of intracranial hemorrhage. The remaining 15 % of cases are hemorrhagic strokes, which are the most deadly. Here, blood vessel rupture or leaks within the artery or on its surface may cause bleeding and damage to the brain. If the vessel breaks inside the brain tissue, it is an intracerebral hemorrhage (ICH), while if it breaks on the brain surface, it is termed subarachnoid hemorrhage. It is usually surgically treated to relieve intracranial pressure caused by bleeding.

Regardless of the stroke typology, a prompt diagnosis is crucial for effective treatment, i.e., an early detection, localization, and classification of the disorder, and subsequent medical care that restore blood flow as soon as possible, mitigating collateral effects. Once the stroke onset and its first symptoms appear, the protocol suggests calling for an ambulance or approaching a medical center/hospital immediately. The medical personnel stabilizes the patient and preliminary assesses his condition. The next 8-hour stage is crucial: clinicians and therapists monitor and evaluate the patient, supporting their prognosis and intervention with imaging-based technologies such as magnetic resonance imaging (MRI) and computerized X-ray tomography (CT). Finally, the patient is hospitalized and monitored due to the high risk of another stroke.

MRI and CT are the current gold standard for medical imaging support, both well-established technologies that deliver highly reliable diagnostic information. However, they have intrinsic drawbacks in terms of portability, cost, time-consume, and safeness (for CT only) that limit their applicability [2, 3]. For instance, an MRI or a CT scanner is bulky and cannot be normally installed in an ambulance or mobile care unit. Moreover, the examination time, the availability of the devices, the costs, and the ionizing radiations (in the CT case), do not allow these technologies to perform real-time monitoring in the emergency department or later at the patient's bedside. In this context, microwave imaging (MWI) has emerged as an attractive non-ionizing, low-intensity, and cost-effective complementary approach, enabling pre-hospital diagnosis of the stroke, bedside brain imaging, and continuous monitoring during the post-acute stage, albeit at the cost of a lower resolution [4–16].

MWI relies on the contrast of the electrical properties (permittivity and conductivity) exhibited by the healthy brain tissues and the stroke-affected ones at microwave frequencies to identify their typology and physio-pathological status. For instance, the electrical properties of the affected area decrease during an IS stroke due to the reduction of the blow flow and, conversely, increase in the case of bleeding. This technique illuminates the body with low-power EM waves in the GHz range of the spectrum, then uses the resultant back-scattered signals to generate images of the internal structures. Hence, MWI devices exploit the non-invasive and harmless nature of microwaves together with the current high-performance microwave hardware and computing power to complement and overcome the limitation of cost, portability, and harmfulness of traditional diagnostic techniques. Summing up, MWI technology:

- exhibits favorable penetration depths that allow imaging of deeper tissues in the human body as compared to other emerging modalities (e.g., optical techniques);
- is entirely harmless since the involved waves are non-ionizing and used in very low doses (low-power EM waves);
- is convenient for the healthcare system, providing mobile and low-cost imaging platforms, backup for advancement in the mobile industry and microwave devices in recent years.

A typical MWI device is made by an array of antennas, which is in charge of probing the region on the body under examination and collect the back-scattered waves data. Moreover, the device must be equipped with a suitable data processing algorithm in charge of pursuing the solution of the non-linear inverse scattering problem to be faced in order to retrieve the EM properties of inspected tissue from the data. In the specific application of brain stroke, several academic [4–12] and industrial prototypes [13–16] have been developed. For example, from the academy stand notable developments as [4–6]. References [4, 5] show a 16-antenna-array-ring system, where the matching medium consists of heat-sealed polyethylene bags filled with a water-glycerine-based liquid, and a two-dimensional (2-D) tomographic approach to resolve the non-linear inversion in Lebesgue spaces. References [6, 7] present a device with an 8-antenna array submerged in glycerin as a matching medium, performing a 2-D non-linear microwave tomography imaging via a distorted Born iterative method (DBIM) with GPU-accelerated Finite-difference time-domain (FDTD). Reference [8] introduces a novel hybrid imaging system using both a radar approach and a frequency approach, testing it with an 8-element array of ultra-wideband Bowtie antenna in a 2-D configuration. Reference [9] poses a 14-antenna-array system, working with a novel polar sensitivity encoding (PSE) algorithm to locate the stroke. Besides, in [10], they propose a flexible-24-antenna cap and three-dimensional (3-D) tomography imaging technique, achieving the location of the stroke in complex scenarios, yet with limited shape retrieval and non-monitoring testing. Moreover, in [11], the authors of this chapter developed a 24-antennas system employing a solid graphite-rubber mixture as a matching medium, and approaching the detection and localization stroke targets via a 3-D imaging algorithm. Additionally, they optimize the system, reducing its dimensions and number of antennas, as well as perform real-time imaging-based monitoring of hemorrhages and ischemia in mimicked lab scenarios [44], which further assessed in this chapter.

On the other hand, the Strokefinder, the EMTensor BrainScanner, and the EMVision Scanner are devices developed by companies. Medfield Diagnostics developed the Strokefinder, a low-complexity and patient-adaptable device made of 8 antennas, aiming to discriminate the stroke typology early in the pre-hospital stage [12, 13]. This first device does not aim to generate images of the stroke. However, it is an excellent example of narrowing the clinical problem and the system functionalities that still retrieves a piece of vital information for doctors and physicians. The second device, EMTensor BrainScanner [14], performs tomographic imaging and approaches in-ambulance and bedside scenarios. However, it uses a highly complex system consisting of 177 loaded and truncated waveguides (radiating elements) in its initial version, increasing the complexity of data processing inherently. The last example, EMVision Scanner [15], is also a tomographic device with encouraging results in the undergoing clinical trials.

This chapter presents the develops and testing of the prototype presented in [11, 44]. In the following, we present the mathematical background behind the image formation and the new system design, which is synthetically tested using different virtual scenarios, providing a preview of the system's capabilities. Second, the whole system is assessed using non-static mimicked clinical scenarios. This part covers the experimental setup, including the manufacturing and characterization of the different elements used during the experimental testing. Finally, all numerical and experimental results are gathered, closing up with the conclusions and perspectives.

2 Microwave-Based Stroke Monitoring

2.1 Imaging Algorithm

This section first discusses the principles and mechanisms associated with MWI monitoring. Then, it describes the algorithm used to deal with the intrinsic non-linearity and ill-posedness of the inverse scattering problem under the imaging-based follow-up task.

In this context, the direct problem involves calculating the scattered electric fields due to the evolution of the pathology when the monitoring device is illuminating the domain of interest (DOI). In other words, the temporal variation of the dielectric properties of the brain tissues in the stroke-affected area causes a change in the scattered field, stating a cause-and-effect relationship. In the inverse problem, the goal is to retrieve the EM properties from the scattered fields. This problem is ill-posed, due to the intrinsic loss of information arising from the fact the sought unknown is not directly probed but assessed through its footprint in the measured scattered field.

To deal with the ill-posedness, the problem needs to be regularized, adding prior information related to the problem's physics [16]. Hence, regularization relies on limiting the family of possible solutions so that the family functions converge to the exact solution, in the case of noise-free data, or to an optimal approximation, in the case of noisy data. In other words, regularization acts with the premise of intentionally using additional information to generate families of approximate solutions consistent with the given image.

Getting into the intended application, the input data are the S-parameters of an array of transceivers placed around the DOI. These parameters sample and describe the illuminating system in terms of the incoming and outgoing port waves, named a_p and b_q, given at the p-th and q-th antenna ports, respectively. Hence, the forward problem can be stated analytically as:

$$\Delta S(\mathbf{r}_p, \mathbf{r}_q; t_0, t_1) = -\frac{j\,\omega\,\varepsilon_b}{2\,a_p\,a_q} \int_{\mathrm{DOI}} \mathbf{E}_p(\mathbf{r}, t_0) \cdot \mathbf{E}_q(\mathbf{r}, t_1)\,\Delta\chi(\mathbf{r}, t_0, t_1)\,d\mathbf{r}, \quad (1)$$

which indicates the scene variation with respect to a constitutive or shape object parameter and the field distribution [17]. In Eq. (1), ΔS is the differential scattering parameter, D is the volume of the domain of imaging, j is the imaginary unit, $\omega = 2\pi f$ is the angular frequency, ε is the complex permittivity distribution, b indicates a reference background scenario, and the symbol "$\cdot$" denotes the dot product. Furthermore, $\mathbf{E}_p$ $(\mathbf{r}, t_0)$ and $\mathbf{E}_q$ $(\mathbf{r}, t_1)$ are the nominal electric field vectors, i.e., the fields radiated by the p-th and q-th probe inside the DOI at the respective state, and

$$\Delta\chi(\mathbf{r}; t_0, t_1) = \frac{\varepsilon(\mathbf{r}, t_1) - \varepsilon(\mathbf{r}, t_0)}{\varepsilon_b(\mathbf{r})}, \quad (2)$$

is the spatial distribution of the normalized contrast, all referenced to the instants t_0 and t_1, respectively.

Notice that Eq. (1) presents a non-linear problem that iterative methods solve by posing it as an optimization task. Examples of these methods are the DBIM and the contrast source inversion (CSI) [19, 45]–see also Chap. "Assessment and Validation of 2-D and 3-D DBIM-TwIST Algorithm for Brain Stroke Detection and Differentiation of this book". However, these methods require to solve the forward problem, that is, the heaviest computational part. Therefore they are not convenient for the targeted real-time application.

Another way to solve the nonlinear problem is to linearize it using the so-called Born approximation. It assumes mild overall perturbation of the electric field in the studied period, which is due to a "weak" and localized perturbation, as in the monitoring problem where the affection is concentrated in an area, and the examination times are user-controlled. Moreover, we approach the problem by applying the distorted version of the approximation considering the morphological variability of the patient. Therefore, the forward formulation is rewritten as:

$$\Delta S(\mathbf{r}_p, \mathbf{r}_q; t_0, t_1) = -\frac{j\,\omega\,\varepsilon_b}{2\,a_p\,a_q} \int_{\text{DOI}} \mathbf{E}_p^{ref}(\mathbf{r}) \cdot \mathbf{E}_q^{ref}(\mathbf{r})\,\Delta\chi(\mathbf{r}, t_0, t_1)\,d\mathbf{r}, \tag{3}$$

where the field distributions at t_0 and t_1 are replaced by nominal reference one, $\mathbf{E}^{ref}$, consisting of a scenario representing the head as a single-tissue organ with the average complex permittivity of a brain. In a compact format, Eq. (3) is expressed as

$$\Delta S = \mathcal{L}\{\Delta\chi\}, \tag{4}$$

where $\mathcal{L}$ is the linear operator from the object space into the image space. The imaging operator kernel describes the system's response to a point source as a function of the observation position, forming the so-called point-spread function (PSF), which is an indicator of the resolution of the system.

To invert Eq. (4), first, the imaging kernel is built by applying a singular value decomposition (SVD) to the discretized integral operator and decomposing it in $\langle [u], [\sigma], [v] \rangle$, where σ_n, u_n and v_n are the n-th singular value, right and left singular vectors, respectively. Then, the differential contrast distribution is retrieved as:

$$\Delta\chi = \sum_{n=1}^{T} \frac{1}{\sigma_n} \langle \Delta S, u_n \rangle\, v_n, \tag{5}$$

where truncation index T acts as a regularizer parameter [16].

In summary, the imaging algorithm uses both an off-line part, compute-intensive but just done once, where the imaging kernel is generated and decomposed, and a real-time (fast) part facing the inversion. Figure 1 illustrates the general scheme of the imaging. Moreover, because the living tissues are linear media whose constitutive parameters do not depend on the field strength, we are dealing with a reciprocal network and ΔS is expected to be symmetric. Thus, in an N-port network, only

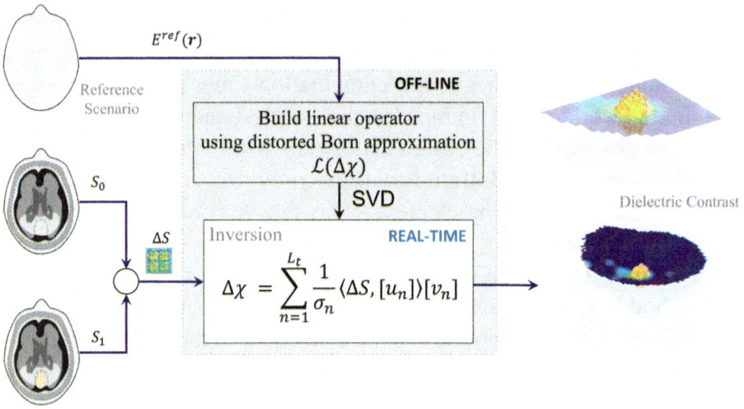

Fig. 1 Differential imaging algorithm scheme

$N(N+1)/2$ S-parameters are independent, representing the upper/lower triangular part of ΔS. Though, we do not employ the reflection self elements for the imaging, as further explained in Sect. 3, thus the number of independent parameters is reduced to $N(N-1)/2$, decreasing both the $\mathcal{L}$ and ΔS dimensions.

2.2 *Electromagnetic Modeling*

To build the imaging kernel (operator), we computed $\mathbf{E}^{ref}$ from numerically modeled scenarios using realistic computer-aided design (CAD) models of the MWI prototype and the head, the nominal dielectric properties of all involved elements, and an electromagnetic (EM) finite element method (FEM) solver. In the specific, the properties of the head tissues are taken from [18] and hemorrhagic or ischemic ones from [20, 21], while others are measured experimentally using the open-coaxial technique, as described in [22]. Besides, the geometrical models came from [23], and in-house FEM implementation is exploited [24].

FEM is a numerical technique used to obtain an approximated solution to boundary-value problems described and governed by either differential or integrodifferential equations, such as Maxwell's equations [25]. It resolves the original problem approximating the solution with a finite number of subdomains in which simple interpolation functions change the unknown function with unknown coefficients. Then, the response is determined by solving the system of equations obtained by applying the Galerkin procedure. FEM is an optimum alternative to approach problems with complex architectures describing scenarios with arbitrary shapes and complex permittivity distributions, where interact, for example, biological tissues, as the head, with close-by radiating elements. In general, it follows the stages listed below:

- **Discretization of subdivision of the domain** This is probably the most important part of FEM procedures because the way in which it is done directly affects the computer storage requirements, the computation time, and the accuracy of the results. In this work, we use tetrahedra, the simplest and best suited for arbitrary-volume domains, such as the tissues and organs within the head.
- **Selection of the interpolation functions** provides an approximation of the unknown solution within the element. We use a first-order polynomial (linear) [25].
- **Formulation of the system of equations** can be done using either the Ritz variational method [25] or the Galerkin one, which is used here.
- **Solution of the system of equations** is the last step of the finite element analysis. In electromagnetics, it is usually associated with problems where exists a source or excitation, e.g., scattering problems.

In this work, the employed FEM uses edge-basis functions for the expansions, the curl-curl formulation for the electric field, and the Galerkin approach for testing [24, 26]. Also, regarding the boundary conditions, first, it applies the Dirichlet condition to the metallic surfaces, such as the radiating element of the antenna and the coaxial connectors, thus stating them as perfect electric conductors (PEC) surfaces. Secondly, it forces absorbing boundary conditions (ABC) on the contour surfaces of the cylindrical volume [26], simulation domain, that holds the whole system. The simulation volume is meshed using a tetrahedral base and subdivided using different density-mesh levels depending on the precision and importance of the discretized volume, e.g., on the port of the antennas. Moreover, the forcing elements of the system, essential components to retrieve the fields and compute the scattering parameters accurately, are included via the coaxial antennas' feeding. Then, each antenna port is modeled as a section of a rigid coaxial cable with lossless Teflon, $\epsilon_r = 2.2$, as a substrate. Then, to excite an antenna is enforced the tangential electric field, $\mathbf{E}^{\text{inc}}$, on the port surface, S_{port}, and proper boundary conditions are applied [24, 25]. The scattering parameter for each antenna pair, p and q, is then evaluated as

$$S_{pq} = \begin{cases} \dfrac{\int_{S_{port}} \mathbf{E}_p \cdot \mathbf{E}_q^{\text{inc}*} \, dS}{\int_{S_{port}} |\mathbf{E}_q^{\text{inc}}|^2 \, dS} & \text{if } p \neq q \\ \dfrac{\int_{S_{port}} \mathbf{E}_q \cdot \mathbf{E}_q^{\text{inc}*} \, dS}{\int_{S_{port}} |\mathbf{E}_q^{\text{inc}}|^2 \, dS} - 1 & \text{if } p = q \end{cases} \tag{6}$$

where $\mathbf{E}_q^{\text{inc}}$ is the electric field forced in the excited port q and $\mathbf{E}_p$ is the electric field evaluated at the port p via the FEM solver. $\mathbf{E}_q^{\text{inc}}$ can be represented with the transverse electromagnetic (TEM) mode of the coaxial cable as

$$\mathbf{E}_q^{\text{inc}} = \frac{V}{\sqrt{2\pi} \ln\left(r_b/r_a\right)} \frac{\hat{\rho}}{\rho}, \tag{7}$$

where r_a and r_b are the internal and external radii of the coaxial cable, and ρ and $\hat{\rho}$ are the radial coordinate and radial unit vector from the center of the coaxial cable in the port plane [25]. The coefficient V is chosen to equal 1 V.

2.3 *Design of a Low-Complexity Microwave Imaging Device*

As mentioned in the introduction, one of the benefits of employing MWI technology is to design portable devices with low cost and suitable imaging capabilities. With these features in mind, we describe the followed design guidelines to design a low-complexity device, taking as reference the outcomes of [27, 28].

2.3.1 Operational Frequency and Matching Medium

Operational frequency and matching medium parameters are some of the most relevant to choose from while developing a MWI prototype for stroke follow-up. Their selection directly influences the incident power penetration and the spatial resolution. For instance, lower frequencies increase the penetration, but decrease the resolution, whereas the opposite occurs at high frequencies. Thus, these parameters are direct degrees of freedom for the device's design since the wave penetration depends on the electrical discontinuity, and the wavelength in the coupling medium dictates the resolution.

To select them, we follow the rigorous procedure proposed in [27] and numerically validated in [29], which guarantees low computation effort. It studies the propagating wave scenario of a multi-tissue head, considering the interaction between the head, the coupling medium, and the electric incident field as a parallel transmission line network. Then, the simplified model assumes the head tissues (e.g., skin, fat, bone, cerebrospinal fluid (CSF), and average brain—grey and white matter—) and the matching medium as line segments, where the brain acts as a load. These segments are characterized by the estimated length of the actual radial thickness and the corresponding impedance, $Z_n = \sqrt{\mu_0/\varepsilon_0\varepsilon_t}$, where ε_t is the complex permittivity of each tissue. At the same time, ε_0 and μ_0 are the free space's permittivity and permeability. Meanwhile, the power source is considered into the matching medium in direct contact with the head.

Once modeled the scenario, the transmittance coefficient T is computed, describing the amount of power penetrating the head, defining it as [30]:

$$T = 1 - |\Gamma|^2, \tag{8}$$

where Γ is the reflection coefficient referenced to the interface plane entering the head. Then, there is a low-transmittance band between 1.5–2.5 GHz for any matching permittivity value from 1 to 80 [27]. Also, the operation in frequencies higher than 2.5 GHz may not be convenient due to 2 cm or less penetration depth, for tissues such as skin, bone, or CSF. Moreover, frequencies below 0.6 GHz reduce the spatial resolution too much, dictated by the background medium wavelength. Finally, the 0.6–1.5 GHz band is identified as an adequate operation frequency band while matching permittivity is higher than 10 and lower than 40, and minimum losses. So, from herein, we choose 1 GHz as the operating frequency. Regarding the matching medium, we decide to employ solid and flexible materials, with adjustable permittivity, which is set around 20.

2.3.2 Optimal Antenna Array Layout

The first intuitive idea while determining an ideal number of antennas is to use as many as possible since having more antennas means more information. However, the selection approach is not straightforward and should consider aspects such as the redundancy of the gathered information, physical constraints, and cost. Thus, we follow the methodology developed and tested in [28, 31], which is an instrument to find a good trade-off between stability and accuracy.

It consists of an SVD-based numerical analysis of the imaging kernel, considering the number of antennas, N_a, as a variable, and the operator condition number (CN), i.e., the ratio between the largest (first) and smallest (last) singular values, as studied parameter. CN indicates the robustness of the inversion to higher-level noise and modeling errors. Hence, the lower CN is more robust the inversion is, but when N_a increases, whatever the antenna polarization, CN increases too. Higher N_a means less distance between antenna elements. Thus, the information gathered by close elements would be strongly dependent, in other words, redundant, while the measurement noise remains independent. This analysis is done in [28] for the head case, showing good compromise when $N_a \approx 24$. In this scenario, CN remains lower than the dynamics and the SNR of a standard measurement system, about 90 dB. Moreover, the perforce does not improve considerably, doubling the antenna number, $N_a = 48$, getting CN$\approx$100 dB, but implying higher complexity, and cost.

N_a and their distribution also directly affect discretization error and imaging accuracy. This error is a metric defined in [31], that considers the normalized difference between a case using a continuous (ideal) scanning surface ($N_a \approx \infty$) and a discretized case using a limited N_a. Moreover, it is worth noting that the system's dynamic also limits the imaging operator's accuracy. So, considering the monitoring application, the actual dynamic range available is around 30 dB, if we estimate the differential scattering as about 70 dB below and the dynamic range of a medium-performance VNA as 100 dB. Therefore, and as [28] shows, working around 20–30 antennas is a good compromise also in terms of imaging accuracy.

Finally, we opt for a 22-antenna configuration placed on a head conformal array that, as shown later in the validation, demonstrates a centimetric resolution (around $\lambda/4$), agreeing with the findings in [28]. This selection also considers the head and antenna dimensions and keeps the low complexity, thus reducing cost.

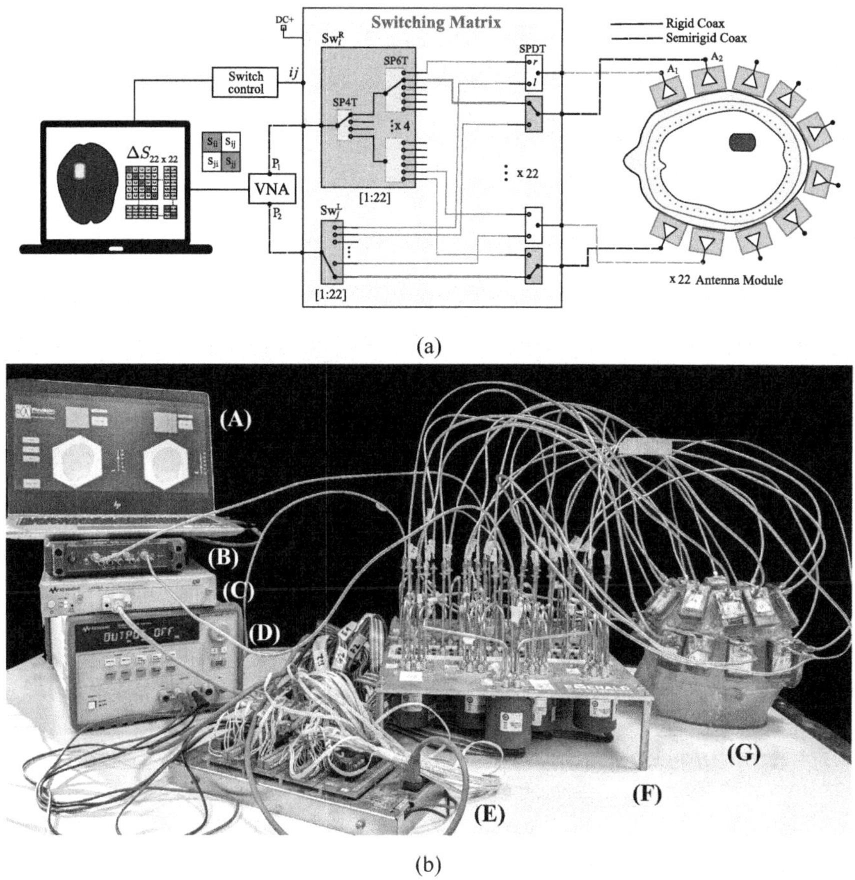

Fig. 2 Microwave brain imaging system prototype. **a** Scheme. **b** Photo: **A** Laptop, **B** vector network analyzer, **C** switch interface card, **D** DC power source, **E** switch control card, **F** switching matrix, **G** head phantom with antenna array on top

3 Microwave Imaging Prototype and Experimental Setup

Considering the design guidelines, we detail the projected MWI system consisting of a 22-antenna array configuration, working with a vector network analyzer and a switching matrix, as shown in Fig. 2.

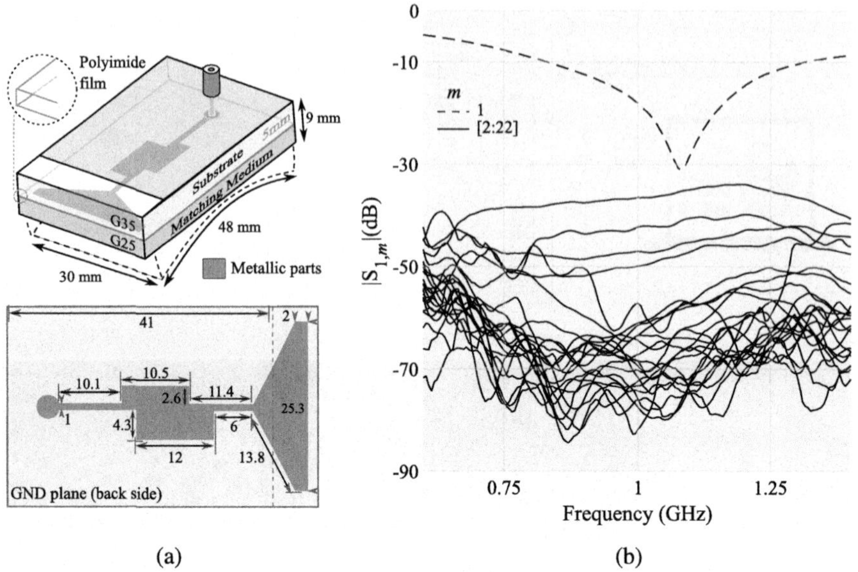

Fig. 3 Flexible custom-made antenna module and measured S-parameters amplitude. **a**: antenna diagram with all dimensions in mm (See the manufactured antenna in Fig. 5c–d). **b**: S-parameters for one antenna of the MWI system

3.1 Antenna Array

As aforementioned, we use a 22-antenna array configuration that is conformal to the head's upper part, reaching a good field coverage (see Sect. 4). Each antenna is positioned directly on the head surface, e.g., the skin in multi-tissue cases, and held by a fixed container in the experimental case.

The antenna element is an optimized back-fed monopole consisting of a triangular radiant part and a double-stub line, as shown in Fig. 3a. It is printed on flexible commercial 50 μm-thickness Polyimide film, placing the ground (GND) plane on the backside. Then, the printed segments are stacked in a two-layer structure, forming a joint antenna module with enough flexibility to adapt to the head curvature. Specifically, each level is made of custom mixtures of urethane rubber and graphite powder adjusted to reach a good trade-off between high permittivity and low losses, as indicated in Table 1, which reports the proportions of used materials and their resulted dielectric properties at 1 GHz, respectively. From back to front, the first layer, dubbed G35, acts as the support substrate of the antenna. While the second one, G25, is a matching medium that minimizes the antenna-to-head mismatch.

The optimized antenna works in a -10 dB band from 0.85 to 1.25 GHz in the head vicinity, with transmission parameters between antenna pairs between -80 and -40 dB (see Fig. 3b). The antennas reach adequate EM wave penetration at 1 GHz into the head with either homogeneous and multi-tissue head models, as shown in

Table 1 Antenna mixtures composition and their permittivity at 1 GHz

	Graphite [%]	Rubber [%]	ϵ_r	σ [S/m]
Matching	25	75	13	0.18
Substrate	35	65	18	0.3

Fig. 4, and cover the upper head domain if we consider a conformal 22-antenna array placed all around the head (see Fig. 2). For penetration analysis, we estimate the power density, $\mathbf{W}_{av}$, associated with the time average Poynting vector (average power density) as in Eq. (9) [30]:

$$\mathbf{W}_{av} = \frac{1}{2}[\mathbf{E} \times \mathbf{H}^*], \tag{9}$$

where **E** and **H**, electric and magnetic fields, respectively, are computed via a full-wave simulation with two lateral conformal placed antennas. The details of the head models are discussed in the numerical validation section. Overall, the antenna module combines the benefits of a flexible antenna with the cheap cost, low complexity, and adaptability of a traditional circuit-printed one, and working within a low-complexity array complies with the requirements demanded by the MWI application.

3.2 Vector Network Analyzer

The vector network analyzer (VNA) is an instrument that measures the constituent wave quantities, e.g., a_1, b_1, or the related phase and magnitude of the S-parameters of passive and active devices. Depending on the application, these measures involve one or more ports, over-swept frequency, or power. For example, we use the swept frequency for characterizing the dielectric properties of the used materials in the experimental part and the MWI monitoring test. Thus, the VNA provides input signals at the different ports while acquiring incident and reflected waves at each port. So, internally it consists of a signal source generator, some receivers, and some directional devices for separating incident and reflected energy [32].

The VNA is an excellent MWI prototyping instrument. It gives an overall design and testing flexibility, allowing us to gather off the needed S-parameter of the imaged situation in multiple configurations and scenarios. Though, its use can be replaced for ad-hoc or customizable devices [33–35]. As the technology's proof-of-concept stage, we opt for the compact Streamline USB Vector Network (P9375A Keysight, 10 MHz-26 GHz) for the presented device in this work [36]. This VNA guarantees gathering the MWI duty information with a noise floor at about −110 dB (at 1 GHz), using an intermediate filter (IF) set 100 Hz, in a single measure, i.e., without averaging.

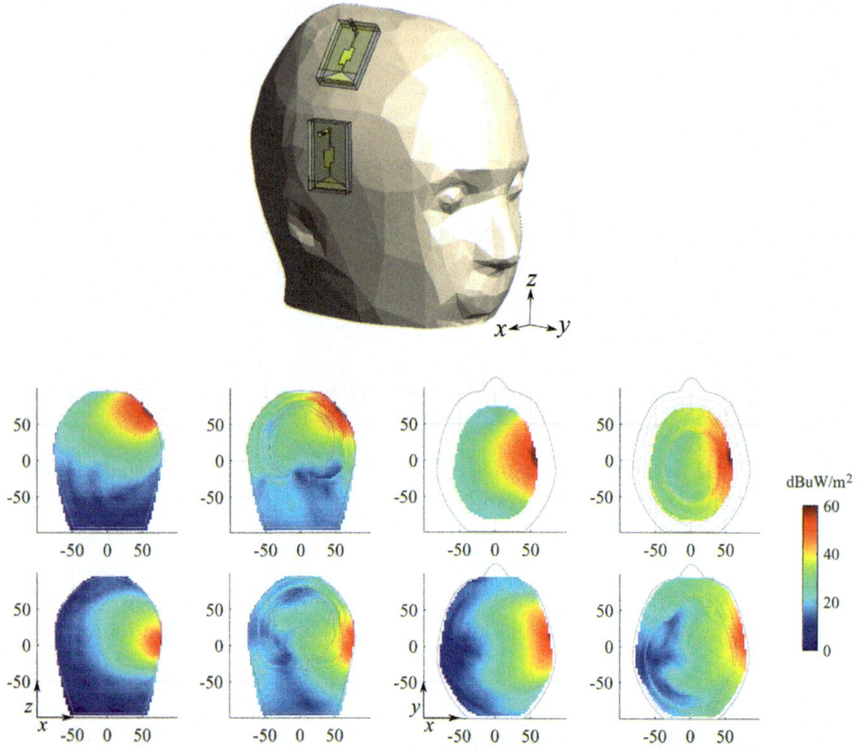

Fig. 4 Sagittal and transverse plane view of the power distribution within the head models for two different lateral antennas. From left-to-right: 1st and 3rd, homogeneous head; 2nd and 4th, multi-tissue head. Top-row: Antenna 1. Bottom-row: Antenna 2. Dimensions in [mm]

3.3 Switching Matrix

The switching matrix is the block of the MWI system that allows for measuring the full S-parameter matrix of the 22×22 antenna array using a 2-port VNA in a multistatic multiview configuration. Then, the switching matrix multiplexes each possible antenna pair, sequentially transmitting and receiving, while completing the full matrix by 2×2 segments. Notice, then, that measure of the matrix is not an instantaneous one. However, considering the target application with much slower variation rates and the controlled condition of the experimental testing, it can be considered instantaneous.

The matrix architecture consists of a two-arm one, where each one-to-twenty arm is connected to a VNA port and to twenty-two selecting switches, which set the operating antenna pair as shown in Fig. 2a. The arms, SW^R and SW^L in the figure, are implemented using a single-pole-four-throw (SP4T) and four single-single-pole-six-throw (SP6T), while selecting ones with pole-double-throw (SPDT), all elec-

(a) (b) (c) (d)

Fig. 5 Manufactured head phantom and antenna array. **a**: Top view, the dashed contour indicates the position of the stroke. **b**: isometric view. **c**: Radiant part of the antenna – top view –. **d**: Lateral view of the antenna

tromechanical coaxial ones, with low insertion (<0.3 dB) loss and high isolation (>90 dB) [37]. Here it is worth noticing that although the advantageous isolation given by the electromechanical switches, they have drawbacks such as slow sweep rate, repeatability, typically no better than a few hundredths to a tenth of a dB at microwave frequencies, and limited lifetime [32]. Thus, updating the switches to solid-state ones would substantially improve the time performance and repeatability, as exhibited in [38], rendering the system real-time responsive. Currently, it spends around five minutes by acquisition set, depending on the VNA setup. Moreover, the connections between the internal sections of the matrix are made with semi-rigid coaxial cable, and the external connections with semi-flexible coaxial ones, and a DC source powers the matrix, and a control card using the VISA/TCPIP (VXI-11) protocol on an Ethernet connection handles it.

3.4 Phantoms

In order to validate the prototype in a meaningful scenario, we employ an anthropomorphic head phantom and a realistic non-static stroke phantom. The former is based on an anthropomorphic 3-D model of an adult head derived by modifying models from [23, 39, 40] as a single-cavity one, considering just the external layer. Moreover, we attach to it the antenna supports as shown in Fig. 5a–b. It is 3-D printed using polyester casting resin, which has dielectric properties similar to plastic $\epsilon_r \approx 3$ and thickness around 3 mm. So, electromagnetically it may compare with air gap with the same thickness, as studied in [24]. For further details on the head phantom design and realization see Chap. Standardized Phantoms of this book. For the stoke a non-static phantom (custom balloon) was developed to mimic the spatial evolution of an early-onset situation, i.e., the phantom can evolve during the monitoring, allowing us to assess the system's capabilities for tracking changes. The shape of this phantom

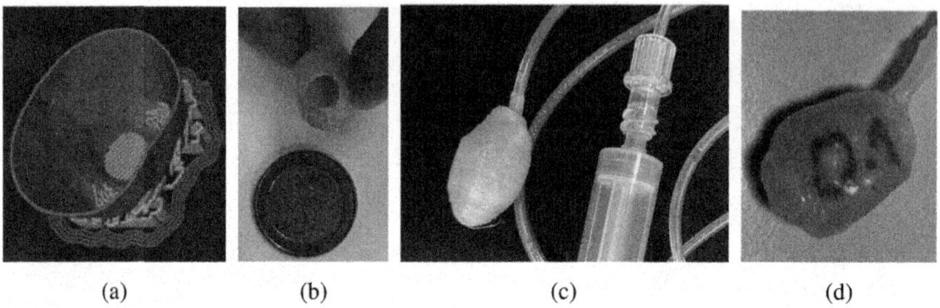

Fig. 6 Stroke phantom manufacturing. a: 3-D model cut view. b: PVA mold. c: Model sealed. d: Stroke phantom before of remove PVA material

Table 2 Mimicking materials composition and permittivity at 1 GHz

	Triton X-100 [%]	Water [%]	Salt [%]	ϵ_r	σ [S/m]
ISC	50	50	5.2	34	0.6
HEM	14	86	9.4	68	1.5
Brain	38	62	5.2	45	0.7

is abstracted by segmenting CT medical[1], and 3-D printed using a PVA 3D filament [41], with the printer's minimum thickness extrusion setting (around 0.1 mm) and the number of layers by thickness set to one as can be seen in Fig. 6a, to reduce the mold material. Then, the printed stroke model, Fig. 5b–c, is coated using a thin layer (less than 1 mm) of slow-cure platinum silicone layer [42], letting cured it as shown in Fig. 5d. The last step exploits the water-soluble nature of the PVA, dissolving the mold by injecting water and then removing the residuals via a small hole, which is used later to attach it to a 3 mm-radius feeding cannula. The result is a resistant, multi-use, and shape-realistic stroke phantom.

The phantoms are filled with Triton-X-based mixtures, mimicking the dielectric properties of the head and the stroke, respectively. The recipes for the employed liquid mixtures are based on [40], and are reported in Table 2, together with the corresponding target dielectric parameters (i.e., relative permittivity, ϵ_r, and conductivity, σ [S/m]). In Table 1, HEM and ISC stand for the hemorrhagic, and ischemic stroke cases, which real reference values are taken from [20], and brain refers to the average dielectric properties of the brain tissues considering a compound of 75% of white matter and 25% of grey matter [40].

The filled head is depicted in Fig. 5a, where the stroke phantom is introduced from the top, and signaled by the white dashed contour. Also, Fig. 7 presents the stroke changing in volume from empty to a maximum of 15 cm^3. These conditions are those used later during the numerical and experimental validation.

[1] Work in collaboration with the Emergency Medicine Dept. of Lariboisière University Hospital, Ilumens Health Simulation Center and Sorbonne University, all in Paris (FR).

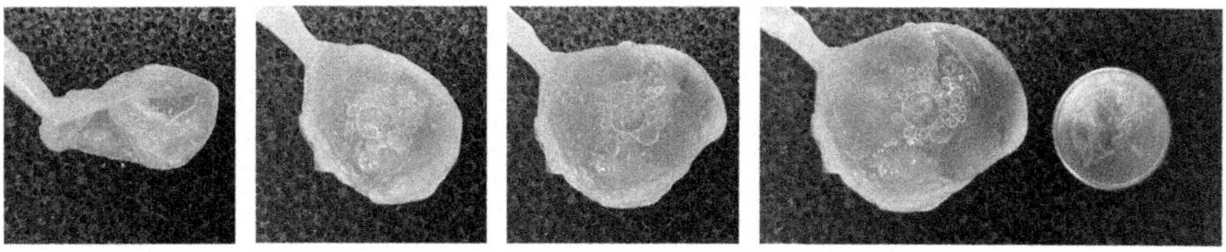

(a) Empty (b) 5 cm^3 (c) 10 cm^3 (d) 15 cm^3

Fig. 7 Non-static stroke phantom growth

4 Numerical Validation

This section presents a thorough numerical analysis of the projected MWI device, working as an instrument of design and validation in fully controlled conditions. It approaches the assessments using realistic full-wave simulations and the imaging algorithm described in the previous section. The analysis focuses first on the imaging operator's capabilities and then on the monitoring imaging capabilities.

4.1 Imaging Operator

The imaging operator $\mathcal{L}$, see Eq. (4), is fundamental to reach outcomes with adequate accuracy, reliability, and quality. It is the framework of the imaging, in which the input differential scattering matrix is projected, and a non-well-conditioned operator might degrade the retrieved images, even in an ideal noiseless condition. Thus, the operator should include a complete and faithful description of the MWI system. With the previous premise in mind, we built the operator of our system using full-wave simulations, including a detailed and highly realistic description of all antennas and their respective layout on top of the head. Though the description of the head is approached using a simplified single-tissue head model ($\epsilon_r = 45$, and $\sigma = 0.7$ [S/m] at 1 GHz), considering that it changes from patient to patient in a real clinical scenario, and the exact morphology of head under test is unknown.

Once built the operator, we analyze it using its decomposed version (see Eq. 5), which is described by Fig. 8. Though, finer details correspond to singular values which are lower in magnitude, whose contribution to the data is most likely over whelmed by measurement noise and model errors. Therefore, considering the actual dynamic range available for the differential signal is around 30 dB, we select an operator truncation (T in Eq. 5) about the first 180 singular values, corresponding to values of not lower than 30 dB the highest one, σ_1, providing stability.

Fig. 8 Imaging kernel singular value behavior

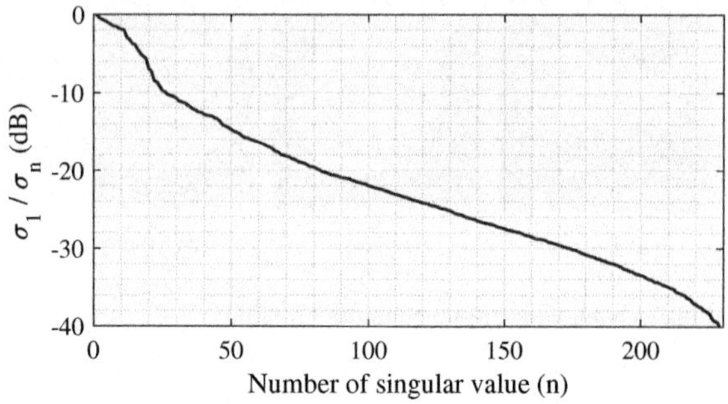

Moreover, to assess the effect of the truncation on spatial resolution, we evaluate the PSF as:

$$[\Pi_{n_T}] = \sum_{n=1}^{n_T} [v_n]\langle \cdot , [v_n]\rangle, \tag{10}$$

for both the full and truncated operator. It represents the orthogonal projection of the contrast functions spanned by the first, n_T, of right singular vectors, $[v_n]$, of $\mathcal{L}$, onto the finite-dimensional space [28]. So, it indicates the contrast that can be retrieved using a specific n_T. In the specific, we used Eq. (10) considering an ideal point-like contrast, i.e., setting $\Delta\chi = 1$ in a 3 mm voxel, while $\Delta\chi = 0$ for all the others elements of DOI.

Finally, Fig. 9 presents the normalized retrieved images, where a minimal variation between cases can be noticed. It confirms that the selected truncation has negligible worsening effects on the system resolution. Then, defining the resolution as the points at -3 dB of the normalized projection, the system shows a centimetric resolution, being around $\lambda/4$, in agreement with [28]. However, the resolution is not symmetrical on all axis, having the major resolution loss along the vertical axis (z-axis), which is reasonable considering the antenna array distribution.

4.2 *Numerical Brain Stroke Monitoring*

In this second stage of numerical validation, we evaluate the imaging capabilities of the system testing in monitoring scenarios of early-stage stroke onset. The stroke is placed between the occipital and parietal zones of the brain (posterior head), miming an actual medical condition that affects the visual functions [43]. Regarding the stroke dimensions, it is projected as temporal dynamic spatial evolution changing its volume from 5 to 15 ċm^3, which is represented in Fig. 10, while considering either hemorrhagic and ischemic typologies of stroke. Moreover, we evaluate each condition using an average single-tissue head model and a multi-tissue one, where

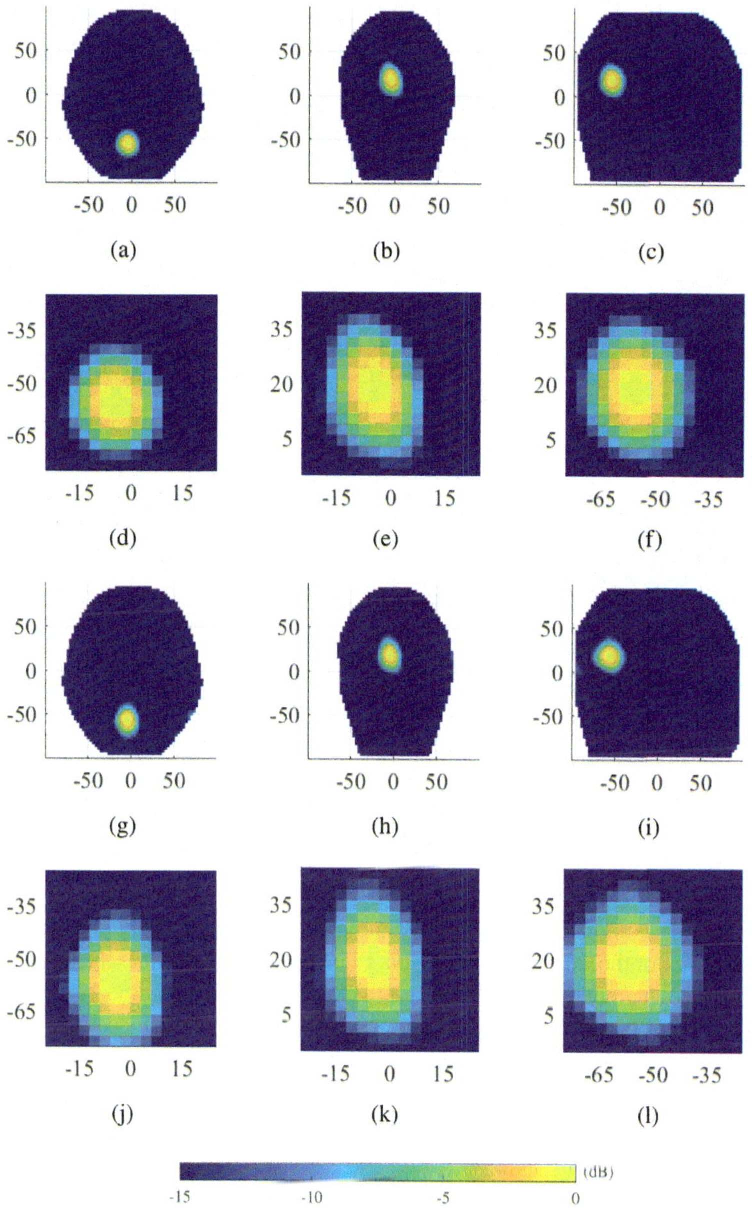

Fig. 9 Normalized projection of 3 mm voxel in the imaging operator. (**a**–**f**): Imaging using the full operator. (**g**–**l**): Imaging truncating the operator at −30 dB. Dimensions in [mm]

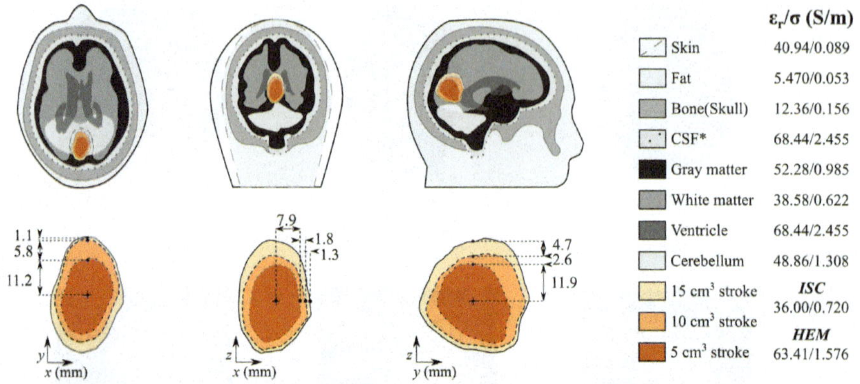

Fig. 10 Numeric mimicked stroke condition considering a multi-tissue head model, and 5, 10 and 15 cm^3 volumetric variations. From left-to-right: transverse, coronal and sagittal plane views cut at the center of the ill-affected area. *CSF stands for cerebro spinal fluid. ϵ_r and σ at 1 GHz

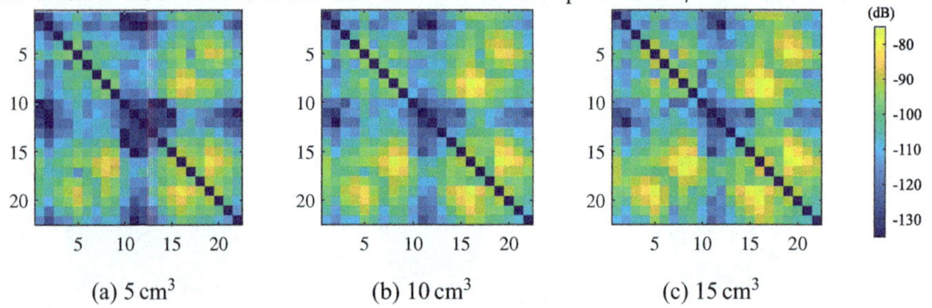

Fig. 11 Simulated differential S-parameters for the different hemorrhagic cases using the homogeneous head. The axis indicate the antenna index

each tissue is homogeneous and takes the electric properties shown in the table in Fig. 10 [18]. For the sake of visualization, Fig. 10 reports the skin as the head contour, though, in the simulated model, it has a thickness of around 3 mm. Also, it is worth highly that the operator in all following tests, even the experimental ones, is computed using the homogeneous head model.

The simulated virtual scenarios provide the scattering matrix of the scenario, s, changing the stroke volume by 5 ċm^3, at each state. Hence, three temporal differential scattering matrices, $\Delta S = S(t_0) - S(t_s)$, are considered. These represent changes from the healthy or without of the stroke to $s = 1$ to 5 cm^3 stroke volume, $s = 2$ to 10 cm^3, and $s = 3$ to 15 cm^3, respectively. Notice, also, that due to the anthropomorphic shape of the stroke, the radial variations from the center of the stroke are not linear between states, as shown in Fig. 10 (bottom row). They depend on the state change and span between 1 to 12 mm in the main axis.

4.2.1 Homogeneous Case

Firstly, we report the differential scattering matrix, $\Delta S_{22\times22}$, at 1 GHz for the three situations in Fig. 11. From these, we null the reflection coefficients (diagonal) mimicking the experimental approach, where these demonstrate a sensibility to the manufacturing tolerance of the antenna that is not numerically considered, introducing a modeling error that harms the imaging. Moreover, notice that the matrices show a non-forced symmetry between the top and bottom triangular matrix, as expected from the transmission coefficients. Thus, the reconstruction algorithm employs either one of the triangular matrices for the imaging. Also, regarding the specific studied cases, it can be glance the matrices presents a high similarity, mainly changing the intensity of certain areas in all three cases. This is expected since they share a common baseline, and the variation is concentrated in a relatively small area.

Applying the imaging algorithm, with the truncation index set at −30 dB, we obtain 3-D electrical contrast maps of the hemorrhagic situation, reporting them in Fig. 12. It depicts the normalized amplitude using a zoomed-up and regular visualization of 2-D and 3-D representations to see the details and indicate the stroke localization. For the 2-D one, we cut at the center of the ill-affected area and plot the transverse, sagittal, and frontal planes, respectively. Instead, for the 3-D one, we plot the voxel with values below −3 dB. Additionally, we include dotted red contours signaling the actual stroke shape on the plane.

Closing the virtual assessment, we also consider the ischemic condition using the same volume variation and changing the dielectric properties of the stroke-affected area. Thus, as expected, and considering we are reporting qualitative images, i.e., we are not discriminating the stroke typology from the images as in the tomographic approaches, the results of the ischemic case are analogous to the hemorrhagic one as shown in Fig. 13, where 0–15 cm^3 scenario is presented. The non-reported cases, compared to the hemorrhagic results, show similar agreement.

4.2.2 Multi-tissue Case

Here we evaluate the algorithm's capabilities to perform in a more realistic clinical scenario, considering first a more complex model of the head and second, the unknown specific morphology of each patient. Hence, we use a homogeneous-head-based operator for the imaging while obtaining the differential scattering parameters using a multi-tissue head model. The pathological evolution of the stroke status, i.e., the three-volume state variation and both typologies, are considered as it was in the homogeneous analysis.

Following the same results scheme, first, we present the differential scattering for the three case variations of a hemorrhagic stroke in Fig. 14. As in the homogeneous case, the matrices present evident resemblances due to the common elements differing in the intensity of some specific areas. Interestingly, the pattern similarity of the multi-tissue scenarios matrices also extends to the homogeneous ones, indicating the effect of the contrast variation of those.

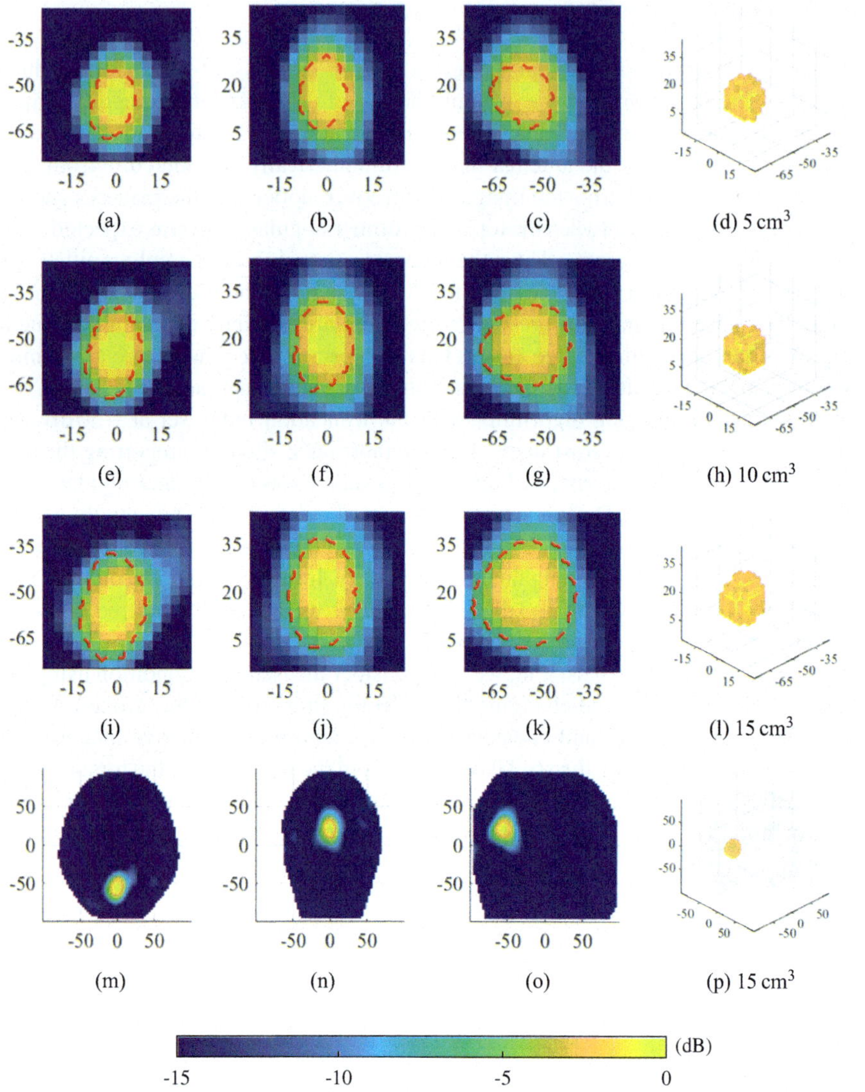

Fig. 12 Normalized amplitude values of the retrieved dielectric contrast for the different hemorrhagic cases using the homogeneous head. (**a–d**) 0–5 cm^3; (**e–h**) 0–10 cm^3; (**i–p**) 0–15 cm^3. The 3-D views indicate the values above −3 dB

Starting with the hemorrhagic case, the figure presents normalized reconstructions for all three studied scenarios. In this case, the overall retrieved contrasts show good stroke localization and volume track capability, which is very promising. However, the shape estimation deteriorated significantly on the y-axis, as seen from the sagittal plane views. Besides, the results present artifacts, nonexistent in the homogeneous case.

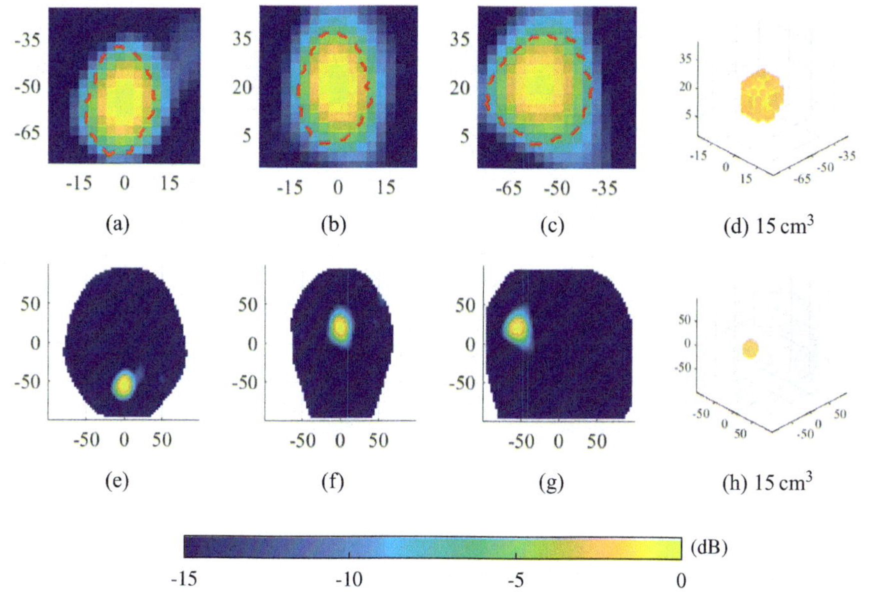

Fig. 13 Normalized amplitude values of the retrieved dielectric contrast for the 0–15 cm^3 ischemic case using the homogeneous head. The 3-D views indicate the values above -3 dB

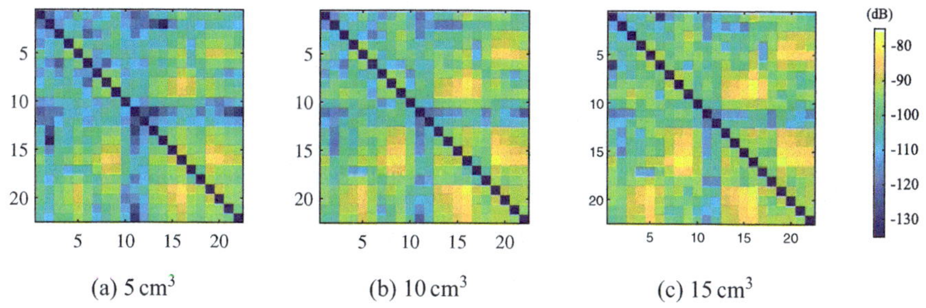

Fig. 14 Simulated differential S-parameters for the different hemorrhagic cases using the multi-tissue head. The axis indicate the antenna index

As in the homogeneous case, the ischemic results demonstrate similar outcomes, showing the same positive aspects and identical drawbacks, as expected. Figure 15 depicts the case for 15 cm^3 (Fig. 16).

In general, the obtained outcomes of all studied cases demonstrate the capabilities to detect, localize and monitor the spatial evolution of the stroke. When the operator describes the situation closely, i.e., the homogeneous case, there is a clear and accurate reproduction, with almost no artifacts. However, the system presents limitations in the most complex and realistic situations, resulting in quality-decreased results due to the distortion of the imaging operator. It is an intrinsic limitation that must be approached before moving to a clinical scenario, though the capabilities to perform

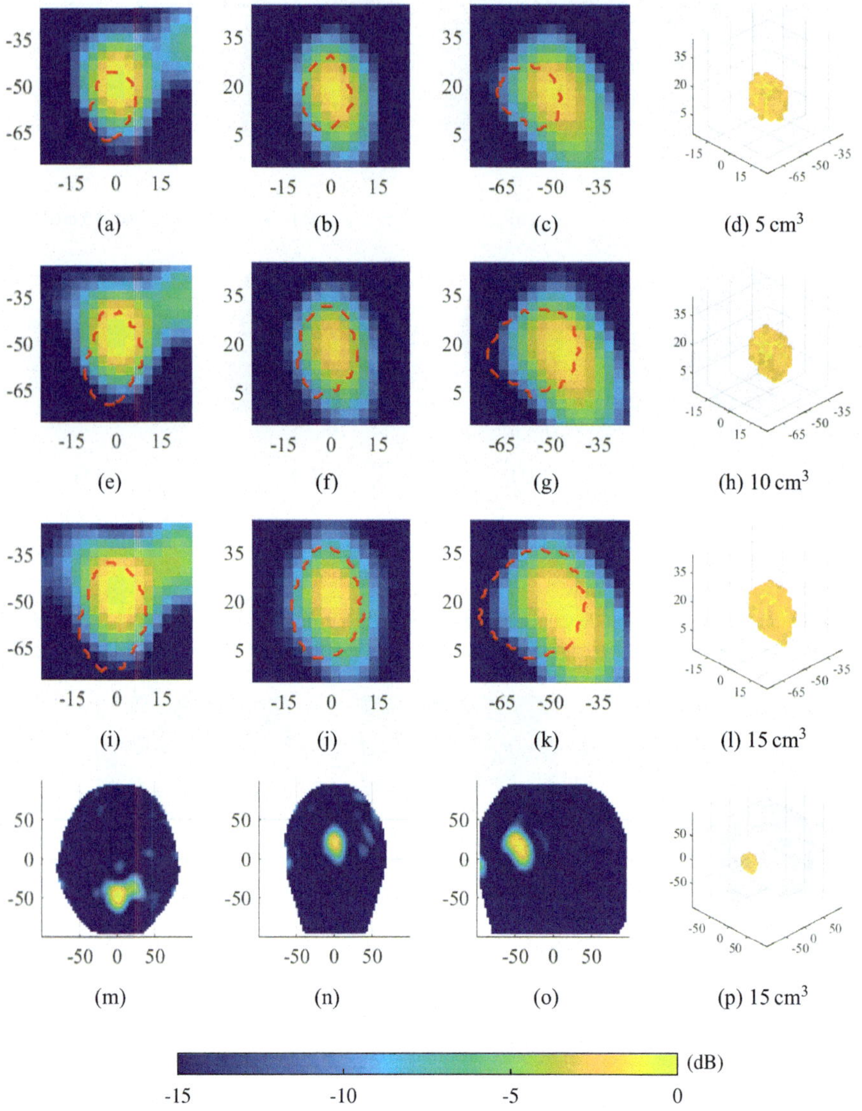

Fig. 15 Normalized amplitude values of the retrieved dielectric contrast for the different hemorrhagic cases using the multi-tissue head. (**a**–**d**) 0–5 cm^3; (**e**–**h**) 0–10 cm^3; (**i**–**p**) 0–15 cm^3. The 3-D views indicate the values above -3 dB

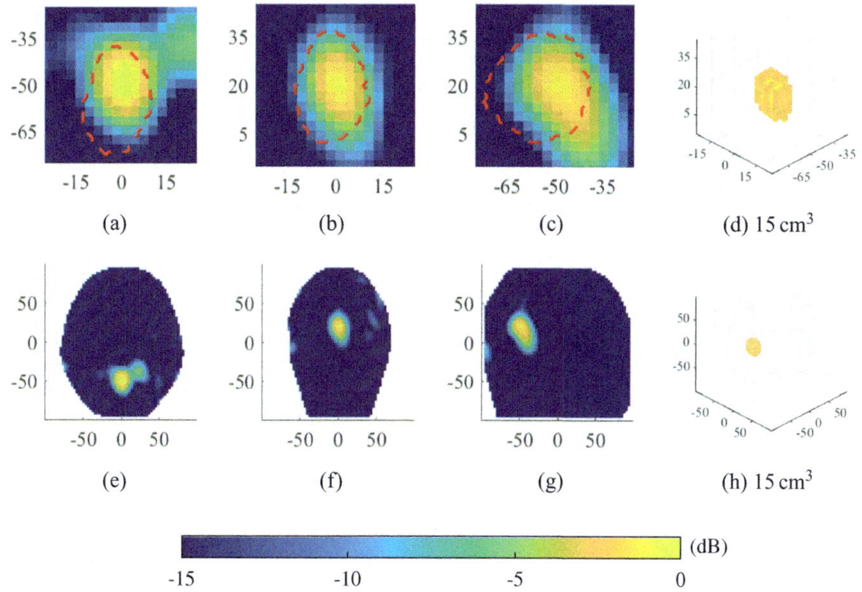

Fig. 16 Normalized amplitude values of the retrieved dielectric contrast for the 0–15 cm^3 ischemic case using the multi-tissue head. The 3-D views indicate the values above −3 dB

precise detection and localization and, to some extent, trace the change of the stroke-affected area is precious medical information.

5 Experimental Validation

The last part of the system validation relies on an experiment mimicking a worsening hemorrhagic condition using the prototype and the phantoms described in Sect. 3. The non-static stroke phantom is positioned in the head, entering from the top, and filled via a cannula and a syringe. With this procedure, we set, one at a time, the different studied states of the stroke, i.e., empty stroke (healthy), 5 cm^3, 10 cm^3, and 15 cm^3, while measuring the scattering parameters. Here, we set the VNA input power to −5 dBm, and the intermediate filter (IF) to 100 Hz, reaching a noise floor around −100 dB at 1 GHz.

Figure 17 shows the differential measured scattering matrices, omitting the reflection terms due to their high dependence on the antenna's manufacturing variability, which is hard to include in the numerical operator. However, the transmission parameters respond similarly to what expected from the numerical analysis, having consistent response patterns but with different intensities depending on the considered case. Moreover, as compared to numerical simulations they present lower intensity values, possibly caused by non-modeled sources of loss. Nevertheless, this is adjustable with

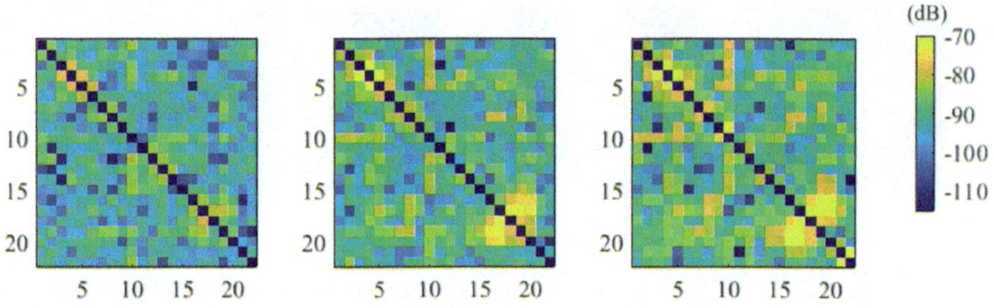

(a) 5 cm^3 (b) 10 cm^3 (c) 15 cm^3

Fig. 17 Measured differential S-parameters for the different hemorrhagic cases using the homogeneous head. The axis indicate the antenna index

calibration if required [22, 44]. Also, the matrices exhibit a mild asymmetry caused by the slight differences in the left and right paths of the switching matrix (see Fig. 2).

Moving to the imaging, we opt for using a truncation index set as -25 dB, retrieving the normalized contrast summarized in Fig. 18, where the main plane views are cut at their maximum value. Moreover, three dotted red circles with radii 10.6, 13.7 and 15.3 mm, corresponding to spheres of volume 5 ,10, and 15 cm^3, act as a visual reference. In agreement with the numerical results, the experiment demonstrate the system's capability to detect and localize the stroke in all cases and estimate and indicate the volume change with a centimetric resolution. Regarding the shape retrieval, it was difficult to determine it with certainty using the tested experimental setup, thus being a point to improve in future testing.

Additionally, we perform both repeatability and false-positive tests as further validation, considering the experimental nature of the data, summarized in Fig. 19. For the repeatability test, we retrieve the 0–5 cm^3 case twice using different measurement sets, achieving equivalent outcomes as shown in Fig. 19a, b. For the false-positive test, the differential matrix is made by subtracting the parameters of two empty cases and normalizing the contrast with the maximum of the previous cases as a reference. As shown in Fig. 19c, the device can discriminate this condition as well.

6 Conclusions

In this chapter, we have presented the design, implementation and experimental validation of a low-complexity microwave-based scanner for brain stroke imaging. The prototype comprises a 22-element array of custom-designed flexible antennas working around 1 GHz, with a differential retrieval algorithm that allows almost real-time imaging. The system is tested in realistic mimicked scenarios, demonstrating its capabilities to detect, localize and trace temporal-spatial variations of both hem-

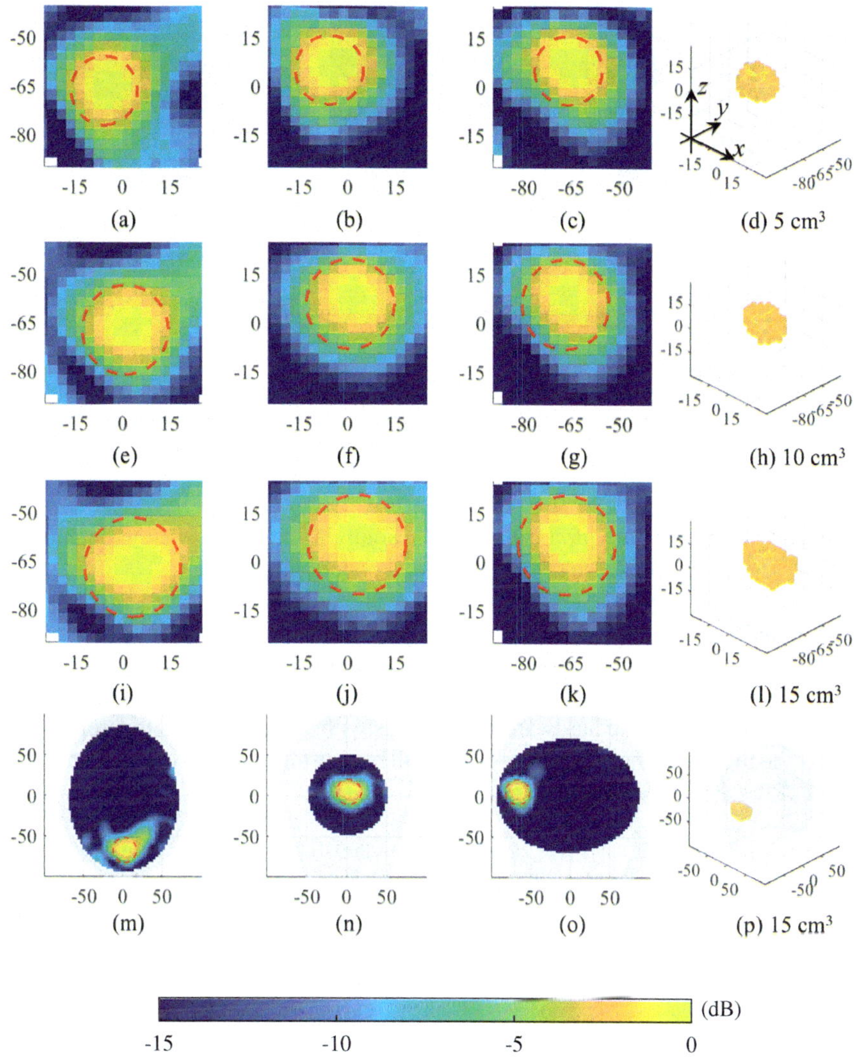

Fig. 18 Experimental test: Normalized amplitude values of the retrieved dielectric contrast for the different hemorrhagic cases using the homogeneous head. (**a**–**d**) 0–5 cm^3; (**e**–**h**) 0–10 cm^3; (**i**–**p**) 0–15 cm^3. The 3-D views indicate the values above -3 dB

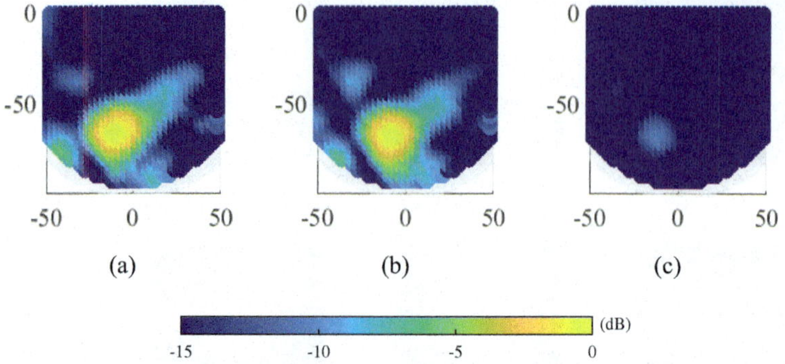

Fig. 19 Transverse views center at the maximum, amplitude normalized of the reconstructed dielectric contrast. (**a**–**b**) repeatability, two 0–5 cm^3 cases using a different set of measurements. **c** false-positive test, difference between two empty cases normalized at maximum value of (**a**)

orrhagia and ischemia with a centimetric resolution. It also shows the potential to operate even in the absence of morphological patient information.

Future work aims to extend the system's validation in other relevant conditions, including tests with experimental multi-tissue phantoms and imaging operators with additional a-priori information, for example, multi-frequency ones, or, operators based on mimicked heads with two or three homogeneous zones. Moreover, it is expected to upgrade the hardware components with ad-hoc devices that replace the VNA and the switching, improving the overall measuring-time performance, a bottleneck of the device.

Acknowledgements We thank Prof. N. Joachimowicz, Prof. H. Roussel and Dr. S. Abedi, Sorbonne University, and Prof. P. Plaisance, Emergency Medicine Dept. of Lariboisière University Hospital, and Eng. V. Lemarteleur, Ilumens Health Simulation Center, for providing the medical images and the fruitful discussions on the medical aspects and on the modeling manufacturing of the phantoms. This work was supported by the Italian Ministry of University and Research under the PRIN project "MiBraScan", and by the European Union's Horizon 2020 Research and Innovation Program under the EMERALD project, Marie Sklodowska-Curie grant agreement No. 764479.

References

1. R. Felberg, A. Naidech, The five Ps of acute ischemic stroke treatment: parenchyma, pipes, perfusion, penumbra, and prevention of complications. Ochsner J. **5**, 5–11 (2003)
2. N. Hu, T. Zhang, Y. Wu, B. Tang, M. Li, B. Song, Q. Gong, M. Wu, S. Gu, S. Lui, Detecting brain lesions in suspected acute ischemic stroke with CT-based synthetic MRI using generative adversarial networks. Ann. Transl. Med. **10**(2), 35 (2022). https://doi.org/10.21037/atm-21-4056
3. K.B. Walsh, Non-invasive sensor technology for prehospital stroke diagnosis: current status and future directions. Int. J. Stroke **14**(6), 592–602 (2019). https://doi.org/10.1177/1747493019866621. (PMID: 31354081)

4. I. Bisio, A. Fedeli, C. Garibotto, F. Lavagetto, M. Pastorino, A. Randazzo, Two ways for early detection of a stroke through a wearable smart helmet: signal processing versus electromagnetism. IEEE Wirel. Commun. **28**(3), 22–27 (2021). https://doi.org/10.1109/MWC.001.2000401
5. A. Fedeli, V. Schenone, A. Randazzo, M. Pastorino, T. Henriksson, S. Semenov, Nonlinear s-parameters inversion for stroke imaging. IEEE Trans. Microw. Theory Tech. **69**(3), 1760–1771 (2021). https://doi.org/10.1109/TMTT.2020.3040483
6. P. Lu, P. Kosmas, Three-dimensional microwave head imaging with GPU-based FDTD and the DBIM method. Sensors **22**(7) (2022). https://doi.org/10.3390/s22072691. https://www.mdpi.com/1424-8220/22/7/2691
7. O. Karadima, P. Lu, I. Sotiriou, P. Kosmas, Experimental validation of the DBIM-TwIST algorithm for brain stroke detection and differentiation using a multi-layered anatomically complex head phantom. IEEE Open J. Antennas Propag. **3**, 274–286 (2022). https://doi.org/10.1109/OJAP.2022.3150100
8. O. Fiser, V. Hruby, J. Vrba, T. Drizdal, J. Tesarik, J. Vrba, D. Vrba, UWB bowtie antenna for medical microwave imaging applications. IEEE Trans. Antennas Propag. 1–1 (2022). https://doi.org/10.1109/TAP.2022.3161355
9. A.S.M. Alqadami, A. Trakic, A.E. Stancombe, B. Mohammed, K. Bialkowski, A. Abbosh, Flexible electromagnetic cap for head imaging. IEEE Trans. Biomed. Circ. Syst. **14**(5), 1097–1107 (2020). https://doi.org/10.1109/TBCAS.2020.3025341
10. A.S.M. Alqadami, A. Zamani, A. Trakic, A. Abbosh, Flexible electromagnetic cap for three-dimensional electromagnetic head imaging. IEEE Trans. Biomed. Eng. **68**(9), 2880–2891 (2021). https://doi.org/10.1109/TBME.2021.3084313
11. J.A. Tobon Vasquez, R. Scapaticci, G. Turvani, G. Bellizzi, D.O. Rodriguez-Duarte, N. Joachimowicz, B. Duchêne, E. Tedeschi, M.R. Casu, L. Crocco, F. Vipiana, A prototype microwave system for 3d brain stroke imaging. Sensors **20**(9) (2020). https://doi.org/10.3390/s20092607. https://www.mdpi.com/1424-8220/20/9/2607
12. S. Candefjord, J. Winges, A.A. Malik, Y. Yu, T. Rylander, T. McKelvey, A. Fhager, M. Elam, M. Persson, Microwave technology for detecting traumatic intracranial bleedings: tests on phantom of subdural hematoma and numerical simulations. Med. Biol. Eng. Comput. **55**, 1177–1188 (2016). PMCID: PMC5544814. https://doi.org/10.1007/s11517-016-1578-6
13. M. Persson, A. Fhager, H.D. Trefná, Y. Yu, T. McKelvey, G. Pegenius, J.E. Karlsson, M. Elam, Microwave-based stroke diagnosis making global prehospital thrombolytic treatment possible. IEEE Trans. Biomed. Eng. **61**, 2806–2817 (2014). https://doi.org/10.1109/TBME.2014.2330554
14. S. Semenov, M. Hopfer, R. Planas, A. Hamidipour, T. Henriksson, Electromagnetic tomography for brain imaging: 3d reconstruction of stroke in a human head phantom. In: *2016 IEEE Conference on Antenna Measurements Applications (CAMA)*. (2016), pp. 1–4. https://doi.org/10.1109/CAMA.2016.7815752
15. D. Cook, H. Brown, I. Widanapathirana, D. Shah, J. Walsham, A. Trakic, G. Zhu, A. Zamani, L. Guo, A. Brankovic, A. Al-Saffar, A. Stancombe, A. Bialkowski, P. Nguyen, K. Bialkowski, S. Crozier, A. Abbosh, Case report: preliminary images from an electromagnetic portable brain scanner for diagnosis and monitoring of acute stroke. Front. Neurol. **12** (2021). https://doi.org/10.3389/fneur.2021.765412
16. M. Bertero, P. Boccacci, *Introduction to Inverse Problems in Imaging* (Institution of Physics, Bristol, U.K., 1998)
17. N.K. Nikolova, Introduction to microwave imaging. *EuMA High Frequency Technologies Series*. (Cambridge University Press, 2017). https://doi.org/10.1017/9781316084267
18. D. Andreuccetti, R. Fossi, C. Petrucci, An internet resource for the calculation of the dielectric properties of body tissues in the frequency range 10 hz – 100 ghz. IFAC-CNR, Florence (Italy), 1997. based on data published by C. Gabriel et al. in 1996. http://niremf.ifac.cnr.it/tissprop/
19. V. Mariano, J.A. Tobon Vasquez, F. Vipiana, A novel discretization procedure in the csi-fem algorithm for brain stroke microwave imaging. Sensors **23**(1), (2023)

20. D. Ireland, M. Bialkowski, Feasibility study on microwave stroke detection using a realistic phantom and the FDTD method. In: *2010 Asia-Pacific Microwave Conference*. (2010), pp. 1360–1363
21. S.Y. Semenov, D.R. Corfield, Microwave tomography for brain imaging: feasibility assessment for stroke detection. Int. J. of Antennas and Propagat. **2008**, 8 (2008). https://doi.org/10.1155/2008/254830
22. D.O. Rodriguez-Duarte, J.A. Tobon Vasquez, R. Scapaticci, G. Turvani, M. Cavagnaro, M.R. Casu, L. Crocco, F. Vipiana, Experimental validation of a microwave system for brain stroke 3-d imaging. Diagnostics **11**(7) (2021). https://doi.org/10.3390/diagnostics11071232. https://www.mdpi.com/2075-4418/11/7/1232
23. S.N. Makarov, G.M. Noetscher, J. Yanamadala, M.W. Piazza, S. Louie, A. Prokop, A. Nazarian, A. Nummenmaa, Virtual human models for electromagnetic studies and their applications. IEEE Rev. Biomed. Eng. **10**, 95–121 (2017). https://doi.org/10.1109/RBME.2017.2722420
24. D.O. Rodriguez-Duarte, J.A.T. Vasquez, R. Scapaticci, L. Crocco, F. Vipiana, Assessing a microwave imaging system for brain stroke monitoring via high fidelity numerical modelling. IEEE J. Electromagn., RF Microw. Med. Biol. **5**(3), 238–245 (2021). https://doi.org/10.1109/JERM.2020.3049071
25. J. Jianming, *The Finite Element Method in Electromagnetic*, 2nd edn. (Wiley, IEEE Press, 2002)
26. E.A. Attardo, A. Borsic, G. Vecchi, P.M. Meaney, Whole-system electromagnetic modeling for microwave tomography. IEEE Antennas Wirel. Propag. Lett. **11**, 1618–1621 (2012). https://doi.org/10.1109/LAWP.2013.2237745
27. R. Scapaticci, M. Bjelogrlic, J.A. Tobon Vasquez, F. Vipiana, M. Mattes, L. Crocco, *Emerging Electromagnetic Technologies for Brain Diseases Diagnostics, Monitoring and Therapy, Chap. 2. Microwave Technology for Brain Imaging and Monitoring: Physical Foundations, Potential and Limitations*. (Springer International Publication 2018), pp. 7–35. https://doi.org/10.1007/978-3-319-75007-1_2
28. R. Scapaticci, J. Tobon, G. Bellizzi, F. Vipiana, L. Crocco, Design and numerical characterization of a low-complexity microwave device for brain stroke monitoring. IEEE Trans. Antennas Propag. **66**(12), 7328–7338 (2018). https://doi.org/10.1109/TAP.2018.2871266
29. R. Scapaticci, L.D. Donato, I. Catapano, L. Crocco, A feasibility study on microwave imaging for brain stroke monitoring. Prog. Electromagn. Res. B **40**, 305–324 (2012). https://doi.org/10.2528/PIERB12022006
30. C. Balanis, *Advanced Enginnering Electromagnetics*. (Willeys, 1989)
31. O.M. Bucci, L. Crocco, R. Scapaticci, G. Bellizzi, On the design of phased arrays for medical applications. Proc. IEEE **104**(3), 633–648 (2016). https://doi.org/10.1109/JPROC.2015.2504266
32. M. Sayed, J. Martens, *Modern RF and Microwave Measurement Techniques, Chap. 5. Vector Network Analyzers*. (Cambridge University Press, 2013), pp. 98–129. https://doi.org/10.1017/CBO9781139567626.006
33. M.R. Casu, M. Vacca, J.A. Tobon, A. Pulimeno, I. Sarwar, R. Solimene, F. Vipiana, A cots-based microwave imaging system for breast-cancer detection. IEEE Trans. Biomed. Circ. Syst. **11**(4), 804–814 (2017). https://doi.org/10.1109/TBCAS.2017.2703588
34. P.M. Meaney, K.D. Paulsen, Theoretical premises and contemporary optimizations of microwave tomography. In: K.H.Y. A. Kishk (ed.) *Microwave Technologies*. (IntechOpen, London, 2022), pp. 1–24. https://doi.org/10.5772/intechopen.103011
35. A.E. Stancombe, K.S. Bialkowski, A.M. Abbosh, Portable microwave head imaging system using software-defined radio and switching network. IEEE J. Electromagn., RF Microwav. Med. Biol. **3**(4), 284–291 (2019). https://doi.org/10.1109/JERM.2019.2901360
36. Keysight Technologies: Keysight 2-port and 4-port PNA network analyzer, N5227A 10 MHz to 67 GHz. Data Sheet and Technical Specifications (2019)
37. J.A. Tobon Vasquez, R. Scapaticci, G. Turvani, G. Bellizzi, N. Joachimowicz, B. Duchêne, E. Tedeschi, M.R. Casu, L. Crocco, F. Vipiana, Design and experimental assessment of a 2D microwave imaging system for brain stroke monitoring. Int. J. Antennas Propag. (Article ID 8065036), 12 (2019). https://doi.org/10.1155/2019/8065036

38. M. Ricci, B. Štitić, L. Urbinati, G.D. Guglielmo, J.A.T. Vasquez, L.P. Carloni, F. Vipiana, M.R. Casu, Machine-learning-based microwave sensing: a case study for the food industry. IEEE J. Emerg. Sel. Top. Circuits Syst. **11**(3), 503–514 (2021). https://doi.org/10.1109/JETCAS.2021.3097699
39. Download visible human project data. Available at https://www.nlm.nih.gov/databases/download/vhp.html
40. N. Joachimowicz, B. Duchêne, C. Conessa, O. Meyer, Anthropomorphic breast and head phantoms for microwave imaging. Diagnostics **85**(8), 1–12 (2018)
41. Esun. Available at https://www.esun3d.net/Products/PVA
42. Smooth-on. Available at https://www.smooth-on.com/product-line/dragon-skin/
43. E.L. Saionz, D. Tadin, M.D. Melnick, K.R. Huxlin, Functional preservation and enhanced capacity for visual restoration in subacute occipital stroke. Brain **143**(6), 1857–1872 (2020). https://doi.org/10.1093/brain/awaa128
44. D.O. Rodriguez-Duarte, C. Origlia, J.A.T. Vasquez, R. Scapaticci, L. Crocco, F. Vipiana, Experimental assessment of real-time brain stroke monitoring via a microwave imaging scanner. IEEE Open J. Antennas Propag. **3**, 824–835 (2022). https://doi.org/10.1109/OJAP.2022.3192884
45. C. Xudong, *Computational Methods for Electromagnetic Inverse Scattering*. (Wiley, IEEE Press, 2018)

The Dielectric Properties of Axillary Lymph Nodes

Matteo Savazzi, Daniela M. Godinho, Niko Ištuk, Tiago Castela, Maria L. Orvalho, Emily Porter, Martin O'Halloran, Carlos A. Fernandes, João M. Felício, and Raquel C. Conceição

Abstract Knowledge of axillary lymph node dielectric properties is fundamental to assess microwave imaging as a potential technique to diagnose axillary lymph nodes in patients affected by breast cancer. The present chapter addresses the characterisation of axillary lymph node dielectric properties with different approaches, as this characterisation is challenged by the axillary lymph node heterogeneity and the clinical restrictions in measuring them. The complex permittivity of human and animal healthy lymph nodes was measured using the open-ended coaxial-probe technique, over the 0.5–8.5 GHz frequency band. It was found that their relative permittivity ranges between 30 and 50 at 4.5 GHz. As a parallel approach, a novel methodology to infer tissue dielectric properties based on magnetic resonance imaging data is proposed as a more accessible method to infer axillary lymph node permittivity. Our results confirm the high variability of healthy axillary lymph node dielectric properties, and suggest that metastasised axillary lymph node properties are 32% (average) higher and more consistent than those of healthy axillary lymph nodes, which motivates studying axillary lymph node microwave imaging. Finally, the effects of

M. Savazzi (✉) · D. M. Godinho · R. C. Conceição
Instituto de Biofísica e Engenharia Biomédica (IBEB), Faculdade de Ciências, Universidade de Lisboa, Campo Grande, 1749-016 Lisbon, Portugal
e-mail: matteo.savazzi@gmail.com

N. Ištuk · E. Porter · M. O'Halloran
National University of Ireland Galway (NUIG), Translational Medical Device Lab, National University of Ireland Galway, H91 ERW1 Galway, Ireland

E. Porter
Department of Electrical and Computer Engineering, The University of Texas at Austin, Austin 78712, USA

T. Castela · M. L. Orvalho
Departamento de Radiologia, Hospital da Luz Lisboa, Luz Saúde, Lisbon, Portugal

C. A. Fernandes · J. M. Felício
Instituto de Telecomunicações, Instituto Superior Técnico (IST), Universidade de Lisboa, 1049-001 Lisbon, Portugal

J. M. Felício
Centro de Investigação Naval (CINAV), Escola Naval, 2810-001 Almada, Portugal

F. Vipiana and L. Crocco (eds.), *Electromagnetic Imaging for a Novel Generation of Medical Devices*, Lecture Notes in Bioengineering,
https://doi.org/10.1007/978-3-031-28666-7_8

freezing and defrosting on the dielectric properties of biological tissues are studied, which is motivated by the possibility of measuring lymph nodes from dead animals, that were previously frozen and defrosted. The results suggest that the measurements of defrosted lymph nodes will not affect their dielectric properties by more than 9% at 4.5 GHz, paving the way for previously frozen lymph node measurements.

Keywords Axillary lymph nodes · Biological tissue dielectric properties · Breast cancer · Medical microwave imaging · Open-ended coaxial-probe

1 Introduction

Breast cancer is the most common cancer worldwide, accounting for more than 2 million new cancer cases in 2020 [1]. The dissemination of breast cancer cells to other parts of the body has a substantial impact on five-year survival rates. In 30% of invasive breast cancer cases, cancer cells can be drained to surrounding Lymph Nodes (LNs), such as the Axillary Lymph Nodes (ALNs) [2]. ALNs are kidney-shaped, encapsulated organs of the lymphatic system, with size ranging from a few millimetres to 1 or 2 cm on their longer axis. ALNs are composed of a surrounding capsule of collagen fibres and an inner part which hosts lymphocytes and macrophages (types of white blood cells). They are located in the axillary region, along the lymphatic vessels, and they serve as a filter for the lymph on its way to the vascular (blood) system. Figure 1 represents a sketch of the lymphatic system, highlighting axillary lymph nodes. The lymph is carried into the node by the afferent lymph vessels, and then leaves the node through the efferent lymphatic vessel which is located in a region called hilum [3, 4].

The number of affected ALNs is considered in breast cancer staging, impacting treatment decisions. Currently, in a first stage, ALN diagnosis is performed using imaging modalities, such as Ultrasound (US) and Magnetic Resonance Imaging (MRI). However, sensitivity and specificity of these imaging modalities are still limited, ranging from 20 to 90% and 40 to 96%, respectively [5]. At the state-of-the-art, the most accurate and widely adopted procedure to diagnose ALNs is the Sentinel Lymph Node Biopsy (SLNB), which consists of the surgical excision and histological analysis of the first ALN receiving drainage from the tumour. SLNB has 100% specificity and around 90% sensitivity [6, 7], yet it is an invasive and time-consuming diagnostic technique. Hence, there is a need for low-cost alternatives to non-invasively detect and diagnose ALNs. Microwave Imaging (MWI) has recently been studied as an alternative imaging modality for this purpose [8–11].

The characterisation of ALN dielectric properties, and the assessment of the level of contrast between healthy and metastasised ALNs is crucial knowledge to validate MWI as a potential technique for ALN screening. However, only a few studies in the literature have reported the dielectric properties of LNs at MicroWave (MW) frequencies. As will be discussed in detail in the next section, such studies are either

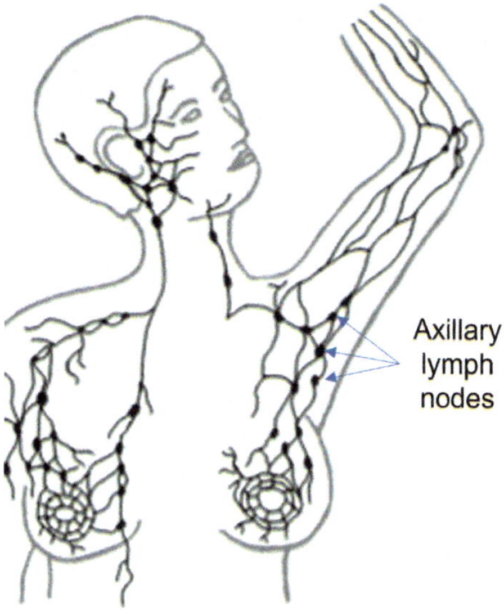

Fig. 1 Sketch of the lymphatic system. The lymphatic system consists of a large network of lymphatic vessels, and lymph nodes (represented as knots in the figure). The position of axillary lymph nodes is indicated in the figure

incomplete or missing important information about the repeatability of the measured data. In this context, the main objective of the present chapter is the characterisation of the dielectric properties of healthy and metastasised ALNs at MW frequencies. The main challenge is the heterogeneity of the tissues under study, which is a critical issue in the application of the state-of-the-art measurement method (i.e., Open-ended Coaxial-Probe, OECP) for tissue dielectric properties. Indeed, after their excision, ALNs are often surrounded by a variable amount of fat, which affects the estimated complex permittivity (generally fat results in lower dielectric constant, due to its low water-content). It is worth emphasising that the properties measured with the OECP method are a weighted average of the properties of the tissues under the sensing volume of the probe. Hence, the influence of the closest tissue to the probe will be predominant in dielectric property estimation, which makes OECP more suitable to measure homogeneous tissues. The second challenge presented by ALN dielectric measurement is related to the common restrictions in human tissue handling. Once ALNs are surgically excised, they are sent to pathology, and further analysed. Given the importance of this analysis, ALNs must be handled very carefully and cannot be cut or punctured. This is a very significant constraint that must be considered since accurate dielectric measurements of ALN content would benefit from having access to the internal cross-section of the node.

Given the above considerations, the present chapter presents different approaches to ALN dielectric characterisation:

- Measurements of freshly excised LNs with the OECP. Human ALNs and animal LNs were measured. Since animal samples do not have any constraints in terms of how they can be handled, valuable information about ALNs may be inferred measuring the dielectric properties of the inner content of animal LNs.

- Development of a novel methodology to infer dielectric properties of structures (specifically, ALNs) from breast MRI exams. The images are processed and segmented in order to apply a dielectric property mapping using known state-of-the-art property curves of breast tissues. The properties of healthy and metastasised ALNs are estimated and compared.
- Given the limited amount of metastasised human ALNs that were measured, one possibility would be to measure LNs excised from dead animals (cats and dogs) which may have been diagnosed with mammary cancer. Our research group has an ongoing collaboration with the Instituto Nacional de Investigação Agrária e Veterinária (INIAV), where animal LNs can be measured—post-mortem—after being frozen and defrosted ahead of their autopsy. It is clear that to infer reliable conclusions regarding LN dielectric properties, it is important to study whether the freezing and defrosting processes affect the dielectric properties of biological tissues, which is the second objective of the present chapter. To this purpose, the dielectric properties of different biological tissues were measured before and after freezing and defrosting them and the results were compared.

The remainder of the chapter is organised as follows: Sect. 2 reviews the existing literature on LN dielectric properties; Sect. 3 reports the measurements performed on freshly excised human and animal LNs; Sect. 4 describes the development of the MRI-based method to estimate the dielectric properties of ALNs; Sect. 5 illustrates the study of freezing and defrosting effects on the complex permittivity of biological tissues.

2 Literature Review

This section reports literature studies that characterised LN dielectric properties.

Joines et al. [12] reported measurements of 2 metastasised LNs from 2 different patients, in the frequency range from 50 to 900 MHz. Measurements were conducted between 1 and 2 h after tissue excision, once the tissue reached room temperature (23–25 °C). For each sample, the measurements were repeated in 3 different locations (6 measurements in total). Results, in terms of average relative permittivity ϵ_r and average conductivity σ, are reported in Fig. 2, using data from [12]. The average (over frequency) Standard Error on the Mean (SEM) was ±7% for relative permittivity and ±9% for conductivity.[1] In that study, no measurement of healthy LNs were conducted, and so differences between healthy and metastasised LNs were not inferred. Additionally, Joines *et al.* did not provide details on the sample preparation nor on the tissue composition under the measurement sites. The latter is critical, due to LN heterogeneity which can affect the measurements. It should also be noted that the anatomical sites of the excised LNs were not specified.

[1] The SEM reported in [12] was computed as an average value among several biological tissue measurements (not only LNs).

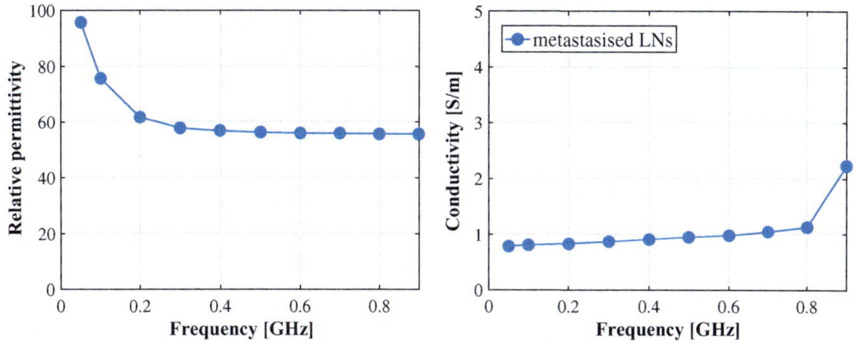

Fig. 2 Metastasised lymph node (LN) dielectric properties measured by Joines et al. [12]. Data are averaged over 6 measurements (2 LNs from 2 different patients at 3 different locations). **(Left)** relative permittivity, and **(right)** conductivity

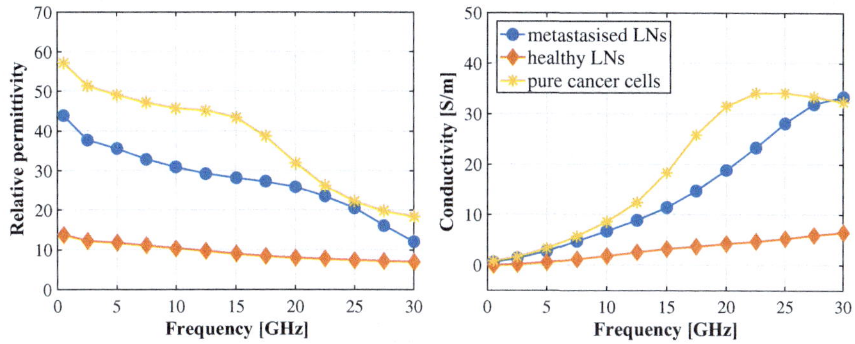

Fig. 3 Dielectric properties of "pure (breast) cancer cells", metastasised and healthy axillary lymph nodes (ALNs), measured by Choi et al. [13]. **Left** real part of relative permittivity; **right** conductivity

Choi et al. [13] measured 27 ALNs (either healthy or metastasised) and 7 "pure (breast) cancer tissue" samples, from 12 patients diagnosed with invasive breast carcinoma. Data were sampled in the range from 0.5 to 30 GHz. The results of the study are plotted in Fig. 3, using data from [13]. The authors concluded that ALN dielectric properties are sensitive to cancer cell presence. In particular, cancer tissue and metastasised nodes exhibit high values of relative permittivity at low-GHz and conductivity at high-GHz frequencies, and pronounced dielectric relaxation between 15 and 30 GHz. However, the problem of fat surrounding ALN was not addressed; this can be critical as results may considerably vary depending on the amount of fat in the sensing volume of the probe. This will be discussed in details in Sect. 3.

Cameron et al. [14] reported 30 measurements of 23 freshly-excised ALNs from 14 patients, in the 0.5–20 GHz frequency range. Interestingly, for some of the analysed ALNs, measurements were done placing a probe both on the external surface

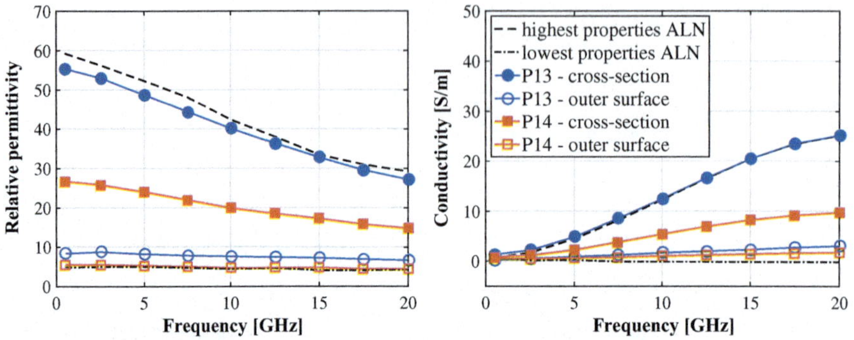

Fig. 4 Axillary lymph node (ALN) dielectric properties measured by Cameron et al. [14]. Black lines show the highest and lowest dielectric property measurements obtained across the 30 samples, indicating the great variability observed in the measurements. Coloured lines represent ALN dielectric properties measured from external and cross-section surfaces. Data refer to 2 samples from 2 patients (P13 and P14). The outer surface measurements correspond to 40% (P13) and 47% (P14) fat under the probe. The cross-sectional measurements correspond to 20% (P13) and 10% (P14) fat content under the probe, with lymphatic tissue becoming the predominant component. **(Left)** relative permittivity, and **(right)** conductivity

of the node and on its cross-section (after slicing it). The results of that study showed wide-range variability of ALN dielectric properties. Figure 4, reports the highest and the lowest dielectric properties (black lines) across the 30 measurements performed in [14]. Cameron et al. concluded that fat tissue tends to dominate the measurements when placing the probe in contact with the exterior of the node, indicating that measurements where the real part of relative permittivity was in the range of 1 to 10 (Fig. 2 in [14]) corresponded to samples that were predominantly fatty. In addition, the authors noticed that the interior of ALNs had higher dielectric properties due to higher water content. Figure 4 (data taken from [14]) also shows the differences between inner (cross-section) and outer (surface) dielectric properties, for two nodes. Lastly, even though results were only presented for one sample, the authors noted that permittivity and conductivity consistently increased with tumor content.

Lu et al. [15–17] measured a total of 219 (178 healthy and 41 metastasised) thoracic LN samples, obtained from lung cancer surgeries of 76 patients. Applying the Mann-Whitney *U* test at single specific frequencies (64, 128, 298, 433 and 915 MHz, and 2.45 GHz), the authors found significant differences (p-value ≤ 0.05) between the dielectric properties of metastasised and healthy LNs at all the analysed frequencies. Independent tests were performed for relative permittivity and conductivity.

3 Measurements of Lymph Nodes Using Open-Ended Coaxial-Probe

This section describes the measurements that were performed with animal LNs and human ALNs at MW frequencies, applying the OECP method. Firstly, dielectric measurement of human ALNs were performed (Sect. 3.2). However, due to tissue handling restrictions, the measurements were limited to the outer surface of the ALNs, allowing limited conclusions to be drawn. Consequently, the study was extended to animal LNs (Sect.3.3), which were sliced in order to enable the measurement of their inner cross-section. Animal LN measurements helped interpreting the measurements performed on human ALNs, which is reasonable as most of the tissue dielectric properties available in the literature [18] are the result of *ex-vivo* animal tissue dielectric measurements. It should be also noted that—despite having different size depending on animal species—animal and human LNs have the same anatomy and physiological function. In addition, some studies in the literature [19, 20] compared human and animal tissues showing good consistency in dielectric properties.

3.1 Measurement Procedure

The measurements were performed using the Keysight slim form probe connected to the Keysight E5063A Vector Network Analyser (VNA) through a right angle SMA-connector. This configuration allowed avoidance of cables, hence excluding a potential source of uncertainty [21]. Measurements were performed between 0.5 and 8.5 GHz, with 30 Hz IF bandwidth and −5 dBm output power. The dielectric properties were retrieved using the Keysight N1500A Materials Measurement software. Prior to measurements, the standard system calibration [22] was performed using the open/short/load method. Deionised water was used as a known load. Figure 5 (left) reports a picture of the measurement setup. Given the sensitivity of the OECP to VNA drift and to probe movements [21, 23], an additional measurement was performed to validate the calibration, which allows to estimate the error of the measurement setup at any given time during a measurement campaign. This validation procedure was originally proposed in [21] and is currently the state-of-the-art procedure to estimate the measurement error, and to minimise the uncertainty of the measurements. In the present study, the validation measurement was performed immediately after setup calibration (prior to tissue measurements) and after some sample measurements (usually within 90 min after the calibration), using a 0.1 molar NaCl solution (with complex permittivity over frequency and temperature documented in [24]). The temperature of the validation solution was recorded to match it to the correct model. When the error was above 10%, the probe calibration and its validation were repeated. Table 1 presents the average percentage error (over frequency and over all measurement sessions) of the validation measurements. The maximum average (±standard deviation) error was 5.50 (±2.13)%, which is acceptable for the purpose of tissue characterisation.

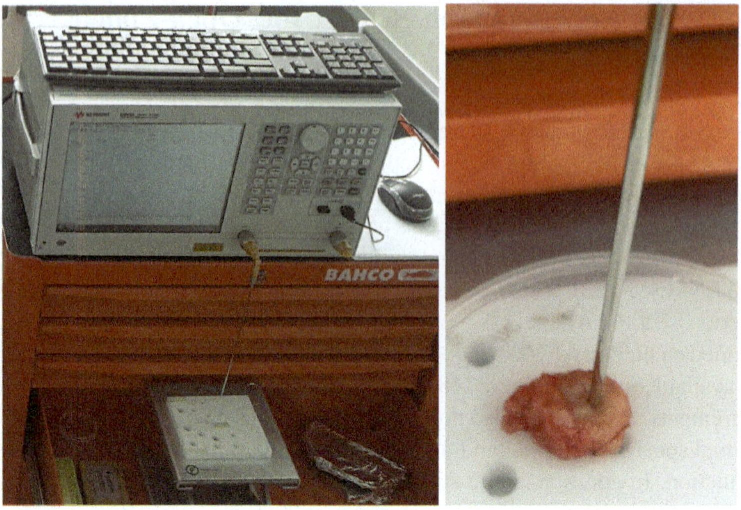

Fig. 5 (**Left**) Dielectric measurement setup: the open-ended coaxial-probe is connected via a right angle SMA-connector to a vector network analyzer during the measurement of the validation load (0.1 molar NaCl water solution). (**right**) Detail of a human ALN under measurement

Table 1 Dielectric measurement validation result: mean percentage difference, considering all frequencies and all measurement sessions, between the permittivity inferred from the validation measurements and the reference values

	Human ALNs		Animal LNs	
	$\Delta\epsilon_r$ [%]	$\Delta\sigma$ [%]	$\Delta\epsilon_r$ [%]	$\Delta\sigma$ [%]
V1	1.06	2.09	0.93	2.93
V2	2.26	2.92	5.50	2.13

Data are separated between pre-measurement validation (V1) and post-measurement validation (V2). The left side of the table refers to human axillary lymph node (ALN) measurements, while the right side of the table refers to animal lymph node (LN) measurements

Regarding tissue measurement, care was taken to apply approximately the same pressure on the samples as the pressure applied with the probe on the tissue can affect the measured dielectric properties [25]. The adopted probe pressure was the least pressure applicable to still ensure full contact between the tip of the probe and the measured tissue, while avoiding air gaps between the two. A detail of a human ALN while being measured is reported in Fig. 5 (right).

The Minimum Information for Dielectric Measurements of Biological Tissues (MINDER), proposed in [26], was used as a guideline to collect experimental data and associated metadata. MINDER specifications indicate a systematic collection of metadata along with dielectric measurement data, in order to support a more informed sharing and re-use of dielectric data. Measurement confounders (e.g., calibration procedure; calibration drift; validation procedure) and clinical confounders (e.g., tissue origin; animal age and weight; time between excision and measurement; tissue

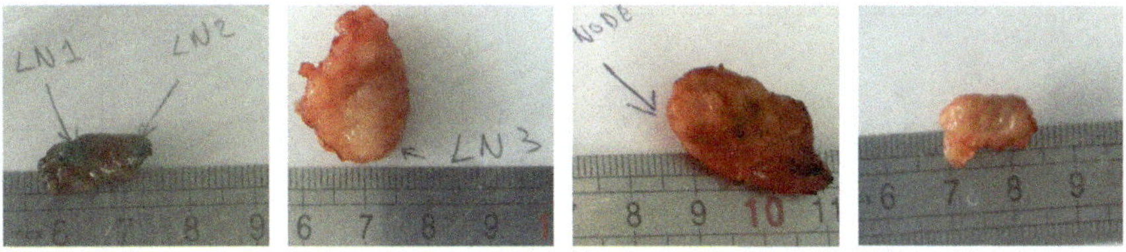

Fig. 6 Examples of five human axillary lymph nodes (ALNs) measured in the present study. The sample shown in the first figure on the left contains two different ALNs, as indicated by the arrows

handling procedures; tissue temperature) can impact the measured data [21, 27, 28]. Reporting confounder-related metadata supports consistent, interpretable, and accurate dielectric data. The measured data and the related metadata are available in the MyWAVE open-access online repository [29].

3.2 *Human Axillary Lymph Node Measurements*

Eleven ALNs from 9 patients diagnosed with breast cancer were measured. Figure 6 reports five of the human ALNs that were measured in this study. All ALNs were excised by a trained surgeon during SLNB scheduled at the University Hospital Galway (Galway, Ireland) [30]. After each ALN excision, the sample was brought to the lab for dielectric property measurement. The sample was measured immediately after receiving it, always within 20 min after excision. The temperature of the sample surface was measured at the same time as the dielectric measurement took place, resulting in an average (±standard deviation) temperature of 20.9 °C (±2.3 °C). The size of ALNs ranged approximately from 5 mm to 2 cm on their longer axis, whereas their thickness varied approximately between 4 mm and 7 mm. These values are greater than the probe's sensing depth and radius, which are estimated to be approximately 0.5 mm and 2.2 mm respectively [31, 32]. The surgeon removed as much fat as possible from the surroundings of the ALNs. It should be also reported that sometimes fat could not be completely removed, due to the risk of cutting or puncturing the ALN, which would compromise the subsequent histopathological examination. However, the surgeons indicated the best measurement point (i.e., the point on the ALN surface that contained the least fat tissue). Between 3 to 5 points of each of the 11 ALNs (depending on the size of the ALN) were measured placing the probe in contact with the outer surface of the sample. A total of 45 measurements from all ALNs was collected. The samples were then given to the pathologist for standard histopathological analysis. All the 11 ALNs were diagnosed as "negative", i.e., no tumor cells were found during the histopathological examination, which represents the main shortcoming of this study. Information from metastasised ALNs should be sought in future work.

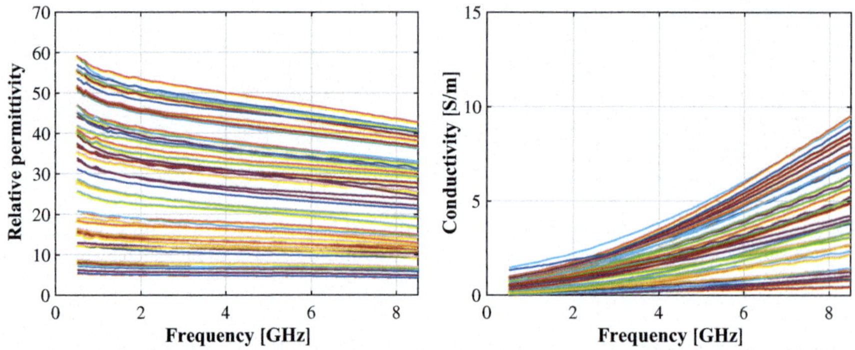

Fig. 7 **(Left)** Relative permittivity and **(right)** conductivity resulting from 45 measurements conducted on 11 ALNs excised from 9 patients. Data refer to measurements on the outer surface of the ALNs

Figure 7 illustrates the dielectric properties of all the human ALNs measured. The results show very high variability of measured ALN dielectric properties, as the relative permittivity ranges from 5.0 to 49.5 and the conductivity from 0.3 S/m to 4.4 S/m at 4.5 GHz. These results are in line with those reported by Cameron et al. [14] (Fig. 4), confirming that the presence of fat on the outer surface of the ALN sample may have a major influence on the permittivity. Due to their high variability, definite conclusions about the permittivity of the content of ALNs cannot be drawn. However, in order to allow the interpretation of the results, data were separated into three different groups, defined as follows:

- $Group_{1-10}$ included measurements where $\epsilon_r(4.5\text{ GHz}) \leq 10$;
- $Group_{10-40}$ included measurements where $10 \leq \epsilon_r(4.5\text{ GHz}) \leq 40$;
- $Group_{40+}$ included measurements where $\epsilon_r(4.5\text{ GHz}) > 40$.

It is worth mentioning that the dielectric properties falling in $Group_{1-10}$ are comparable with the dielectric properties of fat tissue measured in [18] (available on the IT'IS database [33]). It is hence claimed that the lowest values of relative permittivity (i.e., $\epsilon_r < 10$) are a consequence of the presence of a fat layer on the surface of the ALNs and are not of interest to this study. From the remaining values, it may be concluded that ALNs do have higher permittivity than the fatty content that embeds them, indicating that their detection is possible at microwave frequencies. Furthermore, it should be noted that the ALNs, for which the surgeon was able to remove the largest amount of fat from the surface, corresponded to the measurement of the highest permittivity values ($\epsilon_r > 40$ at 4.5 GHz) and showed high consistency across intra-sample measurements. Considering that consistency is a valid indicator of tissue homogeneity, this suggests that $Group_{40+}$ dielectric values effectively correspond to the dielectric properties of the ALN under test.

Lastly, for $Group_{10-40}$ and $Group_{40+}$, the mean permittivity was fitted to a two-pole Debye model, which is commonly adopted in the literature to represent the dispersion of biological tissue permittivity, as follows:

Table 2 Two-pole Debye parameters fitted to the mean permittivity computed on three different groups of data

	ALN $Group_{10-40}$	ALN $Group_{40+}$	Animal LN (inner content)
ϵ_∞	1.00	1.00	9.35
σ_s [S/m]	0.96	1.23	0.04
$\Delta\epsilon_1$	34.17	119.39	211.99
τ_1 [ns]	0.70	1.00	1.65
$\Delta\epsilon_2$	16.22	38.25	45.19
τ_2 [ps]	13.60	10.00	11.25

The first two columns refer to human axillary lymph node data, divided into two different groups, according to their relative permittivity values. The third column refers to all the data acquired when measuring the inner content of ewe lymph nodes

$$\epsilon(\omega) = \epsilon_\infty + \sum_{i=1}^{2} \frac{\Delta\epsilon_i}{1 + j\omega\tau_i} - \frac{\sigma_s}{j\omega\epsilon_0}, \tag{1}$$

where ϵ_∞ is the infinite limit of the permittivity, $\Delta\epsilon_i$ is the change in permittivity due to the ith pole, σ_s is the static ionic conductivity, τ_i is the ith relaxation constant, and 2 is the number of poles. In this study, the Least Squares Method (LSM) [34] was used to minimise the fitting error, which is a widely-adopted approach to retrieve the 2-pole Debye model parameters [35]. The parameters are reported in Table 2. The fitting error, defined as the absolute difference between the average measured data and model, was at maximum 0.4 regarding relative permittivity and 0.1 S/m regarding conductivity, which is negligible for MWI applications.

The high variability observed in the inter-sample analysis (Fig. 7) was often also observed in intra-sample analysis. Figure 8 reports the dielectric properties of several points measured on a single ALN (e.g. ALN 1, shown in Fig. 8). The high variability of the complex permittivity can be justified by the significant amount of surrounding fat. Indeed, as reported in [31], the dielectric properties of heterogeneous tissues depend on the spatial distribution of each material (ALN and fat in this case) within the sensing volume.

Conversely, Fig. 9 shows a case where surgeons were able to better separate the excised ALN (ALN 2 in Fig. 9) from the surrounding fat. In that case, the dielectric properties are more consistent across all measurement sites, which is a valid indicator of tissue homogeneity. The greater homogeneity of the tissue together with the higher dielectric properties measured suggest that the results associated with ALN 2 effectively correspond to the dielectric properties of the node under test.

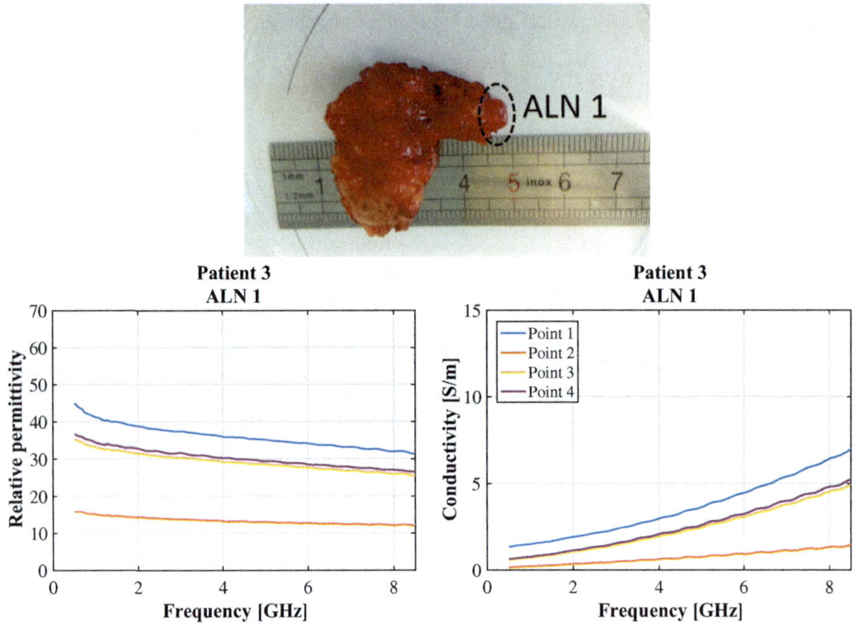

Fig. 8 Example of dielectric measurements of a single axillary lymph node (ALN 1). **(Top)** picture of the sample (the ALN is circled with a dashed line). **(Bottom–left)** ALN 1 relative permittivity; **(bottom–right)** ALN 1 conductivity. Each line corresponds to a measurement site

3.3 Animal Lymph Node Measurements

To overcome the tissue handling limitations of human ALNs, which affect the consistency of measurements, animal LNs were measured. Ex-vivo animal tissues do not have the handling constraints, such as with human tissues. This allowed collection of data from their inner cross-section and more reliable results.

For the purpose of this study, 8 LNs from 2 healthy ewe corpses were measured. Figure 10 (top-line) reports 4 of the ewe LNs that were measured in this study. The ewes were approximately 70 kg weight and about 6 years old. The LNs were excised by a trained veterinary surgeon from the inguinal area, approximately 3 to 4 h after the death of the animal, and placed in a closed container. No preservatives or additives were used as these could impact the dielectric measurements. The surgeon removed as much fat as possible with a scalpel, avoiding puncturing the LN capsule (i.e., a thin layer of connective tissue covering the node [36]). It should be noted that, since there were no handling nor ethical constraints, more fat was removed from around the animal LNs compared to human ALNs. Measurements were performed within 20 min to 4 h after tissue excision.

Firstly, the outer surface of LNs was measured for a direct comparison with human ALNs. The 8 samples were measured in 4 to 6 locations depending on the size of the

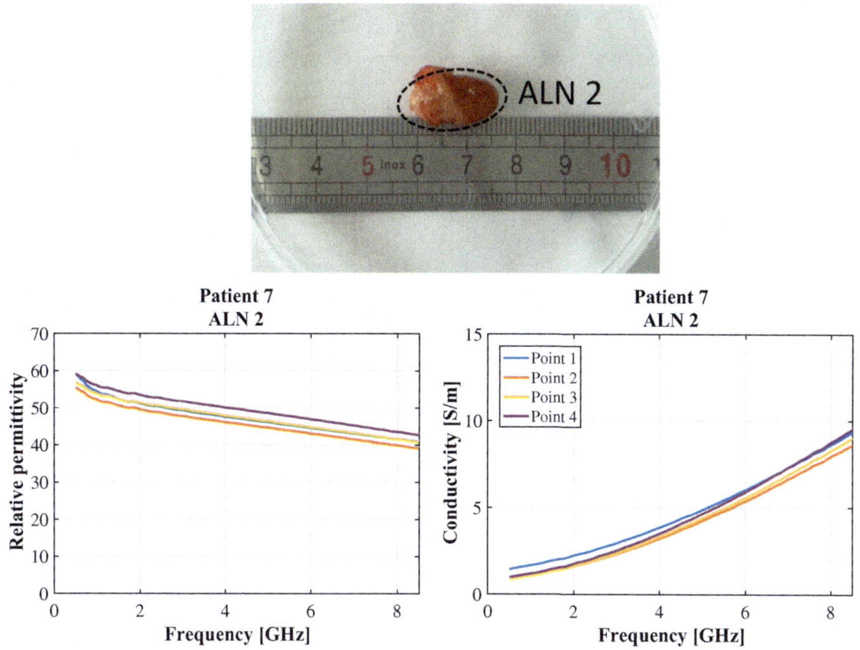

Fig. 9 Example of dielectric measurements of a single axillary lymph node (ALN 2). **(Top)** picture of the sample (the ALN is circled with a dashed line). **(Bottom–left)** ALN 2 relative permittivity; **(Bottom–right)** ALN 2 conductivity. Each line corresponds to a measurement site

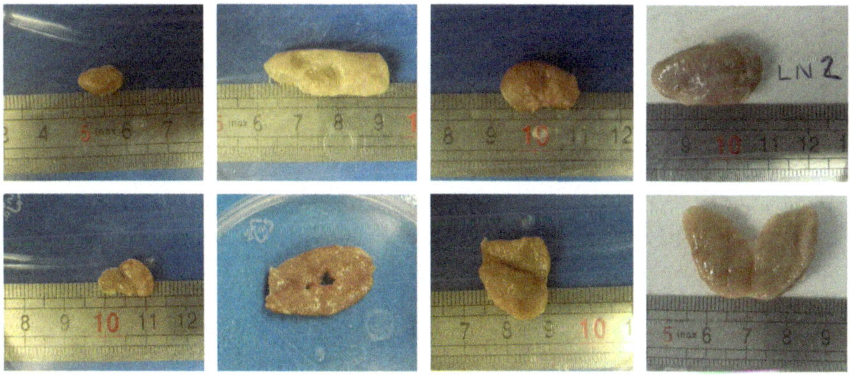

Fig. 10 Example of four ewe inguinal lymph nodes that were measured. (**Top line**) intact ewe LNs; (**bottom line**) same LNs, sliced in half

LN, in a total of 41 measurements, while trying to avoid measurements near regions with residual fat. The average (± standard deviation) temperature of the LN surface was 20.6 °C (± 1.3 °C). The relative permittivity and conductivity are reported in Fig. 11. The maximum permittivity (ϵ_r around 50 at 4.5 GHz) and conductivity (σ

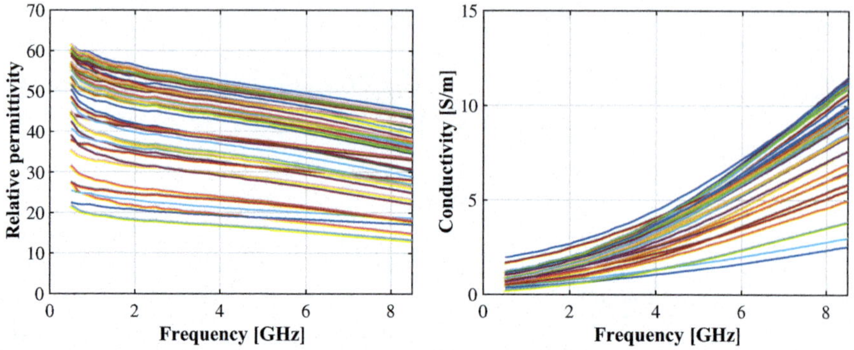

Fig. 11 Dielectric measurements of ewe inguinal LN external surfaces. **(Left)** relative permittivity and **(right)** conductivity resulted from 41 measurements conducted on 8 LNs excised from 2 ewes

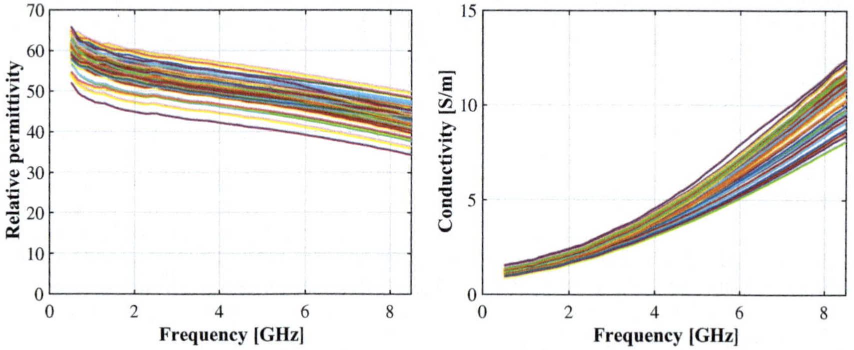

Fig. 12 Dielectric measurements of ewe inguinal LN internal cross-section. **(Left)** Relative permittivity and **(right)** conductivity resulted from 40 measurements conducted on 8 LNs excised from 2 ewes

around 5 S/m at 4.5 GHz) values are similar to those observed in the case of $Group_{40+}$ of human ALNs (Fig. 7). In addition, similarly to the case of human ALNs, there is a significant variability between measurements which is likely to be due to the presence of residual fat.

For each LN, immediately after the outer surface measurement, the LN sample was sliced in half and the probe was placed in contact with the LN cross-section surface. Figure 10 (bottom–line) illustrates 4 ewe LNs after being sliced in half. Again, measurements were taken in 4 to 6 locations for each LN, in a total of 40 measurements. The average ($\pm$standard deviation) temperature of the sample cross-section surface was 21.8 °C ($\pm$0.9 °C). Figure 12 shows the permittivity measured in the core of the ewe LNs.

In contrast to the permittivity measured on the LN surface, measurements in the inner cross-section are consistent with each other due to the homogeneity observed in the interior of LNs. Relative permittivity ranges between 41.7 and 56.7 at 4.5 GHz,

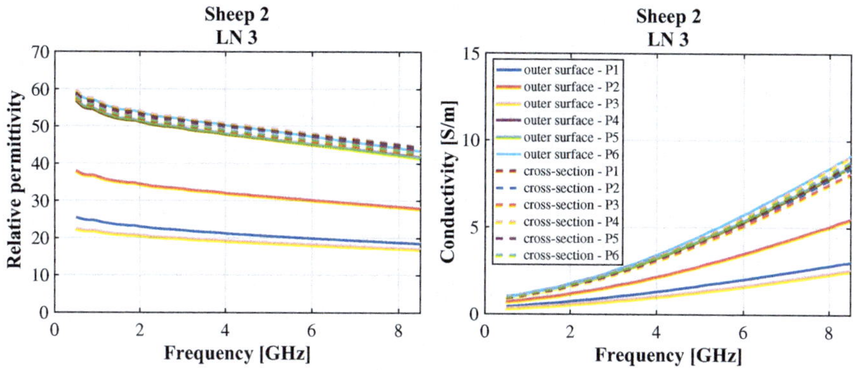

Fig. 13 Dielectric measurements of a single ewe inguinal LN: a comparison between outer surface (continuous line) and cross-section (dashed line) measurements. Each line corresponds to a measurement point. **(Left)** relative permittivity and **(right)** conductivity

while conductivity ranges between 3.6 and 5.3 S/m at the same frequency. These values are similar to the highest permittivity values obtained from the surface measurements of both human ALNs and ewe LNs (see Figs. 7 and 11, respectively). Such high permittivity values are probably due to the very high water-content of the tissue, which is compatible with the physiological function of the organ.[2]

The mean permittivity of all cross-section measurements was fitted to a two-pole Debye model, using the same methodology described for human ALNs. The obtained Debye models' parameters are reported in Table 2. The absolute fitting error was at maximum 0.3 in relative permittivity and 0.1 S/m in conductivity, demonstrating that the model is a good representation of the measured data.

Lastly, it should be reported that similarly to the inter-sample analysis, also the intra-sample analysis shows high consistency and higher dielectric properties in the cross-section measurements. Figure 13 reports the dielectric properties of several points measured on a single ewe LN. A similar behaviour was observed for all the remaining measured 7 LNs.

4 Estimation of Axillary Lymph Nodes Properties from MRI Exams

The limited number of samples and the influence of the ALN heterogeneity in the measurements with OECP motivated the assessment of a novel approach to study the contrast between healthy and metastasised ALNs. This section proposes a methodology to estimate the dielectric properties of tissues from MRI exams. Previous studies have created dielectric properties maps based on MRI exams [37, 38], but no estima-

[2] The inner part of ALNs contains lymph and plasma cells, with high water-content.

tion of properties of biological tissues from this type of exams has been addressed. Only MR-based Electrical Properties Tomography (EPT) has been used to estimate dielectric properties [39]. However, EPT is applied at the Larmor's frequency (up to 300 MHz), which is outside the frequency range of interest for MWI. Our methodology assumes that there is a relationship between voxel intensities, water content, and, therefore, dielectric properties at MW frequencies for specific MR image sequences. A careful analysis of this relationship must be performed as simple MR weighted image sequences can not provide quantitative information. Nonetheless, a relative comparison between the observations is possible if the images are processed correctly. Three main requirements are needed to infer unknown dielectric properties of tissues from MRI exams:

1. A suitable MR image sequence must be selected ensuring that (i) there is sufficient contrast between the tissue of interest and surrounding tissues to allow segmentation, and (ii) there is a direct relationship between the voxel intensities of image, water content and dielectric properties of the different imaged tissues based on state-of-the-art data.
2. The MR image must be pre-processed in order to remove noise and artefacts that might hamper the segmentation of tissues and corresponding dielectric properties estimation.
3. Dielectric properties of the tissues present in the image, besides the tissue of interest, must be known at the frequencies of interest.

This methodology was applied to a dataset of breast MRI exams and estimated the dielectric properties of both healthy and metastasised ALNs. The following sections detail the used dataset, the methodology for image processing and dielectric properties estimation, and the obtained results.

4.1 Methodology

This section presents the methodology used in this study. Sect. 4.1.1 presents the details of the used MRI dataset; Sect. 4.1.2 presents the proposed image processing pipeline that was applied to the MRI exams; and Sect. 4.1.3 presents the methodology to estimate ALNs dielectric properties.

4.1.1 Dataset

This study included 50 breast MRI exams of female patients acquired during regular breast cancer screenings or follow-ups with a 3T clinical MR system (Magnetom Vida, Siemens Healthineers) with an 18-channel dedicated breast coil, at Hospital da Luz Lisboa. The study was approved by the Scientific and Ethical Commission under references CES/44/2019/ME and CES/34/2020/ME, and an informed consent was

Table 3 Demographic data of patients from the MRI database (age and body mass index, BMI). The left side of the table refers to patients with only healthy axillary lymph nodes (ALNs); the right side of the table refers to patients with metastasised ALNs

	Patients with only healthy ALNs (n = 25)			Patients with metastasised ALNs (n = 25)		
	Mean	Std.	Range	Mean	Std.	Range
Age	49	10	34–73	55	13	38–81
BMI	28	6	17–44	27	4	19–36

obtained from all patients. Twenty-five patients with only healthy ALNs and twenty-five patients with one or more metastasised ALNs were included in the study. The demographic data of the patients are shown in Table 3.

Two MRI sequences from the clinical protocol were used: (1) direct coronal two-dimensional (2D) T2-weighted (T2-w) turbo spin echo (TSE) with short-time inversion recovery pulse (STIR); and (2) Direct axial isotropic 3D T1-w fl3D VIBE Dixon image sequence (T1-w Dixon). The 2D T2-w STIR is the most used sequence by radiologists to identify ALNs, since ALNs are usually very well-defined in images with this sequence. However, this 2D image sequence has low spatial resolution in the transversal and sagittal planes ($4 \times 0.75 \times 0.75$ mm^3), meaning an additional sequence must be used to complement the detection of ALNs. The T1-w Dixon image sequence allows a MultiPlanar Reconstruction (MPR) in all anatomical planes without major image artifacts and provides good contrast between internal tissues, such as muscle, adipose, and fibroglandular tissues. This image sequence provides four image sets with different contrasts. For the purpose of this study, the water (W) and fat (F) image contrasts were used. One can assume higher water-content tissues are represented with higher voxel intensity values in T1-w Dixon-W, in order to assume a direct relationship between them (and consequently between dielectric properties). However, this assumption needs to be carefully confirmed for each tissue type individually.

In the T1-w Dixon-W image, there is an increase of voxel intensities from adipose tissue to fibroglandular tissue, muscle and skin. Although muscle and skin are imaged with similar voxel intensities, skin has lower water content (and permittivity) when compared to muscle [35, 40, 41]. This is explained by the skin proximity to the breast MRI coil, which inherently results in higher voxel intensity values. For the remaining tissues, a direct relationship between dielectric properties, water-content and intensity values can be assumed [35, 42]. Therefore, the T1-w Dixon-W image was suitable for this methodology.

Figure 14 shows examples of healthy and metastasised ALNs in T1-w Dixon-W image. The healthy ALN has a large hilum (represented as dark voxel intensities), while the cortex and the remaining ALN structures correspond to the semi-ellipse of light gray voxel intensities. The metastasised ALN has no hilum and is imaged with light gray voxel intensities.

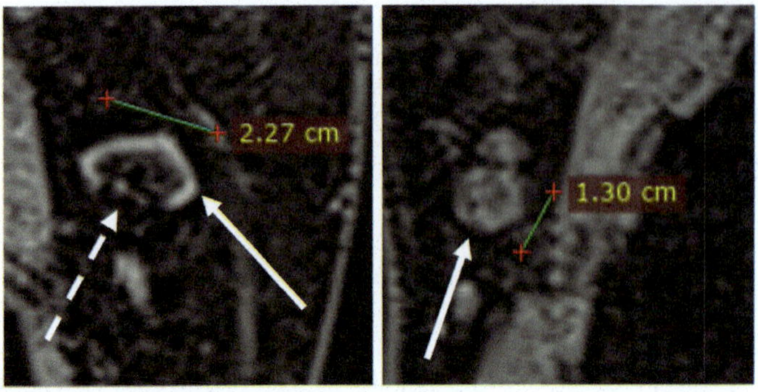

Fig. 14 Examples of axillary lymph nodes (solid arrows) T1-w Dixon-W images. **(Left)** Healthy lymph nodes with a fatty hilum (dashed arrow) and **(right)** metastasised lymph nodes without hilum

4.1.2 Image Processing Pipeline

The pre-processing and semi-automatic segmentation steps towards the estimation of ALNs dielectric properties were fully implemented in Python™ and applied to T1-w Dixon with both Water and Fat contrasts and T2-w STIR. Figure 15 summarises the steps described in the following paragraphs.

The T2-w STIR and T1-w Dixon image sequences have different spatial resolutions and dimensions. In order to correctly superimpose them, they need to be spatially registered to the same spatial reference system. The Insight Toolkit (ITK)'s implementation [43] of an affine registration with linear interpolation was applied to register T2-w STIR to T1-w Dixon. Then, the N4 bias field removal algorithm [44] was applied to the breast MR images. The bias field is an artificial variation of intensities within the tissues of the same type produced during the MRI acquisition due to the inhomogeneities of the magnetic field. The bias field removal is essential to minimise the inhomogeneities within the tissues, and, therefore, improve the segmentation performance and ensure the reliability of the estimated properties.

For each image sequence, the region of interest was selected to avoid including regions of the body of little interest that could compromise the algorithms performance. T1-w Dixon-W images should contain both breasts and axillary regions, as the entire region is needed to complete dielectric property assignment. T2-w STIR should only include the axillary regions, in order to exclude noisy regions that may hamper the segmentation performance. A median filter was applied to both image sequences to remove noise and to smooth the voxel intensity differences within each tissue. Then, a minimum-maximum normalisation was applied to the voxel intensities of each image, which is important for the dielectric property assignment.

Three main segmentation steps were needed before assigning the dielectric properties and estimate ALN properties.

Firstly, the background of the body was segmented by applying an Otsu's thresholding [45] to both T1-w Dixon-W and T1-w Dixon-F images and applying the union

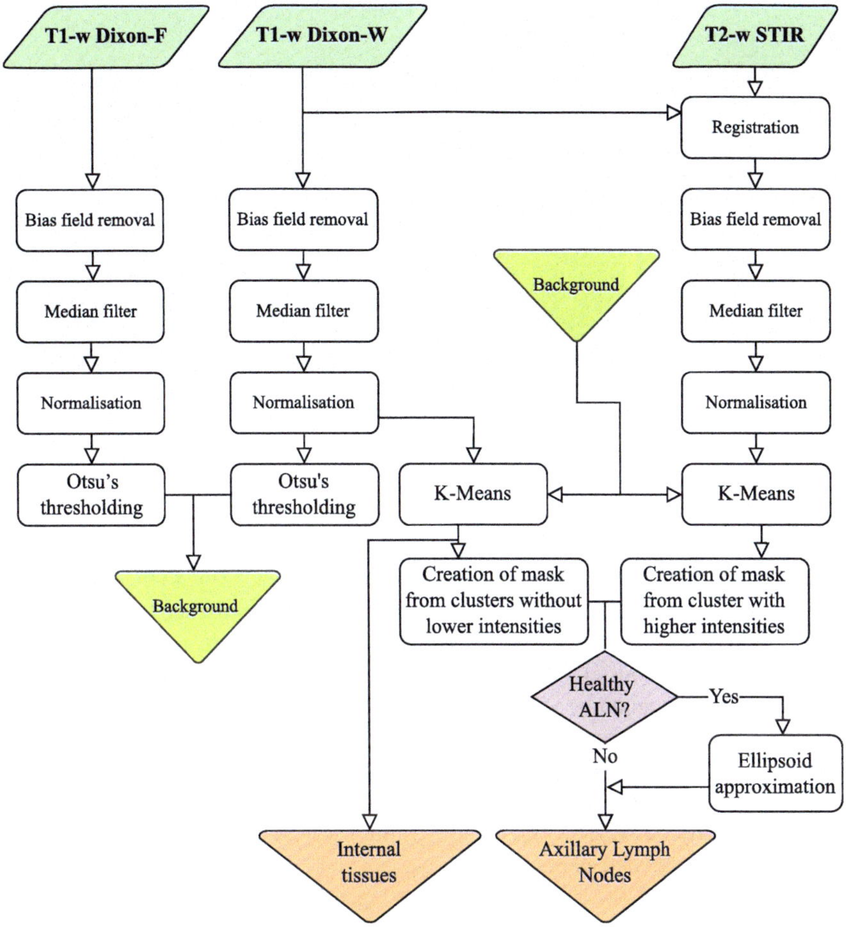

Fig. 15 Flowchart of the processing pipeline (pre-processing and segmentation steps) for the estimation of dielectric properties

of both binarised images. Then, a post-processing step was applied where each axial slice of the resulting mask was scanned from the anterior to the posterior part of the mask, filling any holes within the body region.

Secondly, the internal tissues were segmented using the K-means algorithm. This algorithm separates the tissues in a K number of clusters considering the voxel intensity values. The segmentation results when applying several values of K were compared. The best K value for T1-w Dixon-W images was empirically retrieved considering some qualitative criteria: (1) there is good distinction between fibroglandular, adipose and muscle tissues; (2) ALNs can be identified in more than one cluster but they should be isolated from the surrounding tissues; (3) one single main

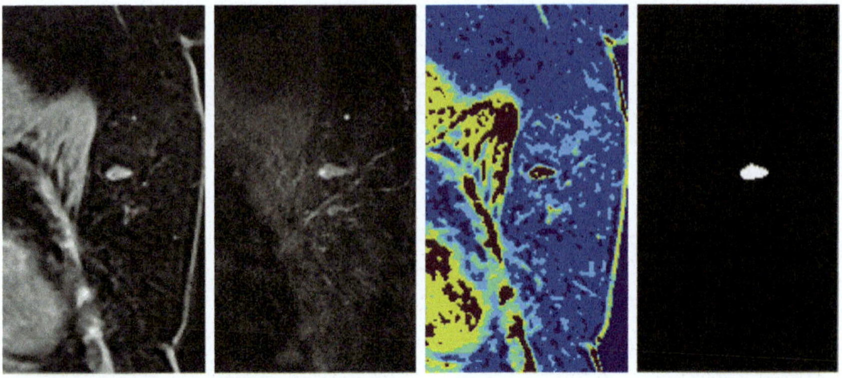

Fig. 16 Examples of MR images and the final segmentation results. The first and second figures are the T1-w Dixon-W and T2-w STIR images, respectively. The third is the resulting segmentation of K-means applied to T1-w Dixon-W and the fourth figure is the resulting segmentation of one ALN

tissue cannot be identified in more than three clusters. For T2-w STIR images, only one criterion was used: all ALNs need to be segmented in only one cluster.

Finally, the ALN segmentation was optimised, as K-means do not usually segment ALNs in only one cluster. To this end, the ALN mask was created by intersecting the resulting segmentations of T1-w Dixon-W and T2-w STIR images. The resulting image is a more accurate representation of the ALN shape and size, due to the higher resolution of T1-w Dixon-W. However, in T1-w Dixon-W, only the ALN cortex has high voxel intensities and was included in the segmentation. Hence, for each detected healthy ALN, the corresponding hilum was included in the segmentation by applying an ellipsoid estimation. More details on this methodology are presented in [42]. Figure 16 shows an example of the original image sequences and resulting segmentations.

4.1.3 Estimation of Dielectric Properties

To estimate the dielectric properties of ALNs, the state-of-the-art dielectric properties of tissues (in particular, adipose and fibroglandular tissues) were, firstly, assigned to the previously processed images. To this end, dielectric property maps were created using a similar approach to those proposed in [37, 38].

Six dielectric property curves were considered for permittivity and conductivity based on the study by Lazebnik et al. [28], as shown in Fig. 17: two curves to limit both fibroglandular and adipose tissues, one minimum and one maximum curve. Choosing these curves will inherently impose a maximum and minimum value of dielectric properties for ALNs. The maximum value corresponds to the value of fibroglandular tissue. However, no other curves (e.g. the water curve) were viable to use in this methodology as no free-water tissues are imaged with these sequences.

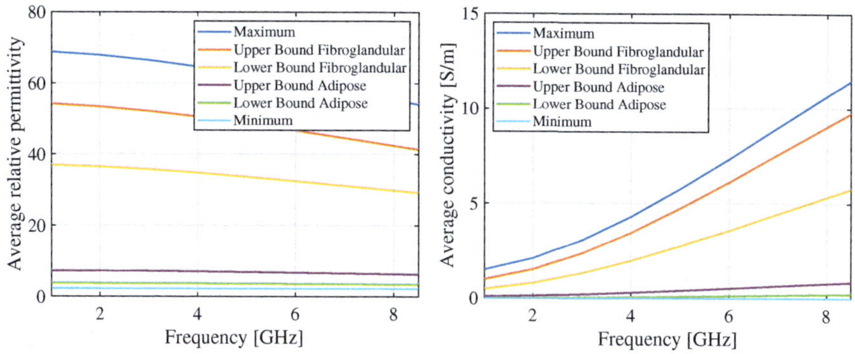

Fig. 17 Dielectric property curves of breast tissues reported in [28]. **(Left)** relative permittivity, and **(right)** conductivity

Then, each cluster obtained from the K-Means segmentation was assigned to an interval between two curves. At each frequency, the minimum and maximum voxel intensities of each cluster were assigned to the dielectric property values of the chosen two curves, as shown in Fig. 18. The remaining voxel intensities were then mapped to a value between the selected curves using a piece-wise linear interpolation:

$$dp(w, f) = w(v) \times c_{\text{upper}}(f) + [1 - w(v)] \times c_{\text{lower}}(f) \tag{2}$$

$$w(v) = \frac{v - v_{\text{min}}}{v_{\text{max}} - v_{\text{min}}} \tag{3}$$

where c_{upper} and c_{lower} are the dielectric property values at the upper and lower bound curves considered for the interpolation, respectively, at a specific frequency f. w is a value between 0 and 1, v is the voxel intensity, and v_{min} and v_{max} are the minimum and maximum voxel intensities of the corresponding cluster, respectively.

After generating the voxelised dielectric property maps for frequencies from 0.5 to 8.5 GHz, with a step of 0.5 GHz, the properties of ALNs can be estimated by superimposing the ALN mask with the dielectric property maps. One ALN from each axillary region was selected for comparison. In exams of patients with metastasised ALNs, one metastasised and one healthy ALNs in opposite axillary regions were compared. This means the entire analysis was performed in a total of 75 healthy ALNs and 25 metastasised ALNs. In order to select a specific ALN, a connected-component labelling method was applied to the ALN mask given the coordinates of a point in the ALN. The dielectric properties of an ALN were obtained by averaging the assigned dielectric properties to each voxel of the ALN per frequency. Then, the first, second, and third quartile curves were calculated for each group of healthy and metastasised ALNs, and the corresponding Debye parameters were obtained by fitting a Debye model using the nonlinear least squares method.

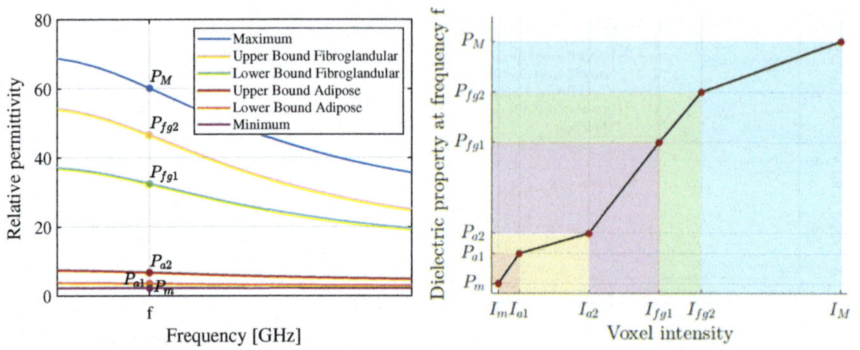

Fig. 18 Interpolation between dielectric property values and voxel intensities at a specific frequency f. **(Left)** shows an example of reference points of relative permittivity curves, and **(right)** shows the piece-wise linear interpolation considering 5 clusters (represented by different colours)

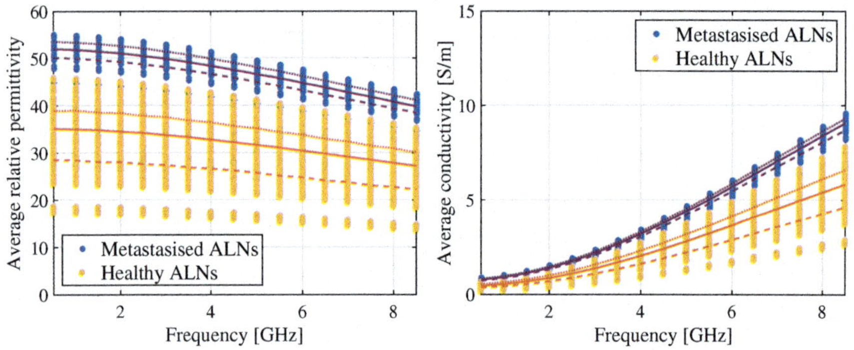

Fig. 19 Average dielectric properties of healthy (yellow dots) and metastasised (blue dots) ALNs over frequency, estimated from MR images: **(left)** relative permittivity, and **(right)** conductivity. The dashed, solid and dotted lines represent the first, second and third quartile of both healthy and metastasised ALNs, respectively

4.2 Results and Discussion

Figure 19 shows the estimated dielectric properties over frequency of the 100 ALNs selected from the MRI exams of the 50 patients included in the study. The curves of the first, second, and third quartiles for both healthy and metastasised ALNs are also shown in Fig. 19, and the corresponding Debye parameters are shown in Table 4.

The dielectric properties of healthy ALNs exhibit high variability: the average relative permittivity (computed, for each healthy ALN, across all the voxels belonging to the ALN) ranges between 16.2 and 42.2 at 4.5 GHz. This variability was also observed in the OECP measurements reported in Sect. 3 of the present chapter, and in the literature [14]. In those studies, the variability was justified by the presence of a fat layer that was covering the ALNs, which introduced a larger heterogeneity of the

Table 4 Debye model parameters for healthy and metastasised ALNs applied to 0.5 to 8.5 GHz frequency range

	Healthy ALNs			Metastasised ALNs		
Quartile	Q1	Q2	Q3	Q1	Q2	Q3
ϵ_∞	9.33	10.94	11.81	14.18	14.49	15.22
σ_s [S/m]	0.36	0.48	0.56	0.75	0.77	0.83
$\Delta\epsilon$	19.37	24.35	27.49	36.14	37.67	38.54
τ [ps]	13.00	13.00	13.00	13.00	13.00	13.00

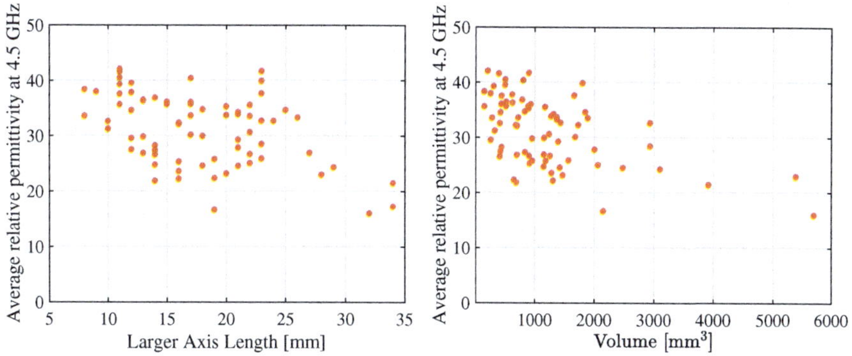

Fig. 20 Estimated relative permittivity at 4.5 GHz for each healthy ALN over **(left)** its larger axis length, and **(right)** volume

sample. In this case, this variability is explained by the size of the hilum (a physiological indentation of the ALN surface that is usually fatty in healthy ALNs), which affects the estimated average dielectric properties. As for metastasised ALNs, the results show higher dielectric properties and lower variability if compared to healthy ALNs: the average relative permittivity ranges from 41.3 to 50.6 at 4.5 GHz. These results can be explained by the fact that metastasised ALNs have a very small hilum or do not have it, which reduces the amount of fat affecting the average properties. The dielectric contrast between the median of both healthy and metastasised ALN groups is 32% at 4.5 GHz.

Figure 20 shows how the average relative permittivity of healthy ALNs changes over their size, specifically over the ALN larger axis length and volume. As the size of the ALNs increases, the lower are the estimated average permittivity values. This is explained by the fact that larger ALNs also have a larger hilum, which means the hilum is contributing more to the average dielectric properties of the ALN.

Figure 21 shows an intra-patient comparison between the estimated average relative permittivity values of the selected ALNs. The values vary between patients, but they are within the same range of values. The dielectric contrast between healthy and metastasised ALNs within the same patient at 4.5 GHz is 35%, which is larger than among healthy ALNs (16%).

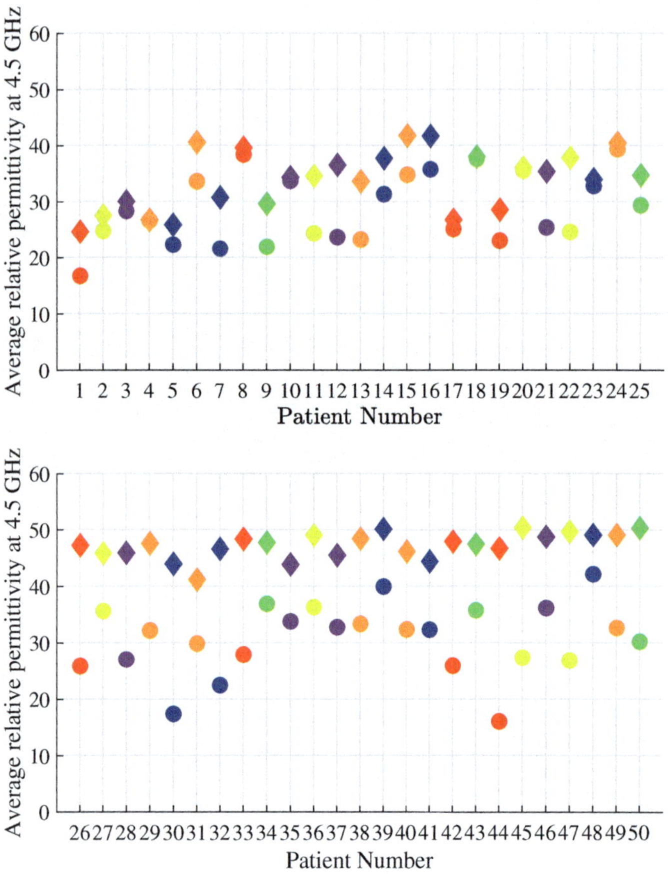

Fig. 21 Comparison of estimated average relative permittivity within each patient for 4.5 GHz. **(top)** shows a comparison between healthy ALNs (both circles and diamonds), and **(bottom)** shows a comparison between healthy (circles) and metastasised (diamond) ALNs

A comparison to actual OECP measurements of the ALNs under study was not possible since no follow-up of the patients was performed. Nevertheless, the obtained results can be compared with the general conclusions drawn in OECP studies. In the case of OECP, the large range of dielectric properties might be due to sample handling and limitations on measuring heterogeneous samples. In contrast, when estimating the properties from MRI, this large range can be explained by the inherent heterogeneity of the ALN and variability intra- and inter-patient. The range of the relative permittivity estimated from MRI varies from 16.2 to 42.2 at 4.5 GHz, which is lower when compared to 5.1 to 53.7 measured with OECP [14]. This may have three explanations: (1) the estimated dielectric properties are limited by minimum and maximum dielectric property curves obtained from measurements of fibroglandular tissues and ALNs could have higher properties than this type of tissue; (2) the voxel intensity values of ALNs can be affected by the distance to the coil even if this effect is reduced by the bias field removal algorithm; (3) OECP measurements may not be sensitive to the heterogeneity of the samples, providing a weighted average of the properties of the measured sensing volume of the ALN, and may be hampered by adipose tissue.

5 Effects of Freezing and Defrosting Processes on Biological Tissue Dielectric Properties

In order to complement the knowledge on ALN dielectric properties, our research group is planning measurements on LNs excised from dead animals (cats and dogs) which may have been diagnosed with mammary cancer. This work will be conducted in collaboration with the Instituto Nacional de Investigacão Agrária e Veterinária (INIAV) [46], where our research group is planning to run a measurement campaign on defrosted LNs from animals that are frozen and defrosted ahead of their autopsy. Therefore, it is important to study whether the freezing and defrosting process can impact the dielectric properties of biological tissue, which is still unknown in the literature. In fact, only very limited number of studies have addressed this subject.

Bengtsson et al. [47] investigated the effect of the freezing process on the dielectric properties of codfish and beef minced meat. In that study, minced meat samples were prepared and frozen at −30°C. Samples were defrosted for dielectric measurement after 1–2 days, after 1–2 weeks, and after 1–3 months. Dielectric measurements were taken at two different frequencies (35 MHz and 100 MHz), at two different temperatures, namely −10°C and +2°C. The authors found no significant effect of frozen storage time for fish dielectric properties, but there was a significant difference for beef at −10°C, where both dielectric constant and loss tangent decreased with prolonged frozen storage.

Bodakian et al. [48] studied the changes of chicken (White Rock; aged 3 to 4 months) muscle dielectric properties after freezing storage, using dielectric spectroscopy over the range 5 Hz – 1 MHz. The researchers divided a chicken breast in two halves: the first one was measured two hours after the animal was slaughtered, while the second one was frozen for four hours, defrosted, and measured. Both samples were measured at 6 °C. The authors noticed that the conductivity substantially increased (approximately 100% increment - corresponding to circa 0.3 S/m - at 1 KHz) by the freezing and defrosting process.

Ley et al. [49] suggested that the storage time in a commercial fridge does not influence the dielectric properties of porcine liver (storage time up to 130:30 h), muscle (storage time up to 56 h), fat (storage time up to 55 h), and blood (storage time up to 77:30 h) in the frequency range between 0.5 and 7 GHz.

The first two studies did not investigate the frequency range of interest to MWI applications, while the third investigated the influence of the storage time at temperatures above 0 °C, at which water molecules do not freeze. Hence, in order to enable dielectric measurements on frozen and defrosted ALNs, the gap observed in the literature should first be bridged. To do so, the effect of freezing and defrosting chicken muscle, bovine liver, and bovine fat on their complex permittivity was investigated. Also, the effect of frozen storage time on permittivity was investigated. The study focuses on the 0.5 – 8.5 GHz frequency band, which comprises the frequencies that could be used to image ALNs with MWI; and it considers frozen storage times up to 14 d, which can be considered a reasonable time for researchers to organise a measurement campaign.

The remainder of this section is organised as follows: Sect. 5.1 illustrates the experimental methodology and evaluation metrics; and Sect. 5.2 presents and discusses the results.

5.1 Methodology and Experimental Setup

This section describes selection and preparation of tissue samples, methodology, and experimental setup adopted during the experiments. Sect. 5.1.1 illustrates the sample preparation and the experimental plan; Sect. 5.1.2 describes the instrumentation, and the measurement technique, including the validation of the measurements; and Sect. 5.1.3 reports the evaluation metric and the statistical hypotheses tests designed for this investigation.

5.1.1 Sample Preparation and Experimental Plan

The study was conducted on chicken muscle (one animal specimen), bovine liver (one animal specimen), and bovine fat tissues (two animal specimens). These tissues were chosen due to their homogeneity, thus minimising the influence of sample heterogeneity, as it acts as a confounder [23]. Additionally, these tissues present relative permittivity of about 4 (fat), 40 (liver), and 50 (muscle) at 4.5 GHz [18], which cover a wide range of biological tissue properties.

Fresh animal tissues were purchased from a local butcher. After the time of death, tissues were stored in a fridge (at approximately 5 °C), and measurement started within 24 h *post mortem*.

On the first day of measurements (day 0), 10 samples (size approximately 3×2 x 1 cm^3) were sliced from each of the 3 tissue types, in a total of 30 samples, and measured using the OECP technique. It should be noted that the size of the samples was sufficiently large, with respect to both sensing radius and sensing depth of the probe according to [31, 32]. Samples were measured only when they reached room temperature. The surface temperature of each sample was recorded with an infrared thermometer at the time of measurement. Within 7 h after the beginning of the measurements, the samples were wrapped in cling film and stored in a sealed container at a temperature of approximately −18 °C.

Table 5 Measured average temperature of the three tissues under study at day 0, $T_s(0)$, and day d, $T_s(d)$, and average absolute difference, $|\Delta T_s(d)|$

	$T_s(0)$ [°C] avg (± std)	$T_s(d)$ [°C] avg (± std)	$\|\Delta T_s(d)\|$ [°C] avg (± std)
Muscle	20.7 (±0.6)	21.3 (±1.4)	0.8 (±0.9)
Liver	20.9 (±0.2)	21.8 (±0.5)	1.0 (±0.5)
Fat	20.7 (±1.1)	21.9 (±1.6)	1.1 (±0.8)

At each subsequent day, one sample of each tissue type was defrosted, and its dielectric properties were measured at approximately the same spot where they had been measured on day 0. Samples were defrosted at room temperature, and measured only once they reached the temperature of the measurements on day 0, which is crucial for a fair comparison between dielectric properties in day 0 and day d. After this final measurement, each sample was discarded. For simplicity, the subscript d will be used in the following text to refer to the number of days each sample was frozen for, where d ranges between 1 and 14.

Table 5 reports the average (± standard deviation) of $|\Delta T_s(d)| = |T_s(d) - T_s(0)|$ for each tissue type, where $T_s(d)$ and $T_s(0)$ are the measured temperatures of sample s on days d and 0, respectively. The average $|\Delta T_s(d)|$ is approximately 1 °C, which is acceptable, as it has been reported that dielectric properties do not vary more than 2% per Celsius degree [50]. For reproducibility purposes, Table 5 also reports the average (± standard deviation) $T_s(d)$ and $T_s(0)$, for each tissue type.

5.1.2 Instrumentation and Measurement Technique

Measurements were performed using an OECP fabricated with an EZ-141 coaxial cable (outer diameter of 3.58 mm). The probe was connected to an Agilent E5071C VNA through a right angle SMA-connector (Fig. 22), in order to eliminate the uncertainty introduced by the use of cables [21]. The reflection coefficient (S_{11}) was measured in 801 equally-spaced frequency points between 500 MHz and 8.5 GHz, with 30 Hz IF bandwidth and −5 dBm output power. The dielectric properties were retrieved implementing "in Matlab" the method proposed in [51]. Such method entails calibrating the probe measuring a short-circuit, an open-circuit, and a known load. Deionised water was used as a known load; its average (± standard deviation) temperature was 21.7 (±0.4) °C.

Following the same reasoning described in Sect. 3.1, validation measurements of 0.1 molar NaCl solution were conducted immediately after setup calibration (prior to tissue measurements) and after some sample measurements (usually after each sample measurement or within 15 min since the last validation measurement), resulting in a total of 99 validation measurements. The temperature of the 0.1 molar NaCl solution was also recorded in order to match it to the correct model. The average

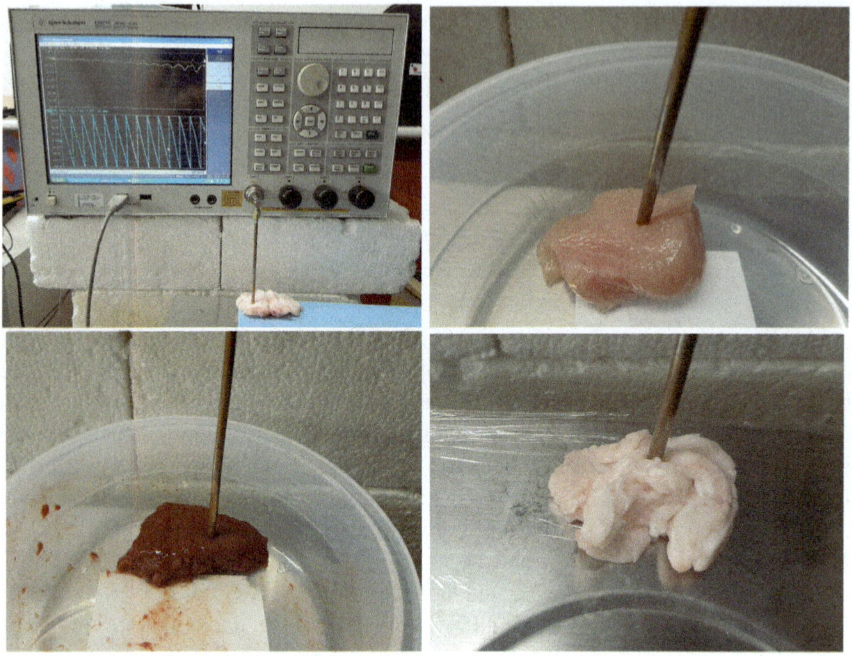

Fig. 22 **(Top–left)** Dielectric measurement setup: open-ended coaxial-probe connected to a right angle SMA-connector, and vector network analyzer during a tissue measurement. **(Top–right)** example of a chicken muscle tissue sample measurement; **(bottom–left)** example of a bovine liver tissue sample measurement; **(bottom–right)** example of a bovine fat tissue sample measurement

($\pm$ standard deviation) temperature of the 0.1 molar NaCl solution was 22.5 ($\pm$0.5) °C. Figure 23 reports the average ($\pm$ standard deviation) relative error, and the average absolute value of the relative error, across all the validation measurements.

Regarding tissue measurement, in order to handle intra-sample dielectric properties variability, between 3 and 7 points (one measurement from each point) were measured on the surface of each sample. It is known that the pressure applied with the probe onto the tissue [23, 25] can affect the measured dielectric properties (acting as a "confounder"). As a result, the least pressure possible was applied, while ensuring (i) full contact between the tip of the probe and the measured tissue, and (ii) that there were no air gaps between the probe and the sample. Figure 22 shows examples of muscle, liver and fat tissue samples while being measured.

5.1.3 Evaluation Metric and Statistical Tests

The relative permittivity and conductivity variations were computed as $\Delta\epsilon_r^s(f, d)$ and $\Delta\sigma^s(f, d)$ respectively, defined as follows:

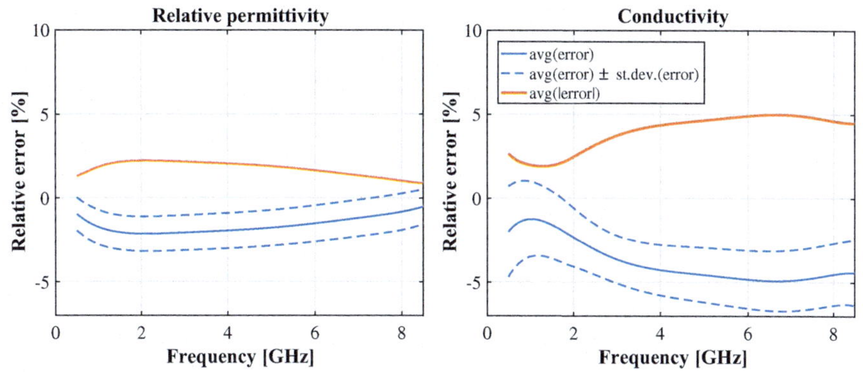

Fig. 23 Dielectric measurement uncertainty: average (±standard deviation) relative error (blue), and average absolute value of the relative error (orange), across all 99 validation measurements. The relative error is computed as the relative [%] difference between the measured and the model (reported in [24]) relative permittivity and conductivity of the 0.1 molar NaCl solution. **(Left)** relative permittivity, and **(right)** conductivity

$$\Delta\epsilon_r^s(f, d) = \epsilon_r^s(f, d) - \epsilon_r^s(f, 0) \quad (4a)$$

$$\Delta\sigma^s(f, d) = \sigma^s(f, d) - \sigma^s(f, 0) \quad (4b)$$

where $\epsilon_r^s(f, d)$ and $\epsilon_r^s(f, 0)$ are the de-embedded relative permittivity of sample s at frequency f measured on days d and 0, respectively. A similar metric is used to compute the conductivity variation $\Delta\sigma^s(f, d)$.

Finally, the following two hypotheses were investigated:

- Hypothesis 1 (H1): the dielectric properties of defrosted tissues do not correlate with frozen storage time. To evaluate H1, the trend of $\Delta\epsilon_r^s(f, d)$ and $\Delta\sigma^s(f, d)$ over d is analysed by fitting the data with a linear regression model (slope m). The p-value for m is computed to evaluate its statistical significance (significance level $\alpha = 0.05$).
- Hypothesis 2 (H2): the processes of freezing and defrosting do not significantly affect the dielectric properties of biological tissues, with respect to fresh tissues. To evaluate H2, two sets of paired data are defined: (i) the dielectric properties measured on day 0, and (ii) the dielectric properties measured in the following days (d). For each tissue type, each of the two datasets comprises a total of 10 measurements. The Wilcoxon signed-ranked test[3] is used to assess whether the two sets of data significantly differ. It should be noted that the investigation of H2 does not take variable d into account, hence this hypothesis is only tested if H1 is accepted. The sample size (N) was calculated so that the Wilcoxon signed-ranked test could

[3] The Wilcoxon signed-rank test is a non-parametric statistical hypothesis test suited for paired samples with non-normally distributed differences. The Wilcoxon signed-rank test is used to test the null hypothesis that the median of the paired differences of the two samples is 0.

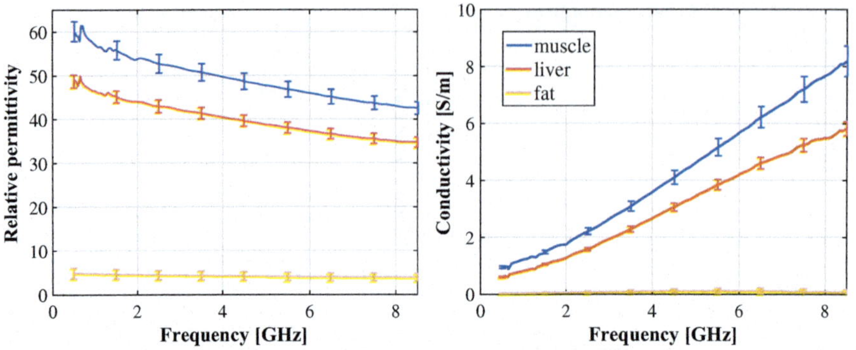

Fig. 24 Day 0 average dielectric properties of the three tissues analysed: **(left)** relative permittivity, and **(right)** conductivity. The results are obtained by averaging the dielectric properties measured on all the 10 measured samples for each tissue. The standard deviation across the 10 samples is reported with 1 GHz intervals via a whisker plot

detect, with power[4] 0.8 and significance level $\alpha = 0.05$ (commonly adopted in the literature [52]), a minimum difference of approximately 2–3 in the relative permittivity, and 0.2–0.3 in the conductivity for muscle and liver, and a minimum difference of approximately 1.5 (relative permittivity), and 0.1 (conductivity) in fat tissue. The Effect Size (E) corresponding to the above indicated differences was estimated ≥ 1 (E was defined as per [53, 54]). As a result, it was observed that N = 10 samples was sufficient to assess the sought differences.

5.2 *Results and Discussion*

This section presents the dielectric properties results and discusses their variability with freezing process over time. Figure 24 shows the day 0 dielectric properties results of the 3 tissues analysed. The whisker bars in Fig. 24 represent the inter-sample variability which is kept below 5% for the three tissues. This is an important indicator of the consistency of the measurements, as it allows to compare the dielectric properties variation measured on different samples. Also, it should be emphasised that the measured dielectric properties are similar to those reported in [55] (for bovine liver), [56] (for chicken muscle), and [28] (for freshly excised human breast fat), which makes our measurements more reliable.

[4] Statistical power is defined as $1 - \beta$; β is the Type-II error rate [52].

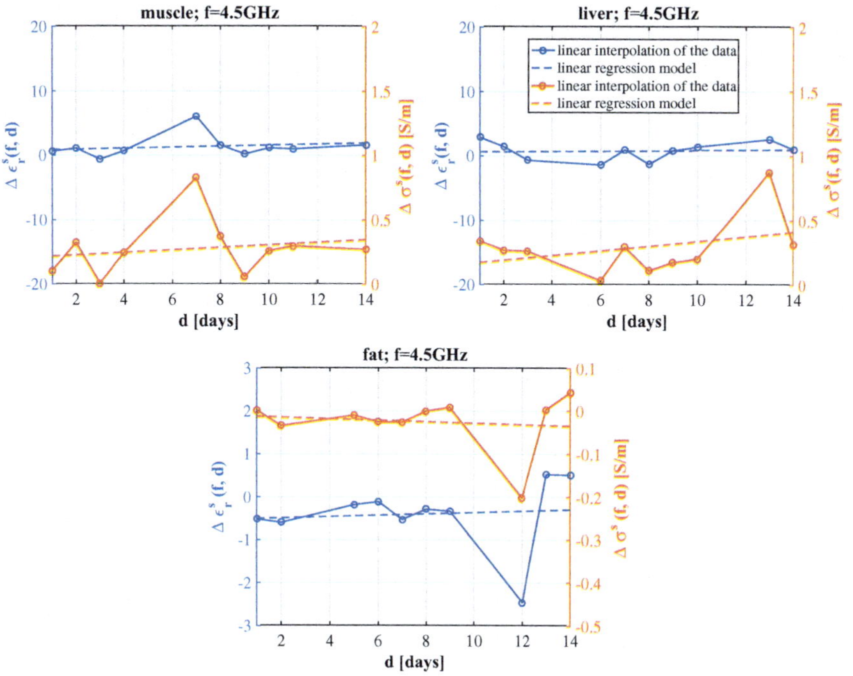

Fig. 25 Investigation of H1: dielectric properties variation, $\Delta\epsilon_r^s(f, d)$ and $\Delta\sigma^s(f, d)$, versus frozen storage time d, at 4.5 GHz. **a** muscle, **b** liver, and **c** fat. Note that each sampling point univocally corresponds to one tissue sample (measured both on day 0 and day d)

5.2.1 Investigation of Dependence of Frozen Storage Time (H1)

Figure 25 reports $\Delta\epsilon_r^s(f, d)$ and $\Delta\sigma^s(f, d)$, at $f = 4.5$ GHz over d, computed for all tissues. The data points were fitted with a linear regression—their slopes m are reported in Table 6.

The slope of the linear regression fitted to $\Delta\epsilon_r^s(f, d)$ and $\Delta\sigma^s(f, d)$ indicates a marginal increase of the dielectric constant with the increase of frozen storage time d. In addition, the p-values, also presented in Table 6, estimated for the slopes m indicate that there is no sufficient evidence that a non-zero correlation exists between dielectric properties variation and frozen storage time. Similar analyses were performed at $f = 2.5$ GHz and $f = 6.5$ GHz, and comparable conclusions were drawn.

It can be concluded that the dielectric properties of the defrosted tissues do not significantly depend on the frozen storage time (for frozen storage times up to 14 d). As such, H2 may be investigated while disregarding any dependency on variable d.

Table 6 Linear regression model parameters, fitted for the three tissue types at 4.5 GHz: slope m, and corresponding p-value (in brackets) for the fitted models

	$\Delta\epsilon_r^s(f, d)$ m (p-value)	$\Delta\sigma^s(f, d)$ m (p-value)
Muscle	0.08 (0.59)	0.01 (0.61)
Liver	0.03 (0.84)	0.02 (0.31)
Fat	0.02 (0.81)	0.00 (0.76)

The models are defined as $\widehat{\Delta\epsilon_r^s}(f, d) = m * d + m_0$, where m is the slope, and m_0 is the vertical intercept. (A similar definition holds for $\widehat{\Delta\sigma^s}(f, d)$)

5.2.2 Investigation of Effect of Freezing and Defrosting (H2)

Based on the results discussed in Sect. 5.2.1, H2 can be investigated. Figure 26 illustrates $\Delta\epsilon_r^s(f)$ and $\Delta\sigma^s(f)$ at $f = 2.5,\ 4.5,\ 6.5$ GHz, in a box-plot. At the centre frequency of $f = 4.5$ GHz, after the tissue is frozen and defrosted, the relative permittivity difference increases in the range between +0.8 (first quartile, Q1) and +2.0 (third quartile, Q3) in the case of chicken muscle (which corresponds to a percentage variation between 1.2% and 3.2%); while it varies within Q1 = 0.1 S/m (2.6%) and Q3 = 0.3 S/m (7.6%), in conductivity. For bovine liver, the relative permittivity varies within Q1 = -0.7 (−1.7%) and Q3 = +1.4 (+3.7%), while the conductivity varies within Q1 = +0.2 S/m (+5.6%) and Q3 = +0.3 S/m (+10.3%) As for bovine fat, the median relative permittivity decreases by −0.3 (−7.6%); while the median conductivity variation is below 10^{-2}. It should be noted that the conductivity shows a larger percentage increase than the relative permittivity, which can be partially explained by the fact that conductivity is more susceptible to error associated to the OECP technique. In order to assess the significance of the observed differences, the Wilcoxon signed-ranked test was used (it was first verified that $\Delta\epsilon_r^s(f)$ and $\Delta\sigma^s(f)$ do not follow a Normal distribution by applying the Kolmogorov-Smirnov test). The statistical powers, computed *post-hoc*, are reported in Fig. 26 for each test.

Significant differences (p-value ≤ 0.05) were found for muscle (both relative permittivity and conductivity), for liver conductivity, and for liver relative permittivity at 2.5 GHz, as marked by the stars in Fig. 26. Regarding the cases where significant differences were not found, it should be emphasised that it may be due to the low power (< 0.8) of the corresponding tests (the observed effect size is considerably lower than the one estimated *a-priori*). This implies that if differences existed, their value would be below the threshold that was defined as meaningful. As a result, investigating the significance of such small differences with higher power is not of interest for most of the applications referred above.

The results indicate that the characterisation of dielectric properties of biological tissues can be performed on previously frozen low water-content tissues. Regarding high water-content tissues, defrosted samples may be used for the estimation of the dielectric properties if one can accept a slight overestimation, depending on the purpose of their study.

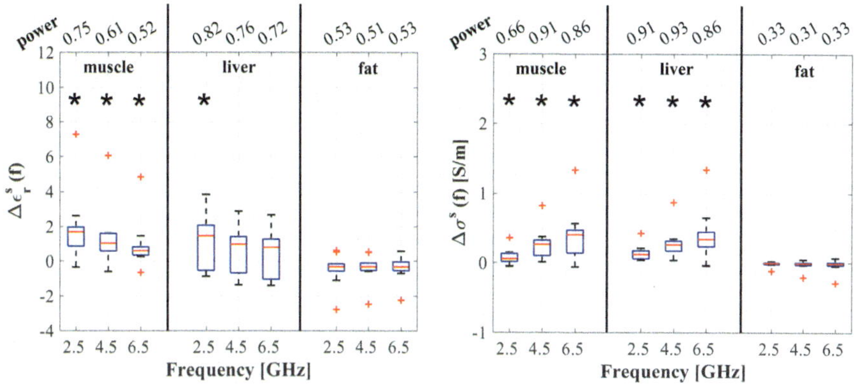

Fig. 26 Investigation of H2: the box-plots report the distribution of $\Delta\epsilon_r^s(f)$ and $\Delta\sigma^s(f)$ measured—for each tissue at 2.5, 4.5, and 6.5 GHz. Red line: median; bottom and top edges of the box: lower and upper quartiles respectively; black whiskers: maximum and minimum values excluding outliers; '+' symbol: outliers (i.e., values that are 1.5 times more than the interquartile range away from the edge of the box). The stars above indicate significant differences according to the Wilcoxon signed-ranked test. The power of the test is computed *post-hoc* and presented at the top

6 Conclusions

The present chapter reviewed the literature studies on LN dielectric properties, identified the main challenges in measuring them, attempted the estimation of ALN dielectric properties with two parallel approaches, and paved the way for the measurement of defrosted samples for tissue dielectric characterisation.

The chapter first reviewed the literature studies on LN dielectric properties, including both animal and human studies (within the axillary region or in other anatomical regions). Such analysis suggested uncertainties in the current knowledge of human ALN dielectric properties at MWI frequencies (approximately 1–10 GHz), and lack of information about differences between healthy and metastasised ALNs, motivating further studies.

The first proposed approach, to complement literature knowledge on LN dielectric properties, consisted of applying the state-of-the-art method for dielectric measurements (i.e., the OECP) to both human ALNs and animal LNs. Even if no statistical evidence was carried out, the results indicated an approximation of the dielectric properties of ALNs. For instance, the relative permittivity of healthy ALNs is approximately in the range of 30 to 50 at 4.5 GHz, while the conductivity ranges between 3 and 4 S/m at the same frequency. Despite the challenges related to tissue heterogeneity, the estimation of ALN dielectric properties was possible due to the measurements performed of the inner cross-section of animal (ewe) LNs. In addition, when measuring human ALNs from the outer surface, consistency across measurements sites (same ALN sample) was considered as a valid indicator of the homogeneity of the tissue being measured, helping in the discrimination between actual ALN tissue measurements and measurements affected by fat presence. In contrast, low dielectric properties were considered as an indicator that fat was being measured. However, given the limited availability of metastasised ALNs, it was not possible to investigate differences between healthy and pathological ALNs, which is recognised as the main limitation of this part of our study.

The second approach to ALN dielectric characterisation employed a novel methodology which correlated voxel intensity in MR images with water-content of the tissues (and hence dielectric properties). The results confirmed the high variability of the dielectric properties of healthy ALNs (75 samples), with relative permittivity varying in between 16.2 and 42.2 at 4.5 GHz. The results also showed the different composition of ALNs (proportion between cortex and hilum) influences the estimated properties. In addition, based on the analysis of 25 samples, this study suggested that the dielectric properties of metastasised ALNs are higher (relative permittivity between 41.3 and 50.6 at 4.5 GHz) and more consistent than those of healthy ALNs. An average dielectric contrast of 32% between healthy and metastasised ALNs was found at 4.5 GHz, which is a good indicator to pursue the development of ALN-MWI systems.

Lastly, the effect of freezing and defrosting processes on biological tissues dielectric properties was studied. This study was motivated by the fact that our research group will have the possibility of measuring previously frozen LNs to complement the partial conclusions that were drawn after measuring freshly excised tissues. The effect of the freezing and defrosting processes on the dielectric properties of biological tissues was investigated in the 0.5–8.5 GHz frequency band, by measuring chicken muscle, bovine liver and bovine fat (relative permittivity ranging between 4 and 50 at 4.5 GHz), using the OECP technique. Firstly, the dependency of electrical properties on frozen storage time (up to 14 d) was assessed. The results showed that there is no significant dependency on the frozen storage time. In other words, once the samples are frozen, they may be preserved for at least 14 d without a significant impact on the dielectric properties. Secondly, the variation of complex permittivity with freezing/defrosting processes was investigated. For high water content tissues, the real part of relative permittivity (median value at 4.5 GHz) increased by 2.1% and 2.6% for muscle and liver respectively, while the imaginary part of relative permittivity increased by 6.4% and 8.9%. Regarding fat tissue, the observed variations were −0.3 and −0.02 (median value at 4.5 GHz) in the real and imaginary parts respectively, which are negligible for most applications. It was concluded that frozen samples may be suitable for high water content tissues (e.g. LNs) measurements if one accepts a slight overestimation of the properties, which is particularly useful in light of the measurement that is being performed on defrosted LNs from cats and dogs.

The presented studies contributed to the knowledge of ALN dielectric properties, addressing the gaps in the literature. However, more studies are encouraged, mainly to validate the contrast between healthy and metastasised ALNs. Considering the results of the third study, frozen ALNs can be measured without compromising the viability of the results. The effect of the heterogeneity of the samples should also be evaluated, as OECP only measures a limited sensing volume that is influenced by the closest structures to the probe. Finally, it should be emphasised how important it is to report metadata regarding how samples are prepared and preserved, and hope other researchers will be encouraged to do so.

Acknowledgements This work was supported by the EMERALD project funded from the European Union's Horizon 2020 research and innovation programme under the Marie Skłodowska-Curie grant agreement No. 764479. This work was also supported by Fundação para a Ciência e a Tecnologia-FCT and European Social Fund under the fellowships 2021.05385.BD and SFRH/BD/129230/2017, FCT/MEC (PIDDAC) under the Strategic Programme UIDB/00645/2020, and co-funded by FEDER-PT2020 partnership agreement under project UIDB/50008/2020.

References

1. The global cancer observatory—world health organization, breast cancer fact sheets, (2020). Accessed on 05/2021. [Online]. Available: http://gco.iarc.fr/today
2. C.E. DeSantis, J. Ma, A. Goding Sauer, L.A. Newman, A. Jemal, Breast cancer statistics, 2017, racial disparity in mortality by state. CA: Cancer J. Clin. **67**(6), 439–448 (2017)
3. S. Standring, Blood, lymphoid tissues and haemopoiesis, in *Gray's Anatomy: The Anatomical Basis of Clinical Practice*, 41st ed. (Elsevier Limited, 2016), pp. 68–80
4. National Cancer Institute, Lymph Nodes (2018). [Online]. Available: https://training.seer.cancer.gov/anatomy/lymphatic/components/nodes.html
5. S.A. Valente, G.M. Levine, M.J. Silverstein, J.A. Rayhanabad, J.G. Weng-Grumley, L. Ji, D.R. Holmes, R. Sposto, S.F. Sener, Accuracy of predicting axillary lymph node positivity by physical examination, mammography, ultrasonography, and magnetic resonance imaging. Ann. Surg. Oncol. **19**(6), 1825–1830 (2012)
6. U. Veronesi, G. Paganelli, G. Viale, A. Luini, S. Zurrida, V. Galimberti, M. Intra, P. Veronesi, C. Robertson, P. Maisonneuve, G. Renne, C. De Cicco, F. De Lucia, R. Gennari, A randomized comparison of sentinel-node biopsy with routine axillary dissection in breast cancer. N. Engl. J. Med. **349**(6), 546–553 (2003)
7. D.N. Krag, S.J. Anderson, T.B. Julian, A.M. Brown, S.P. Harlow, T. Ashikaga, D.L. Weaver, B.J. Miller, L.M. Jalovec, T.G. Frazier, R.D. Noyes, A. Robidoux, H.M.C. Scarth, D.M. Mammolito, D.R. Mccready, P. Eleftherios, Technical outcomes of sentinel-lymph-node resection and conventional axillary-lymph-node dissection in patients with clinically node-negative breast cancer: results from the NSABP B-32 randomised phase III trial. Lancet Oncol. **8**(10), 881–888 (2007)
8. D.M. Godinho, J.M. Felício, C.A. Fernandes, R.C. Conceição, Experimental evaluation of an axillary microwave imaging system to aid breast cancer staging. IEEE J. Electromagnet., RF, Microwaves Med. Biol. **6**(1), 68–76 (2022)
9. D.M. Godinho, J.M. Felício, C.A. Fernandes, R.C. Conceição, Optimisation of artefact removal algorithm for microwave imaging of the axillary region using experimental prototype signals, in *15th European Conference on Antennas and Propagation (EuCAP)*. (Düssoldorf, Germany, 2021)
10. M. Savazzi, J.R. Costa, C.A. Fernandes, J.M. Felício, R.C. Conceição, Numerical assessment of microwave imaging for axillary lymph nodes screening using anthropomorphic phantom, in *15th European Conference on Antennas and Propagation (EuCAP)*. (Düssoldorf, Germany, 2021)
11. M. Savazzi, O. Karadima, J.M. Felício, C.A. Fernandes, P. Kosmas, R.C. Conceição, Effect of varying prior information in axillary 2D microwave tomography, in *16th European Conference on Antennas and Propagation (EUCAP)*. (Madrid, Spain, 2022)
12. W.T. Joines, Y. Zhang, C. Li, R.L. Jirtle, The measured electrical properties of normal and malignant human tissues from 50 to 900 MHz. Med. Phys. **21**(4), 547–550 (1994)

13. J.W. Choi, J. Cho, Y. Lee, J. Yim, B. Kang, K.O. Ki, H.J. Woo, J.K. Hee, C. Cheon, H.D. Lee, Y. Kwon, Microwave detection of metastasized breast cancer cells in the lymph node; potential application for sentinel lymphadenectomy. Breast Cancer Res. Treat. **86**(2), 107–115 (2004)
14. T.R. Cameron, M. Okoniewski, E.C. Fear, "A preliminary study of the electrical properties of healthy and diseased lymph nodes, in *International Symposium on Antenna Technology and Applied Electromagnetics & the American Electromagnetics Conference (ANTEM-AMEREM)*. (Canada, Ottawa, 2010), pp. 1–3
15. X. Yu, Y. Sun, K. Cai, H. Yu, D. Zhou, D. Lu, S.X. Xin, Dielectric properties of normal and metastatic lymph nodes ex vivo from lung cancer surgeries. Bioelectromagnetics **41**(2), 148–155 (2020)
16. D. Lu, H. Yu, Z. Wang, Z. Chen, J. Fan, X. Liu, J. Zhai, H. Wu, X. Yu, K. Cai, Classification of metastatic and non-metastatic thoracic lymph nodes in lung cancer patients based on dielectric properties using adaptive probabilistic neural networks. Front. Oncol. **11**, 376 (2021)
17. D. Lu, J. Peng, Z. Wang, Y. Sun, J. Zhai, Z. Wang, Z. Chen, Y. Matsumoto, L. Wang, S.X. Xin, K. Cai, Dielectric property measurements for the rapid differentiation of thoracic lymph nodes using XGBoost in patients with non-small cell lung cancer: a self-control clinical trial. Transl. Lung Cancer Res. **11**(3), 342 (2022)
18. S. Gabriel, R. Lau, C. Gabriel, The dielectric properties of biological tissues: II. Measurements in the frequency range 10 Hz to 20 GHz. Phys. Med. Biol. **41**(11), 2251 (1996)
19. T. Karacolak, R. Cooper, E.S. Unlu, E. Topsakal, Dielectric properties of porcine skin tissue and in vivo testing of implantable antennas using pigs as model animals. IEEE Antennas Wirel. Propag. Lett. **11**, 1686–1689 (2012)
20. P. Stauffer, F. Rossetto, M. Prakash, D. Neuman, T. Lee, Phantom and animal tissues for modelling the electrical properties of human liver. Int. J. Hyperth. **19**(1), 89–101 (2003)
21. C. Gabriel, A. Peyman, Dielectric measurement: error analysis and assessment of uncertainty. Phys. Med. Biol. **51**(23), 6033 (2006)
22. Keysight Tecnologies, Inc., Keysight 85070E, Dielectric Probe Kit 200 MHz to 50 GHz, (2017). Accessed on 06/2020. [Online]. Available: https://www.keysight.com/us/en/assets/7018-01196/technical-overviews/5989-0222.pdf
23. A. La Gioia, E. Porter, I. Merunka, A. Shahzad, S. Salahuddin, M. Jones, M. O'Halloran, Open-ended coaxial probe technique for dielectric measurement of biological tissues: challenges and common practices. Diagnostics **8**(2), 40 (2018)
24. A. Peyman, C. Gabriel, E. Grant, Complex permittivity of sodium chloride solutions at microwave frequencies. Bioelectromagnetics: J. Bioelectromagnetics Soc. Soc. Phys. Regul. Biol. Med., Eur. Bioelectromagnetics Assoc. **28**(4), 264–274 (2007)
25. G. Maenhout, T. Markovic, I. Ocket, B. Nauwelaers, Effect of open-ended coaxial probe-to-tissue contact pressure on dielectric measurements. Sensors **20**(7), 2060 (2020)
26. E. Porter, A. La Gioia, S. Salahuddin, S. Decker, A. Shahzad, M.A. Elahi, M. O'Halloran, O. Beyan, Minimum information for dielectric measurements of biological tissues (MINDER): a framework for repeatable and reusable data. Int. J. RF Microwave Comput.-Aided Eng. **28**(3), e21201 (2018)
27. A. Campbell, D. Land, .: Dielectric properties of female human breast tissue measured in vitro at 3.2 GHz. Phys. Med. Biol. **37**(1), 193 (1992)
28. M. Lazebnik, L. McCartney, D. Popovic, C.B. Watkins, M.J. Lindstrom, J. Harter, S. Sewall, A. Magliocco, J.H. Booske, M. Okoniewski, S.C. Hagness, A large-scale study of the ultrawide-band microwave dielectric properties of normal breast tissue obtained from reduction surgeries. Phys. Med. Biol. **52**(10), 2637–3656 (2007)
29. MyWAVE COST Action: Open-access Repository of Dielectric and Thermal Properties of Biological Tissues, (2022). Available from 10/2022. [Online]. Available: https://www.um.edu.mt/projects/mywave/data-repository/
30. University Hospital Galway (Galway, Ireland). Accessed on 06/2020. [Online]. Available: https://saolta.ie/hospital/university-hospital-galway
31. A. La Gioia, M. O'Halloran, A. Elahi, E. Porter, Investigation of histology radius for dielectric characterisation of heterogeneous materials. IEEE Trans. Dielectr. Electr. Insul. **25**(3), 1064–1079 (2018)

32. P.M. Meaney, A.P. Gregory, J. Seppälä, T. Lahtinen, Open-ended coaxial dielectric probe effective penetration depth determination. IEEE Trans. Microwave Theory Tech. **64**(3), 915–923 (2016)
33. P. Hasgall, F. Di Gennaro, C. Baumgartner, E. Neufeld, B. Lloyd, M. Gosselin, D. Payne, A. Klingenböck, N. Kuster, *IT'IS Database for Thermal and Electromagnetic Parameters of Biological Tissues*. itis.swiss/database, version 4.0 (May 2018)
34. W.D. Hurt, Multiterm Debye dispersion relations for permittivity of muscle. IEEE Trans. Biomed. Eng. **1**, 60–64 (1985)
35. S. Gabriel, R.W. Lau, C. Gabriel, The dielectric properties of biological tissues: III parametric models for the dielectric spectrum of tissues. Phys. Med. Biol. **41**(11), 2271–2293 (1996)
36. W. Pawlina, M.H. Ross, *Histology: A Text and Atlas: with Correlated Cell and Molecular Biology*. (Lippincott Williams & Wilkins, 2018)
37. E. Zastrow, S.K. Davis, M. Lazebnik, F. Kelcz, B.D.V. Veen, S.C. Hagness, Development of anatomically realistic numerical breast phantoms with accurate dielectric properties for modeling microwave interactions with the human breast. IEEE Trans. Biomed. Eng. **55**(12), 2792–2800 (2008)
38. A.H. Tunçay, I. Akduman, Realistic microwave breast models through T1-weighted 3-D MRI data. IEEE Trans. Biomed. Eng. **62**(2), 688–698 (2015)
39. N. Mori, K. Tsuchiya, D. Sheth, S. Mugikura, K. Takase, U. Katscher, H. Abe, Diagnostic value of electric properties tomography (EPT) for differentiating benign from malignant breast lesions: comparison with standard dynamic contrast-enhanced MRI. Eur. Radiol. **29**(4), 1778–1786 (2019)
40. H.Q. Woodard, D.R. White, The composition of body tissues. Brit. J. Radiol. **59**(708), 1209–1219 (1986)
41. L.W. Thomas, The chemical composition of adipose tissue of man and mice. Q. J. Exp. Physiol. Cognate Med. Sci. **47**, 179–188 (1962)
42. D.M. Godinho, J.M. Felício, T. Castela, N.A. Silva, M.D.L. Orvalho, C.A. Fernandes, R.C. Conceição, Development of MRI-based axillary numerical models and estimation of axillary lymph node dielectric properties for microwave imaging. Med. Phys. **48**(10), 5974–5990 (2021)
43. Z. Yaniv, B.C. Lowekamp, H.J. Johnson, R. Beare, Simple ITK image-analysis notebooks: a collaborative environment for education and reproducible research. J. Digit. Imaging **31**(3), 290–303 (2018)
44. N.J. Tustison, J.C. Gee, N4ITK: Nick's N3 ITK implementation for MRI bias field correction. Insight J. **9**, 1–8 (2009)
45. N. Otsu, A threshold selection method from gray-level histograms. IEEE Trans. Syst., Man, Cybern. **9**(1), 62–66 (1979)
46. Instituto Nacional de Investigação Agrária e Veterinária (INIAV), Lisbon, Portugal. Accessed on 10/2020. [Online]. Available: http://www.iniav.pt/
47. N. Bengtsson, J. Melin, K. Remi, S. Säoderlind, Measurements of the dielectric properties of frozen and defrosted meat and fish in the frequency range 10–200 MHz. J. Sci. Food Agric. **14**(8), 592–604 (1963)
48. B. Bodakian, F. Hart, The dielectric properties of meat. IEEE Trans. Dielectr. Electr. Insul. **1**(2), 181–187 (1994)
49. S. Ley, S. Schilling, O. Fiser, J. Vrba, J. Sachs, M. Helbig, Ultra-wideband temperature dependent dielectric spectroscopy of porcine tissue and blood in the microwave frequency range. Sensors **19**(7), 1707 (2019)
50. M. Lazebnik, M.C. Converse, J.H. Booske, S.C. Hagness, Ultrawideband temperature-dependent dielectric properties of animal liver tissue in the microwave frequency range. Phys. Med. Biol. **51**(7), 1941 (2006)
51. J.S. Bobowski, T. Johnson, *Permittivity Measurements of Biological Samples by an Open-Ended Coaxial Line*, *arXiv preprint* arXiv:2007.07343 (2012)
52. T.W. Beck, The importance of a priori sample size estimation in strength and conditioning research. J. Strength Conditioning Res. **27**(8), 2323–2337 (2013)
53. J. Cohen, Statistical Power Analysis for the Behavioral Sciences. (Academic Press, 2013)

54. Heinrich-Heine-Universität Düsseldorf, G*Power 3.1 Manual, Chapter 22. Accessed on 04/2021. [Online]. Available: https://www.psychologie.hhu.de/fileadmin/redaktion/Fakultaeten/Mathematisch-Naturwissenschaftliche_Fakultaet/Psychologie/AAP/gpower/GPowerManual.pdf
55. H. Fallahi, J. Sebek, P. Prakash, Broadband dielectric properties of ex vivo bovine liver tissue characterized at ablative temperatures. IEEE Trans. Biomed. Eng. **68**(1), 90–98 (2020)
56. P. Basaran, N. Basaran-Akgul, B. Rasco, Dielectric properties of chicken and fish muscle treated with microbial transglutaminase. Food Chem. **120**(2), 361–370 (2010)

SAFE—Microwave Imaging Device for Breast Cancer Early Screening and Diagnostics

Aleksandar Janjic, Ibrahim Akduman, Mehmet Cayoren, Onur Bugdayci, and Mustafa Erkin Aribal

Abstract Microwave Imaging has emerged as a promising technique that can contribute to the breast lesion detection and classification. Using harmless electromagnetic waves, MWI is providing relevant diagnostic information without resorting to X-rays. Due to its harmless nature, frequent scanning is accessible for women of all age. In this chapter, we discuss about the clinical implementation of our novel MWI device, namely SAFE, which utilizes the variance in the electromagnetic properties of healthy and cancer affected breast tissue in order to provide information regarding women breast tissue health. As the device is not requiring physical compression, it can also detect the lesions that are near to the thoracic wall. SAFE detection and localization outcomes were evaluated based on the clinical reports provided by radiologists. A clinical study was conducted on 59 patients, including 32 with benign and 27 with malignant pathology outcomes. By using machine learning (ML) model,

A. Janjic (✉) · I. Akduman · M. Cayoren · M. E. Aribal
Mitos Medical Technologies, ITU Ayazaga Ari Teknokent 2-B Block 2-2-E, 34469 Maslak, Istanbul, Turkey
e-mail: aleksandarjanjic@mitosmedikal.com

I. Akduman
e-mail: akduman@itu.edu.tr

M. Cayoren
e-mail: cayoren@itu.edu.tr

M. E. Aribal
e-mail: erkin.aribal@acibadem.com

A. Janjic · I. Akduman · M. Cayoren
Faculty of Electrical and Electronics Engineering, Istanbul Technical University, 34469 Maslak, Istanbul, Turkey

O. Bugdayci
Department of Radiology, School of Medicine, Marmara University, 34899 Pendik, Istanbul, Turkey
e-mail: onur.bugdayci@marmara.edu.tr

M. E. Aribal
Department of Radiology, Breast Health Center, Altunizade Hospital, Acibadem M.A.A. University, 34684 Atasehir, Istanbul, Turkey

F. Vipiana and L. Crocco (eds.), *Electromagnetic Imaging for a Novel Generation of Medical Devices*, Lecture Notes in Bioengineering,
https://doi.org/10.1007/978-3-031-28666-7_9

more specifically Stochastic Gradient Descent (SGD), SAFE correctly detected 81% of the lesions present in the breasts tissue, from which 83% of them were localized correctly. Results indicate that SAFE is capable of detecting and localizing high percentage of lesions present in the dataset, regardless of breast density, lesion size or participants age. In this chapter, we elaborate on our approach and achieved results and discuss SAFE ongoing and intended future work.

Keywords Microwave imaging · Breast lesion detection · Breast lesion localization · Machine learning · SGD

1 Introduction

Breast cancer is one of the most common cancers in the female population and accounts for approximately 25% of total diagnosed cases in women worldwide, with around 2.3 million new cases per year [1–5]. The American Cancer Society reports a 5-year survival rate for localized and regional stages of breast cancer to be 99% and 86%, respectively, contrary to the advanced stage with the rate of 28%, emphasizing the correlation between early breast cancer detection and patient survival.

Mammography (MMG) is the primary modality for breast cancer screening and diagnostics but is limited by several risk factors such as false-positive examinations, ionizing radiation exposure and false negativity in dense breasts [6–12]. Ionizing radiation also limits the possibility of frequent or repeated examinations. In addition to MMG, ultrasonography (US) or magnetic resonance imaging (MRI) can be used for breast cancer imaging. US is operator-dependent and therefore susceptible to errors. Furthermore, US can result in an increased number of unnecessary biopsies, thus increasing the hospital costs [13–15]. MRI, although increasingly used for screening of high-risk patients, is not considered practical for population-wide screening, due to its costs and special facility requirements [16–18].

Many efforts have been conducted to provide a breast imaging device that will not resort to harmful ionizing radiation, will perform better in dense breasts compared to MMG, and where risk and cost–benefit would justify its clinical implementation. Additionally, a device allowing painless, fast, and frequent examinations would significantly increase the number of breast check-up tests, especially in the younger population, consequently increasing the number of detected early-stage cancers, making a positive impact on the patient's long-term survival rate. In this context, microwave imaging (MWI) has emerged as a promising technique with the protentional to contribute to breast cancer early screening and diagnostics.

MWI represents an imaging modality where the unknown features of the object-of-interest (OI), such as shape, location, or dielectric properties (dielectric permittivity and conductivity), are reconstructed based on the measured field OI scatters. The technology itself utilizes the difference in dielectric properties of healthy and cancerous breast tissue to detect the presence of anomalies inside the breast. It is proven that the cancerous tissue is characterized by higher dielectric permittivity compared to the healthy one [19–27]. One of the most appealing features of the MWI is the non-ionizing nature of the electromagnetic waves used to penetrate the breast tissue. The non-ionizing radiation used allows safe and more frequent examinations of all women, regardless of their age or condition (pregnancy, specific illness, etc.). This can have a major impact on the younger population, especially on the women with hereditary genetic mutations, who are at a considerable risk of developing breast cancer and where there is a necessity for examinations from an early age.

However, MWI implementation is not straightforward due to the non-linear and ill-posed nature of the electromagnetic problem at hand. To cope with the non-linearity and ill-posedness, and find a stable solution of the problem, various inverse imaging techniques were, and are still being developed. These techniques can be of qualitative or quantitative nature. Qualitative methods intend to evaluate an indicator function holding the information about the shape and location of the OI (cancerous tissue). On the other hand, quantitative methods intend to retrieve the dielectric properties of OI. Within the scope of this chapter, only qualitative imaging methods will be considered, as the main goal of the work presented is to identify (localize) the existence of the anomaly, not analyze its properties. To do the reconstruction based on the MWI projection data acquired at many different angles around the patient breast, qualitative algorithms such as Linear Sampling Method (LSM) [28–35] and Factorization Method (FM) [28–30, 36, 37], were implemented. The reason why qualitative methods are convenient for real-life applications is they do not entail any approximation, and they can produce robust reconstructions with reduced computational time.

Up to this point, a variety of MWI clinical evaluations have been demonstrated in the literature [38–60]. A comprehensive overview was given by [61–63]. A considerable variance in design, patient interface, and reconstruction algorithms exist between the aforementioned clinically tested prototypes. The number of patients included in the study varies from 2, reported in [56] up to the 200 reported by the University of Bristol and Micrima, UK [46]. Besides Micrima's MARIA studies, studies with a significant number of patients were reported by Dartmouth College, USA, which studies included 150 patients, Microwave Vision SA, France, with the WAVELIA studies of 51 respondents, Umbria Bioengineering Technologies, Italy, which MAMMOWAVE studies involved 58 patients and Mitos Medical Technologies, Turkey, with the SAFE studies that counted 115 participants. In all of the studies reported, patients were required to lie prone on the table, making the examinations more comfortable for the women involved. The systems and clinical results of the aforementioned studies have not been comparatively discussed in the literature, in particular: the complexity of the hardware, the effect of the hardware and implemented algorithms on the scanning time and image reconstruction, influence of the hardware design on clinical use, and the clinical findings of the studies correlated with the participant's age, breast density scores, breast sizes, or pathological diagnosis outcomes.

In this work, as a part of the SAFE (Scan and Find Early) clinical study, we implement machine learning (ML) approach which helps us to detect the presence of anomaly inside patient's breast, and we analyze the applicability of the aforementioned qualitative imaging algorithms by reconstructing the MWI images used for present anomaly localization. SAFE is the microwave tomography device developed by the joint work of Mitos Medikal Technologies A.S. and the Medical Device Research, Development, and Application Laboratory of Istanbul Technical University. The device itself is intended for breast cancer early screening and diagnostics. The SAFE patent is registered in Turkey and USA, while the patent application for the EU market is currently pending approval. The device uses non-ionizing radiation and does not require breast compression, providing a painless and harmless examination procedure. The appealing feature of the SAFE is that it may be employed in screening rather than in differential diagnostics. This suggests that the ability of the SAFE to detect malignant lesions as a screening methodology may add worth in triaging true negative patients, thus reducing the amount of unnecessary mammographic examinations.

2 Materials and Methods

2.1 Device Description

SAFE employs low power microwave signals in the range of 1–8 GHz. This range was chosen to acquire the suitable balance between the absorption rate and resolution Device contains one transmitting (TX) antenna used to illuminate the breast with electromagnetic waves, and one receiving (RX) antenna used to measure scattered electromagnetic fields from different angular positions around the breast. Antennas are connected to the 2-port VNA (Keysight 2-port usb vector network analyzer). For each TX position, there is a N number of RX positions. At each measurement point, S_{21} parameter, which corresponds to the electromagnetic field emerged from the interaction between the illuminated wave and breast, is recorded. Total number of measurements per scan is 1296. The S_{21} parameters are recorded in the frequency range of 1–8 GHz, with the 200 MHz step.

The examination time is around 15 min for both breasts. Due to that time, patient is required to lie prone on the table with one breast inserted in the breast cup which is a part of coupling cylinder (Fig. 1a). Coupling medium (Zirconia: $\epsilon = 8, \sigma = 0.12$) is implemented for impedance matching. TX and RX are immersed in the coupling medium blocks, and are circling around coupling cylinder. Since breast size differs between patients, size adjustable cups are available. Prior to each scan, technician performing the examination decide which cup suits patient the best. Compression is not applied on the breast during examination.

Fig. 1 SAFE imaging system in clinics: **a** scanning configuration, **b** clinical design

2.2 Data Acquisition

SAFE clinical study was approved by the Ethics committee of the Marmara University School of Medicine and was done in collaboration with the Radiology Department of the same institution. Patient participation was voluntary, and written informed consent was required prior to participation. All protocols and procedures were in accordance with both institutional and national ethical standards in research and with the World Medical Association Declaration of Helsinki. Only the patients intended for biopsy after routine imaging were considered for the study. SAFE scanning procedure was done prior to the biopsy, and was performed by the medical staff of Mitos Medical Technologies.

Information about patient's breast health, as well as the location of the anomaly present, was acquired after the radiologists reviewed conventional exams for each patient involved in the study. In this context, conventional exams used on site were MMG (Mammomat Inspiration, Siemens Healthcare, Erlangen, Germany), US (Toshiba Aplio 400, Canon Medical Systems Corporation, Tochigi, Japan), and MRI (3T scanner, Magnetom Verio, Siemens Healthcare, Erlangen, Germany). Patient's breast were classified in two groups: healthy (no anomalies were detected inside the breast tissue), and affected (presence of anomaly detected with the conventional imaging). Localizaton of the anomalies present were done in accordance with clock positions, quadrants and ICD-O codes of the breast (Fig. 2).

Anomalies inside the breast were classified as benign or malignant according to the histopathological results. Breast density was provided based on the BI-RADS density classification, where the breasts were categorized in two groups: non-dense (type A and B) and dense (type C and D) [64]. According to the age, participants were divided in younger group, including the patients of age 18–39, and older group, with the patients above 40 years of age. Lesions present in the breast tissue were divided in three groups according to their size: T_1—size of the lesion is $\leq$ 20 mm, T_2—lesion size 20–50 mm, and T_3—lesions with the size greater than 50 mm. Patients undergoing bilateral biopsies were not included in the study.

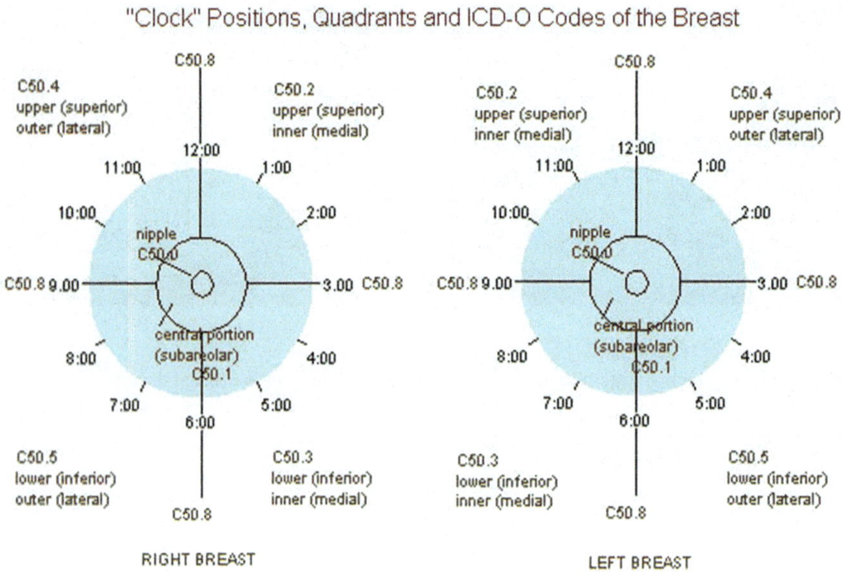

Fig. 2 Clock positions, quadrants and ICD-O codes of the breast used to compare conventional imaging and MWI outcomes

2.3 Data Evaluation

Data collected was evaluated by the joint work of the engineering department of Mitos Medical Technologies and radiological department of Marmara University School of Medicine. Two step evaluation was performed. Firstly, we used S_{21} parameters (raw data) to decide which breast is affected, and afterwards we use inverse scattering algorithms to create the image of the affected breast and provide information about the location of the affected area.

2.3.1 Detection

Proposed approach is only considering the real-part of the S_{21} complex matrix to decide which of the patient's breast is affected and which one is healthy. Machine learning (ML) method, namely Stochastic Gradient Descent (SGD) was used to distinguish backscattered signals of healthy breast from the backscattered signals of affected ones. Information regarding the breast health (healthy or affected tissue) collected through conventional imaging were used to train the SGD model. Probability threshold of 50% was used for predictions. More specifically, if we consider the affected cases as ones and healthy as zeros, if the probability that the case is one is higher than 50%, SGD model will label the breast as affected one. If the probability is lower than 50%, the case will be labeled as healthy one.

Due to the limited dataset, stratified fivefold cross validation was used to test the performance of SGD model. Stratified fivefold cross validation considers splitting the dataset in 5 subsets, where training is performed on the n-1 number of sets ($n = 5$), while the remaining subsets (in this case 1) are used for testing. As the number of folds is 5, we iterate 5 times with the different subset reserved to test the model. Additionally, two-sample t-test was performed to evaluate if the real S_{21} are suitable for detecting breast anomalies. Test null hypothesis (H_0) assumes that the two sets of real S_{21} (healthy and affected) have equal means. This means that the two sets of S_{21} can be employed for the detection if t-test rejects H_0 and accept alternative hypothesis (H_α). The H_α assumes that two sets of real S_{21} have unequal means (existence of significant statistical difference between the sets). Significance level chosen for rejecting the H_0 was $p < 0.05$. The outcome of the test, where $p = 2.9 \times 10^{-5}$, reject the H_0 and accepts H_α, hence indicates that the two sets of real S_{21} have unequal means and are suitable for detecting breast anomalies.

2.3.2 Localization

Non-blinded evaluation intended to confirm the approximate location of the lesion of interest was performed. A-priori information about the lesion location was provided by conventional imaging. Clock positions, quadrants and ICD-O codes of the breast were used for location labeling. LSM and FM were used to reconstruct MWI images. Correctly localized affected area was considered as an area in which, by the clock positioning standard, location of the lesion in both (FM and LSM) reconstructed MWI images is in concordance with the location reported by conventional imaging. If the location of MWI and the one reported by conventional imaging are not in concordance, this was considered as negative outcome. Small tolerances regarding to lesion location were accepted, as the positioning of the patient (breast twisting inside the cup) can produce small mismatch in the location reported by conventional imaging and MWI. In the imaging phase, differential MWI approach was used [65]. The approach requires reference measurement to suppress the background. In this specific case, we use one of the breast measurements as the reference in order to obtain the image and evaluate if location of a present lesion matches the one reported by conventional imaging.

Acquired images are pseudo-color images, where the colormap intensities are in correlation with the dielectric properties of the breast tissues. Higher permittivity corelates with higher colormap intensity. Threshold filtering was performed on the image in order to eliminate the noise. Each image was created using MATLAB graphical user interface, with the processing time of around 1–2 s per image.

Qualitative Inverse Scattering Algorithms

Qualitative inverse scattering methods intend to reconstruct the shape and provide information about the location of inaccessible targets based on the field these objects

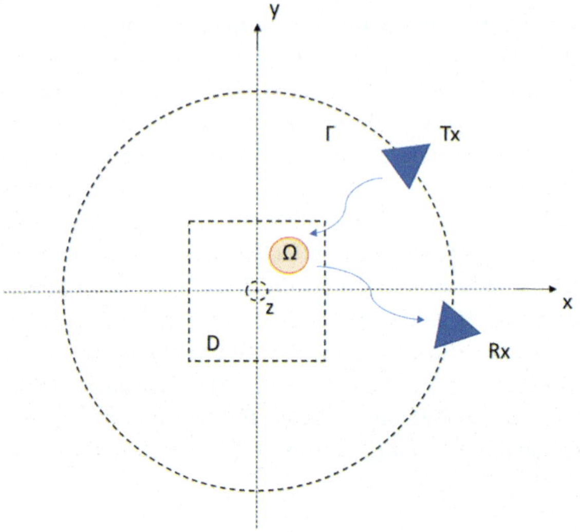

Fig. 3 Configuration of inverse problem: D—domain embedding the scatterer Ω, Γ—excitation-measurement curve, Tx\Rx—transmitter\receiver

scatters. Two closely related methods of such a formulation are LSM and FM. In this subsection, we give a brief explanation of these methods. Formulation is based on the work [66].

Let us consider the 2D inverse scattering problem (Fig. 3). Let D be the domain which embeds the scatterer Ω. The scatterer Ω is probed by the incident field radiated by transmitting antenna T_x located on the closed curve Γ having the radius R. Interaction between incident field and scatterer give raise to scattered electric field E^{sct}, which is measured by the receiving antenna R_x located on Γ. The scattered field can be expressed in the form:

$$E^{sct}(\underline{R}, \Theta) = k_b^2 \int_D G(\underline{R}, \underline{r}')\chi(\underline{r}')E^{tot}(\underline{r}', \Theta)d\underline{r}' \quad \underline{R} \in \Gamma \tag{1}$$

$$E^{tot}(\underline{r}', \Theta) - E^{inc}(\underline{r}', \Theta) = k_b^2 \int_D G(\underline{r}, \underline{r}')\chi(\underline{r}')E^{tot}(\underline{r}', \Theta)d\underline{r}' \quad \underline{r} \in D \tag{2}$$

where we consider Θ as generic incident direction. $\underline{R} = (R, \phi)$, ϕ being a generic observation direction, $\underline{r} = (x, y)$, k_b is the wave-number of surrounding medium, E^{inc}, E^{tot}, are incident and total field, respectfully, $\chi(\underline{r}') = k_b^2(\underline{r}) - k_0^2(\underline{r})$ represents contrast function, and $G(\underline{r}, \underline{r}')$ is the scalar Green's function given by:

$$G(\underline{r}, \underline{r}') = -\frac{j}{4}H_0^2(k_0|\underline{r} - \underline{r}'|) \tag{3}$$

where H_0^2 being a zero order second kind Hankel's function.

The inverse scattering problem is then expressed as retrieving the contrast function $\chi\left(\underline{r} \in D\right)$ from the measured scattered field $E^{sct}\left(\underline{R} \in \Gamma\right)$ for the known incident field E^{inc}.

Linear Sampling Method

LSM provides estimation of the target's shape and location by solving an auxiliary linear ill-posed inverse problem instead of the non-linear problem presented in (1) and (2) [67]. The linear inverse problem can be expressed as:

$$\mathcal{F}[\xi] = \int_{\Gamma} E_s(R, \Theta)\xi(r_s, \Theta)d\Theta = G(R, r_s) \tag{4}$$

where ξ is the unknown to be determined, $r_s \in D$ is a point in the grid which samples the domain D, and $\mathcal{F}$ is the far-field operator [67]. Due to the ill-posed nature of (4), regularization is necessary to acquire the stable solution. Exploiting singular value decomposition (SVD) [68], and Tikhonov regularization [68], solution can be obtained in the form:

$$\xi\left(\underline{r_s}, \Theta\right) = \sum_{n=1}^{\infty} \frac{\lambda_n^2}{\lambda_n^2 + \alpha^2} \left\langle G\left(\underline{R}, \underline{r_s}\right), u_n\left(\underline{R}\right)\right\rangle_{\Gamma} v_n(\Theta) \tag{5}$$

where α is Tikhonov regularization parameter, λ_n stands for singular values of $\mathcal{F}$, u_n and v_n are left and right singular vectors of $\mathcal{F}$, and $\langle \cdot, \cdot \rangle$ represents scalar product on Γ. As the L^2 norm of regularized solution ξ assumes low values when $\underline{r_s}$ belongs to targets and large elsewhere, estimation of the target is achieved by plotting the LSM indicator over the domain of interest D and associating the sampling points with low indicator to unknown objects.

Explicit expression of the indicator can be achieved from (5) in the form:

$$\Upsilon\left(\underline{r_s}\right) = \int_{\Gamma} \left|\xi\left(\underline{r_s}, \Theta\right)\right|^2 d\Theta = \sum_{n=1}^{\infty} \frac{\lambda_n^2}{\lambda_n^2 + \alpha^2} \left\| G\left(\underline{R}, \underline{r_s}\right), u_n\left(\underline{R}\right)\right\|_{\Gamma}^2 \tag{6}$$

where the computation of $\mathcal{F}$ is done only ones, as the kernel of linear system is same for every sampling point. For further discussion on the theoretical background of LSM we refer reader to [29].

Factorization Method

Similarly, to LSM, FM requires to sample the domain of interest D into an arbitrary grid of points and at each sampling point r_s solve linear equation:

$$\left(\mathcal{F}^{*}\mathcal{F}\right)^{1\backslash 4} g = \tilde{G} \tag{7}$$

where $\mathcal{F}$ is a far-field operator and $\tilde{G}$ stands for Green's function defined in (3). The Eq. (7) has a finite solution only if the sampling point $r_s \in D$ coincides with the object Ω [69]. Indicator function can then be expressed as:

$$I(r_s) := \left[\sum_{m=1}^{M} \frac{\left| \left\langle \tilde{G}(\cdot, r_s), \psi_m(\cdot) \right\rangle \right|^2}{|\lambda_m|} \right]^{-1} > 0 \tag{8}$$

Here, $\lambda = \{\lambda_1, \lambda_2, \ldots, \lambda_N\}$ and $\psi = \{\psi_1, \psi_2, \ldots, \psi_N\}$ are the sets of eigenvalues and their corresponding eigenvectors of the far-field operator $\mathcal{F}$. Additional theoretical background of FM can be found in [69].

3 Results

Fifty-nine patients were included in the study. Patient age ranged from 17 to 71 years (± 13, mean age: 48 years). Dataset included 32 benign and 27 malignant lesions. Histopathological results of all cases involved is presented in Table 1. Twenty-five patients had non-dense breasts, 14 patients had dense breasts, while for 20 patients, density was unknown, as US imaging was only available. Lesion size ranged between 6 and 120 mm with the average mass size of 26 mm (± 20 mm, median lesion size: 20 mm).

Table 1 Histopathology of the patients included in the study

Histopathology	Number of cases
Intraductal papilloma	4
Fibroepithelial lesion	3
Adenosis	10
Columnar cell change	2
Fibroadenoma	5
Lymphoma	1
Invasive ductal carcinoma	19
Acute and chronic inflammatory changes	4
Invasive lobular carcinoma	2
Fat necrosis	1
Fibrosis	3
Ductal carcinoma in situ	3
Sclerosing adenosis	1
Pseudo-angiomatous hyperplasia	1

3.1 Detection

Considering the whole dataset, SAFE was able to distinguish between healthy and affected breasts in 47/59 patients achieving an accuracy of 81%. Sensitivity achieved was 81%, while specificity was 83% (Fig. 4).

Device accuracy at 85% was higher when malignant lesions were involved 78%. Sensitivity was considerable higher in malignant group compared to the one achieved for benign group (85% vs. 75%). Specificity at 85% was slightly higher in malignant group compared to the specificity at 81% achieved for benign group (Fig. 5).

Density dataset showed that SAFE was better in distinguishing between healthy and affected breasts when dense breasts were involved, achieving the accuracy of 85%, contrary to the accuracy of 79% achieved for the non-dense breasts. Sensitivity of 79% and specificity of 90% were also higher in dense breasts, compared to the sensitivity of 73% and specificity of 83% acquired in non-dense breasts (Fig. 6).

Considering the patient's age, SAFE achieved same accuracy, sensitivity and specificity of 77% when younger patient's data was analyzed. When older patients were considered, device achieved accuracy of 83%, while sensitivity and specificity were 80% and 85%, respectively (Fig. 7).

The effect of the lesion size on detection capabilities of SAFE is presented in Fig. 8. Number of lesions per group was: $T_1 = 30$, $T_2 = 23$, $T_3 = 6$. Due to the small dataset, the T_3 group was not analyzed. Accuracy (86%) and specificity (90%) were higher in the T_2 group than the accuracy and specificity achieved for T_1 group, which were 80% and 77%, respectively. Contrary, sensitivity was slightly higher in T_1 group then in T_2 group (83% vs. 81%). Smallest lesion detected was 6 mm.

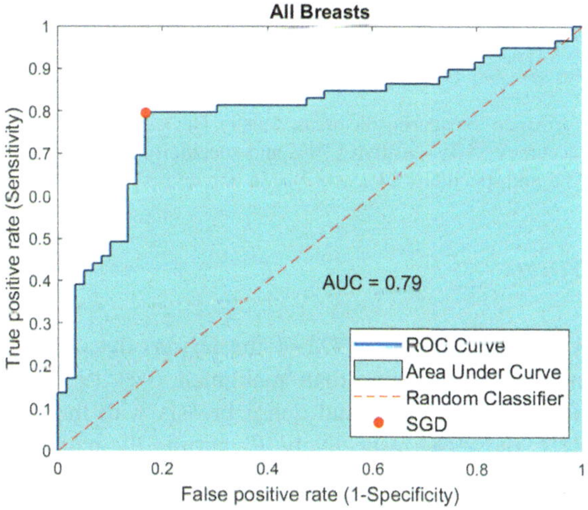

Fig. 4 Detection results considering all samples involved. SAFE achieved accuracy of 81% (47/59 correctly detected cases), with the sensitivity of 81% and specificity of 83%

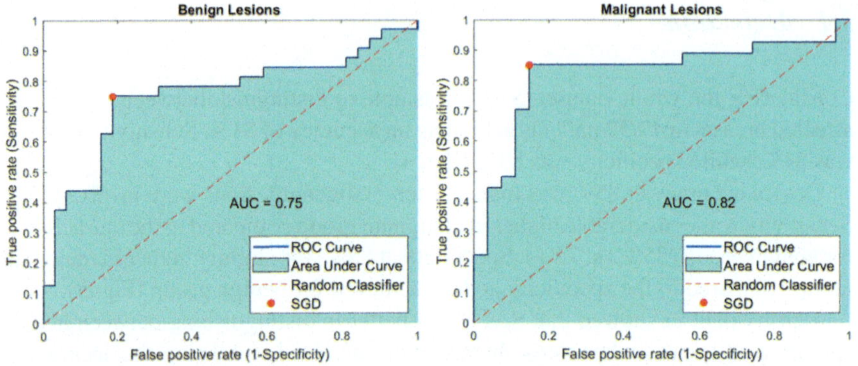

Fig. 5 Device performance based on the breast pathology. Device detected higher percentage of malignant lesions (85%) then benign ones (78%). Considering sensitivity and specificity, they were also higher in malignant group (85% and 85%, respectively), compared to the benign group (75% and 81%, respectively)

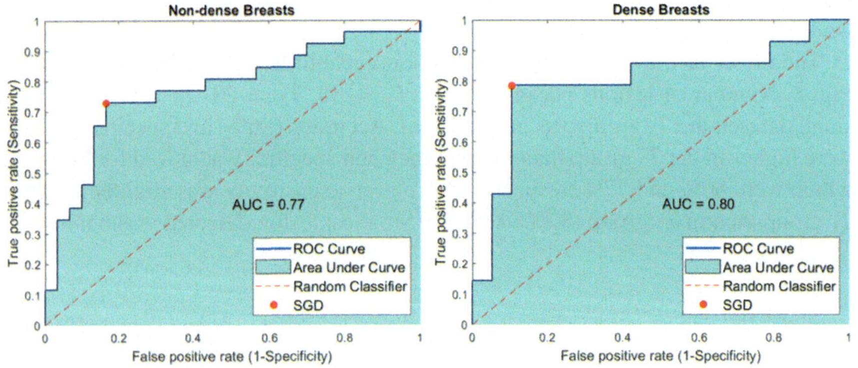

Fig. 6 SAFE performance considering the breast density. Device performed better in dense breasts achieving higher accuracy 85%, sensitivity 79% and specificity 90%, compared to the accuracy 79%, sensitivity 73% and specificity 83% achieved in non-dense breasts

3.2 *Localization*

SAFE correctly localized 83% (39/47) of the lesions detected. Device localized considerably more benign lesions than malignant ones (92% vs. 74%). SAFE performed similarly in non-dense and dense breasts with the correctly localized lesion rate of 82% and 79%, respectively. Regarding the patients age, the device performed better in the younger population, where it correctly localized 90% of the cases, whereas it correctly localized 84% of cases in the older population. Considering the T_1 lesion size group, the device correctly localized 84% of the cases, while in T_2 group this rate was 82%. As were the cases in the detection analysis, due to the limited dataset, the T_3 group was not analyzed. All results are presented in Table 2.

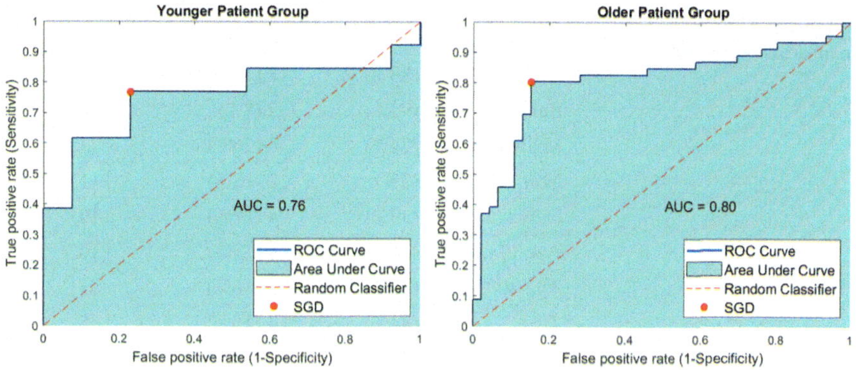

Fig. 7 SAFE performance based on participant's age. Better performance was noticed when older patients were analyzed with device accuracy of 83% compared to the 77% achieved for the younger participants group. Sensitivity 80% and specificity 85% were also higher when older patients were involved, compared to the sensitivity 77% and specificity 77% achieved for the younger participants group

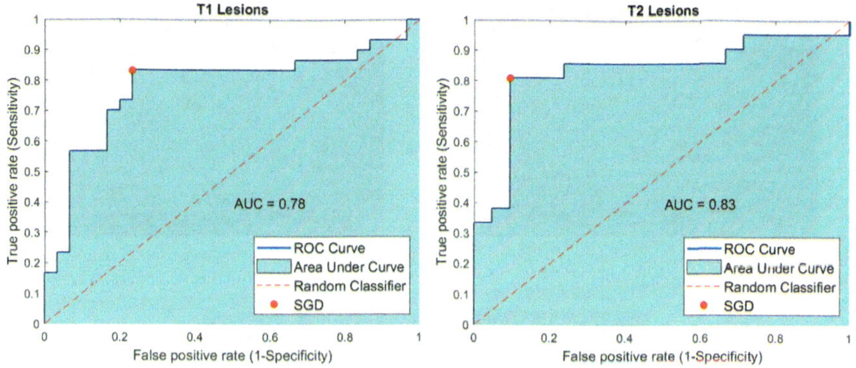

Fig. 8 Results considering lesion size effect on SAFE detection. As expected accuracy and specificity increased with the lesion size (T_1: accuracy 80%, specificity 77%; T_2: accuracy 86%, specificity 90%). Contrary to expectations, sensitivity decreased with the lesion size (T_1: sensitivity 83%; T_2: sensitivity 81%)

Examples of SAFE's correctly localized imaging outcomes, together with corresponding conventional imaging outcomes, can be seen in Figs. 9 and 10. Here we present two cases: one benign case with MRI showing the lesion at 1 o'clock position, and the other where US was used to detect the malignant lesion at 12 o'clock in patient's right breast. LSM and FM images were reconstructed for both cases (Figs. 9a, b and 10a, b) where the locations of high-contrast regions, indicating the presence of anomaly inside the breast, are concordant with the locations reported by conventional imaging (Figs. 9c and 10c). In the second case, a small mismatch between the location reported by MWI and conventional imaging is noticeable, but

Table 2 SAFE results for main areas of interest

Reads	Positive outcome	Negative outcome	Total cases
All	39 (83%)	8 (17%)	47
Benign	22 (92%)	2 (8%)	24
Malignant	17 (74%)	6 (26%)	23
Non-dense	9 (82%)	2 (18%)	11
Dense	15 (79%)	4 (21%)	19
Age 17–39	9 (87%)	1 (13%)	10
Age 40–71	31 (84%)	6 (16%)	37
T_1	21 (84%)	4 (16%)	25
T_2	14 (82%)	3 (18%)	17

the tolerance is acceptable as the mismatch can arise due to the breast twisting inside the cup.

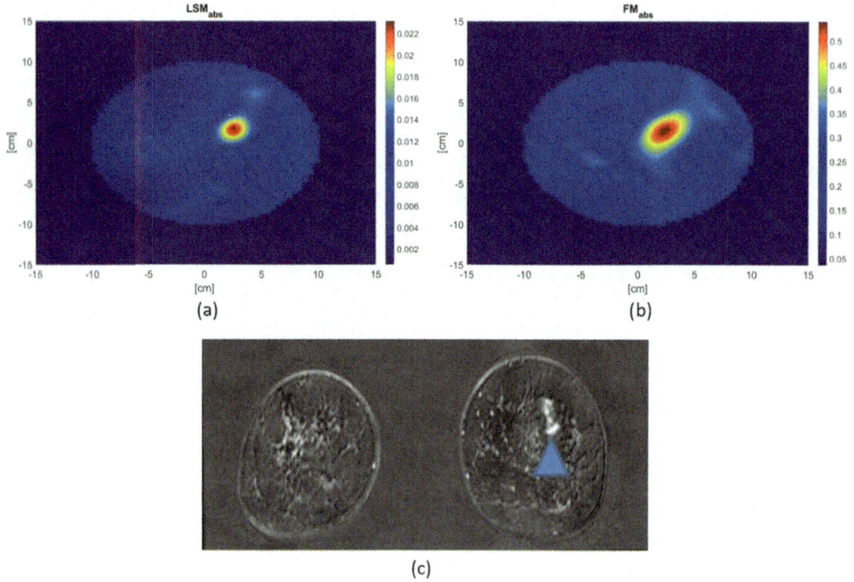

Fig. 9 Comparison of MWI outcomes with conventional imaging. **a**, **b** LSM and FM reconstructions showing the high-contrast region (present breast anomaly) at 1 o'clock, **c** MRI scan showing the presence of the anomaly at 1 o'clock inside patient left breast. MWI imaging outcome is concordant with MRI

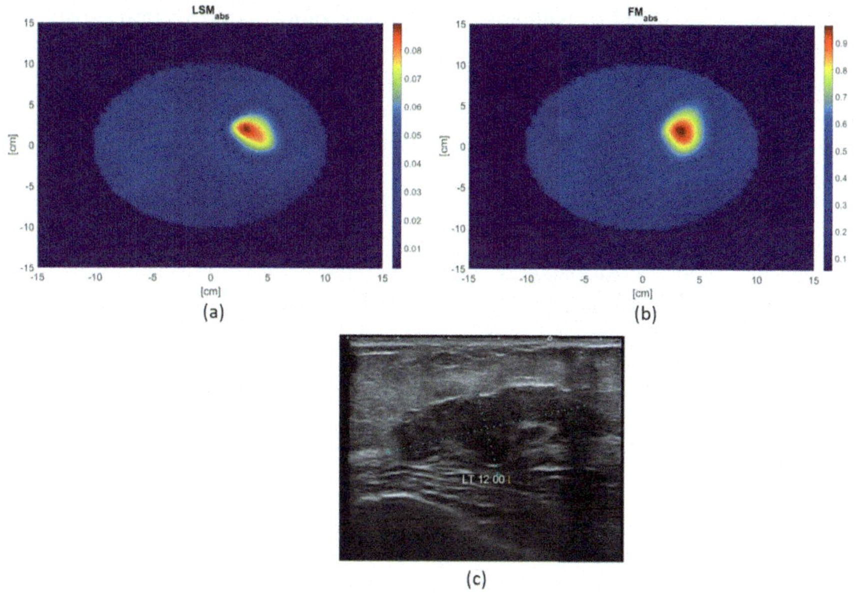

Fig. 10 Comparison of MWI outcomes with conventional imaging. **a**, **b** LSM and FM reconstructions showing the high-contrast region (present breast anomaly) between 12 and 2 o'clock, **c** US scan showing the presence of the anomaly at 12 o'clock inside patient right breast. MWI imaging outcome is concordant with MRI. Small mismatch between MWI and US reported location is acceptable as position can slightly vary due to the breast twisting inside the cup

4 Discussion

The MWI prototype for breast cancer early screening and diagnostics was developed and tested in clinical environment. The device uses harmless non-ionizing electromagnetic waves to illuminate the breast and does not require breast compression. The examination time of around 15 min is comparable with mammography examination.

These early results demonstrate that SAFE could be employed to distinguish between healthy and affected breasts with an accuracy of 81%, sensitivity of 81% and specificity of 83%. For this purpose, machine learning model, namely Stochastic Gradient Descent (SGD) was trained using the real parts of S_{21} complex matrix as the features to detect the presence of anomaly in one of the patient's breasts. This study also demonstrated that SAFE is capable of correctly localizing 83% of detected anomalies.

Pathology had an impact on the device performance, as accuracy achieved for benign and malignant group of patients differed slightly (78% vs. 85%). Furthermore, sensitivity of the device was considerable affected by the lesion pathology. Contrary to the higher values for malignant groups considering the lesion detection, the rate of correctly localized lesions at 74% was considerably lower in malignant lesions than the rate at 92% achieved in benign lesions. Further investigation will be necessary to understand the effect of pathology on the localization and detection capabilities of the SAFE.

Breast density is considered to be one of the most important factors affecting mammography, as its sensitivity can vary between 60% in dense breasts and 89% in non-dense ones [70, 71]. Favorable performance of SAFE in dense breasts, where the detection accuracy at 85%, sensitivity at 79% and specificity at 90%, is encouraging. For this group of patients, the device correctly localized 79% of detected cases. SAFE accuracy at 79%, sensitivity at 73% and specificity at 83% in non-dense breasts suggest that SAFE performance is slightly affected by the breast density. This can also be concluded based on the rate of correctly localized lesions which was 82%.

Favorable performance of SAFE in the younger population (< 40 years of age), where the accuracy of the test was 77%, gives it an enviable advantage over MMG which is not recommended for screening in this group of patients due to its harmful effects of ionizing radiation. This opens up the possibility to provide harmless, fast and cost-effective solution for younger women, especially the ones with hereditary genetic mutations, where there is a need for screening from an early age. The device successfully localized 90% of detected lesions in the younger group. SAFE performed better in older patient's group (≥ 40 years of age) where the accuracy of 83%, sensitivity of 80% and specificity of 85% were achieved, and where device correctly localized 84% of detected cases. These results suggest that device is also not considerable affected by the patient's age.

In this study, SAFE successfully detected and localized the lesion as small as 6 mm. This suggests that SAFE is capable of detecting the lesion in its early stage, which can have a major impact on planning further treatments. As expected, accuracy increased with the lesion size, where device accuracy for T_1 group was 80% compared to the 86% achieved for the T_2 group. Contrary to expectations, T_1 group had higher sensitivity than T_2 group, which will be further investigated.

The study has its own limitations. The dataset is considerable low to produce any statistically significant conclusions. The participants included in the study were only the ones intended for biopsy. This means that a-priori information about the patient health is already available and device performance in screening patients is still unknown. Additionally, device performance could be affected if post-biopsied or post-operational patients would be included [72]. Due to the subjective assessment of medical staff regarding the suitability of patient's breast and intended breast cup, mismatch error can occur. The effect of menstrual cycle on the device performance is unknown, as the information was not available during analysis. As the T_3 group did not have enough samples, effect of the largest lesions on the device performance could not be considered. In 20 patients, density was unknown as these patient's only underwent US examination.

5 Conclusions

MWI is an emerging technology with the potential to contribute to the breast cancer imaging. Due to its harmless non-ionizing nature and cost-effectiveness, it can be considered as an attractive alternative to already existing conventional imaging tools. The study showed that device is not considerable affected by the participant's age, lesion size or breast density. Favorable performance when younger patients, and in dense breasts are involved is encouraging due to the well-known MMG limitations in both groups. Even though the study involved limited dataset, results encourage further clinical trials that will provide better insight into the clinical role of our MWI breast cancer device.

Acknowledgements This work was supported by the EMERALD project funded from the European Union's Horizon 2020 research and innovation programme under the Marie Skłodowska-Curie grant agreement No. 764479, and by the Scientific and Technology Research Council of Turkey (TUBITAK) grant number 120N388.

References

1. H. Sung et al., Global cancer statistics 2020: GLOBOCAN estimates of incidence and mortality worldwide for 36 cancers in 185 countries. CA Cancer J. Clin. **71**(3), 209–249 (2021). https://doi.org/10.3322/caac.21660
2. J. Ferlay et al., Estimating the global cancer incidence and mortality in 2018: GLOBOCAN sources and methods. Int. J. Cancer **144**(8), 1941–1953 (2019)
3. M. Malvezzi et al., European cancer mortality predictions for the year 2019 with focus on breast cancer. Ann. Oncol. **30**(5), 781–787 (2019)
4. R.L. Siegel, K.D. Miller, A. Jemal, Cancer statistics, 2020. CA Cancer J. Clin. **70**(1), 7–30 (2020)
5. R.L. Siegel, K.D. Miller, A. Jemal, Cancer statistics, 2019. CA Cancer J. Clin. **69**(1), 7–34 (2019)
6. C.H. Neal, M.A. Helvie, Overdiagnosis and risks of breast cancer screening. Radiol. Clin. North Am. **59**(1), 19–27 (2021)
7. N.F. Boyd et al., Mammographic density and the risk and detection of breast cancer. N. Engl. J. Med. **356**(3), 227–236 (2007)
8. J.N. Wolfe, Breast patterns as an index of risk for developing breast cancer. Am. J. Roentgenol. **126**(6), 1130–1137 (1976)
9. M. von Euler-Chelpin, M. Lillholm, I. Vejborg, M. Nielsen, E. Lynge, Sensitivity of screening mammography by density and texture: a cohort study from a population-based screening program in Denmark. Breast Cancer Res. **21**(1), 1–7 (2019)
10. N. R. Council and others, *Mammography and Beyond: Developing Technologies for the Early Detection of Breast Cancer* (National Academies Press, 2001)
11. N.F. Boyd et al., Mammographic breast density as an intermediate phenotype for breast cancer. Lancet Oncol. **6**(10), 798–808 (2005)
12. P.T. Huynh, A.M. Jarolimek, S. Daye, The false-negative mammogram. Radiographics **18**(5), 1137–1154 (1998)
13. P.E. Freer, P.J. Slanetz, Breast density and screening for breast cancer. https://www.uptodate.com/contents/breast-density-and-screening-for-breast-cancer. Accessed Dec 01, 2021

14. V. Shambhavi, P.J. Slanetz Priscilla, Breast imaging for cancer screening: mammography and ultrasonography (2020). https://www.uptodate.com/contents/breast-imaging-for-cancer-screening-mammography-and-ultrasonography. Accessed Dec 01, 2021
15. J.M. Lee et al., Performance of screening ultrasonography as an adjunct to screening mammography in women across the spectrum of breast cancer risk. JAMA Intern. Med. **179**(5), 658 (2019). https://doi.org/10.1001/jamainternmed.2018.8372
16. L.J. Esserman, B.N. Joe, Diagnostic evaluation of women with suspected breast cancer. Accessed Dec 01, 2021 [online]. Available https://www.uptodate.com/contents/diagnostic-evaluation-of-suspected-breast-cancer
17. J.G. Elmore, C.I. Lee, Screening for breast cancer: strategies and recommendations. Accessed Dec 01, 2021 [online]. Available https://www.uptodate.com/contents/screening-for-breast-cancer-strategies-and-recommendations
18. J.G. Elmore, C.I. Lee, Screening for breast cancer: evidence for effectiveness and harms. Accessed Dec 01, 2021 [online]. Available https://www.uptodate.com/contents/screening-for-breast-cancer-evidence-for-effectiveness-and-harms
19. S. Gabriel, R.W. Lau, C. Gabriel, The dielectric properties of biological tissues: III. Parametric models for the dielectric spectrum of tissues. Phys. Med. Biol. **41**(11), 2271 (1996)
20. S. Gabriel, R.W. Lau, C. Gabriel, The dielectric properties of biological tissues: II. Measurements in the frequency range 10 Hz to 20 GHz. Phys. Med. Biol. **41**(11), 2251 (1996)
21. A.M. Campbell, D.V. Land, Dielectric properties of female human breast tissue measured in vitro at 3.2 GHz. Phys. Med. Biol. **37**(1), 193 (1992)
22. C. Gabriel, S. Gabriel, YE. Corthout, The dielectric properties of biological tissues: I. Literature survey. Phys. Med. Biol. **41**(11), 2231 (1996)
23. M. Lazebnik et al., A large-scale study of the ultrawideband microwave dielectric properties of normal breast tissue obtained from reduction surgeries. Phys. Med. Biol. **52**(10), 2637 (2007)
24. S.S. Chaudhary, Dielectric properties of normal and malignant human breast tissues at radiowave and microwave frequencies. Indian J. Biochem. Biophs. **21**, 76–79 (1984)
25. A.J. Surowiec, S.S. Stuchly, J.R. Barr, A. Swarup, Dielectric properties of breast carcinoma and the surrounding tissues. IEEE Trans. Biomed. Eng. **35**(4), 257–263 (1988)
26. M. Lazebnik et al., A large-scale study of the ultrawideband microwave dielectric properties of normal, benign and malignant breast tissues obtained from cancer surgeries. Phys. Med. Biol. **52**(20), 6093 (2007)
27. D. Andreuccetti, R. Fossi, *Dielectric Properties of Human Tissues: Definitions, Parametric Model, Computing Codes*, p. 34 (2000) [online]. Available http://scholar.google.com/scholar?hl=en&btnG=Search&q=intitle:Dielectric+Properties+Of+Human+Tissues:+Definitions,+Parametric+Model,+Computing+Codes#0%5Cn, http://niremf.ifac.cnr.it/tissprop/document/tissprop.pdf
28. O. Scherzer, *Handbook of Mathematical Methods in Imaging* (2015)
29. F. Cakoni, D. Colton, *A Qualitative Approach to Inverse Scattering Theory* (2014) [online]. Available http://www.springer.com/series/34
30. M.N. Akıncı, T. Çağlayan, S. Ozgur, U. Alkaşı, M. Abbak, M. Çayören, Experimental assessment of linear sampling and factorization methods for microwave imaging of concealed targets. Int. J. Antennas Propag. **2015** (2015). https://doi.org/10.1155/2015/504059
31. M.N. Akıncı et al., Qualitative microwave imaging with scattering parameters measurements. IEEE Trans. Microw. Theory Tech. **63**(9), 2730–2740 (2015)
32. I. Catapano, L. Crocco, An imaging method for concealed targets. IEEE Trans. Geosci. Remote Sens. **47**(5), 1301–1309 (2009)
33. I. Catapano et al., On quantitative microwave tomography of female breast. Progr. Electromagn. Res. **97**, 75–93 (2009)
34. I. Catapano, L. Crocco, T. Isernia, On simple methods for shape reconstruction of unknown scatterers. IEEE Trans. Antennas Propag. **55**(5), 1431–1436 (2007). https://doi.org/10.1109/TAP.2007.895563
35. I. Catapano, L. Crocco, T. Isernia, Linear Sampling Method: Physical interpretation and guidelines for a successful application. Progr. Electromagn. Res. Symp. **2**, 1139–1143 (2008). https://doi.org/10.2529/piers071005053840

36. S. Coşğun, E. Bilgin, M. Çayören, Microwave imaging of breast cancer with factorization method: SPIONs as contrast agent. Med. Phys. (2020)
37. L. Crocco, L. di Donato, I. Catapano, T. Isernia, The factorization method for virtual experiments based quantitative inverse scattering. Progr. Electromagn. Res. **157**, 121–131 (2016). https://doi.org/10.2528/PIER16072809
38. P.M. Meaney, M.W. Fanning, D. Li, S.P. Poplack, K.D. Paulsen, *A Clinical Prototype for Active Microwave Imaging of the Breast* (2000)
39. P.M. Meaney et al., Initial clinical experience with microwave breast imaging in women with normal mammography. Acad. Radiol. **14**(2), 207–218 (2007). https://doi.org/10.1016/j.acra.2006.10.016
40. S.P. Poplack et al., Electromagnetic breast imaging: results of a pilot study in women with abnormal mammograms. Radiology **243**(2), 350–359 (2007). https://doi.org/10.1148/radiol.2432060286
41. A. Janjic et al., SAFE: a novel microwave imaging system design for breast cancer screening and early detection—clinical evaluation. Diagnostics **11**(3), 533 (2021)
42. M. Klemm et al., *Clinical Trials of a UWB Imaging Radar for Breast Cancer* (2010)
43. T. Henriksson et al., Clinical trials of a multistatic UWB radar for breast imaging, in *2011 Loughborough Antennas & Propagation Conference*, Nov 2011, pp. 1–4. https://doi.org/10.1109/LAPC.2011.6114004
44. M. Shere, I. Lyburn, R. Sidebottom, H. Massey, C. Gillett, L. Jones, MARIA® M5: a multicentre clinical study to evaluate the ability of the micrima radio-wave radar breast imaging system (MARIA®) to detect lesions in the symptomatic breast. Eur. J. Radiol. **116**, 61–67 (2019)
45. A.W. Preece, I. Craddock, M. Shere, L. Jones, H.L. Winton, MARIA M4: clinical evaluation of a prototype ultrawideband radar scanner for breast cancer detection. J. Med. Imaging **3**(3), 33502 (2016)
46. N. Ridley, M. Shere, I. Lyburn, P. Bannister, Cancer detection in dense tissue using radiofrequency imaging—a clinical evaluation. Dec 2017
47. N. Ridley et al., Automatic labelling of lesions using radiofrequency feature discrimination. Dec 2017
48. H. Massey, N. Ridley, I. Lyburn, S. Taylor, M. Schoenleber-Lewis, P. Bannister, Radio-wave detection of breast cancer in the symptomatic clinic—a multi-centre study. Dec 2017
49. E.C. Fear, J. Bourqui, C. Curtis, D. Mew, B. Docktor, C. Romano, Microwave breast imaging with a monostatic radar-based system: a study of application to patients. IEEE Trans. Microw. Theory Tech. **61**(5), 2119–2128 (2013). https://doi.org/10.1109/TMTT.2013.2255884
50. B.R. Lavoie, M. Okoniewski, E.C. Fear, Estimating the effective permittivity for reconstructing accurate microwave-radar images. PLoS One **11**(9) (2016). https://doi.org/10.1371/journal.pone.0160849
51. C. Curtis, B.R. Lavoie, E. Fear, An analysis of the assumptions inherent to near-field beamforming for biomedical applications. IEEE Trans. Comput. Imaging **3**(4), 953–965 (2017). https://doi.org/10.1109/tci.2017.2756022
52. D. Kurrant, J. Bourqui, E. Fear, Surface estimation for microwave imaging. Sensors **17**(7) (2017). https://doi.org/10.3390/s17071658
53. H. Song et al., Detectability of breast tumor by a hand-held impulse-radar detector: performance evaluation and pilot clinical study. Sci. Rep. **7**(1) (2017). https://doi.org/10.1038/s41598-017-16617-6
54. E. Porter, M. Coates, M. Popovic, An early clinical study of time-domain microwave radar for breast health monitoring. IEEE Trans. Biomed. Eng. **63**(3), 530–539 (2016). https://doi.org/10.1109/TBME.2015.2465867
55. E. Porter, H. Bahrami, A. Santorelli, B. Gosselin, L.A. Rusch, M. Popovic, A wearable microwave antenna array for time-domain breast tumor screening. IEEE Trans. Med. Imaging **35**(6), 1501–1509 (2016). https://doi.org/10.1109/TMI.2016.2518489
56. Y. Kuwahara, Microwave imaging for early breast cancer detection, in*New Perspectives in Breast Imaging*, InTech (2017).https://doi.org/10.5772/intechopen.69562

57. A. Fasoula et al., Pilot patient study with the Wavelia Microwave Breast Imaging system for breast cancer detection: clinical feasibility and identified technical challenges, in *14th European Conference on Antennas and Propagation, EuCAP 2020*, Mar 2020. https://doi.org/10.23919/EuCAP48036.2020.9135549
58. B.M. Moloney et al., Microwave imaging in breast cancer—results from the first-in-human clinical investigation of the Wavelia system. Acad Radiol (2021). https://doi.org/10.1016/j.acra.2021.06.012
59. L. Sani et al., Breast lesion detection through MammoWave device: empirical detection capability assessment of microwave images' parameters. PLoS One **16**(4) (2021). https://doi.org/10.1371/journal.pone.0250005
60. A. Vispa et al., UWB device for breast microwave imaging: phantom and clinical validations. Measurement **146**, 582–589 (2019). https://doi.org/10.1016/j.measurement.2019.05.109
61. D. O'Loughlin, M. O'Halloran, B.M. Moloney, M. Glavin, E. Jones, M.A. Elahi, Microwave breast imaging: clinical advances and remaining challenges. IEEE Trans. Biomed. Eng. **65**(11), 2580–2590 (2018). https://doi.org/10.1109/TBME.2018.2809541
62. S. Kwon, S. Lee, Recent advances in microwave imaging for breast cancer detection. Int. J. Biomed. Imaging **2016** (2016). https://doi.org/10.1155/2016/5054912
63. R. Benny, T.A. Anjit, P. Mythili, *An Overview of Microwave Imaging for Breast Tumor Detection* (2020)
64. ACR BI-RADS® ATLAS-MAMMOGRAPHY
65. A. Abbosh, B. Mohammed, K. Bialkowski, Differential microwave imaging of the breast pair. IEEE Antennas Wirel. Propag. Lett. **15**, 1 (2015). https://doi.org/10.1109/LAWP.2015.2512260
66. L. Crocco, I. Catapano, L. di Donato, T. Isernia, The linear sampling method as a way to quantitative inverse scattering. IEEE Trans. Antennas Propag. **60**(4), 1844–1853 (2012). https://doi.org/10.1109/TAP.2012.2186250
67. F. Cakoni, D. Colton, P. Monk, The linear sampling method in inverse electromagnetic scattering. Soc. Ind. Appl. Math. (2011). https://doi.org/10.1137/1.9780898719406
68. D. Colton, R. Kress, *Inverse Acoustic and Electromagnetic Scattering Theory*, vol. 93 (Springer, New York, NY, 2013). https://doi.org/10.1007/978-1-4614-4942-3
69. A. Kirsch, N. Grinberg,*The Factorization Method for Inverse Problems* (Oxford University Press, 2007).https://doi.org/10.1093/acprof:oso/9780199213535.001.0001
70. P.E. Freer, Mammographic breast density: impact on breast cancer risk and implications for screening. Radiographics **35**(2), 302–315 (2015). https://doi.org/10.1148/rg.352140106
71. W.A. Berg et al., Diagnostic accuracy of mammography, clinical examination, US, and MR imaging in preoperative assessment of breast cancer. Radiology **233**(3), 830–849 (2004)
72. A. Fasoula, L. Duchesne, J.D.G. Cano, B.M. Moloney, S.M.A. Elwahab, M.J. Kerin, Automated breast lesion detection and characterization with the wavelia microwave breast imaging system: methodological proof-of-concept on first-in-human patient data. Appl. Sci. **11**(21) (2021). https://doi.org/10.3390/app11219998

Microwave Ultra-Wideband Imaging for Non-invasive Temperature Monitoring During Hyperthermia Treatment

Alexandra Prokhorova, Ondrej Fiser, Jan Vrba, and Marko Helbig

Abstract Microwave medical imaging can become a good alternative for common imaging approaches in the very near future since it is safe due to non-ionizing radiation, cost-efficient and to this end promising for clinical applications. This chapter is devoted to microwave imaging for non-invasive temperature monitoring during hyperthermia therapy. To ensure a constant desired temperature level in the cancerous tissue and to prevent damage of healthy cells, accurate temperature control is necessary during thermal treatment. We present a temperature estimation approach based on the ultra-wideband M-sequence radar technology developed at the Technische Universitaet Ilmenau. The methodology is based on the knowledge of temperature dependencies of tissue physical parameters and on ongoing ultra-wideband measurements, followed by imaging and estimation of dielectric properties which are converted to temperature values. The prototype components from both sensing and heating parts of the system are investigated numerically so that suitable configurations of the antenna array can be defined. Furthermore, the system is experimentally validated on a neck phantom filled with corresponding tissue mimicking materials, which well imitate the dielectric properties of the specific tissues. Exemplary results of these developments are presented in this chapter.

A. Prokhorova (✉) · M. Helbig
Institute of Biomedical Engineering and Informatics, Technische Universität Ilmenau, Ilmenau, Germany
e-mail: alexandra.prokhorova@tu-ilmenau.de

M. Helbig
e-mail: marko.helbig@tu-ilmenau.de

O. Fiser
Department of Biomedical Technology, Czech Technical University in Prague, Prague, Czech Republic
e-mail: ondrej.fiser@fbmi.cvut.cz

J. Vrba
Department of Electromagnetic Field, Czech Technical University in Prague, Prague, Czech Republic
e-mail: vrba@fel.cvut.cz

F. Vipiana and L. Crocco (eds.), *Electromagnetic Imaging for a Novel Generation of Medical Devices*, Lecture Notes in Bioengineering,
https://doi.org/10.1007/978-3-031-28666-7_10

Keywords Ultra-wideband imaging · Non-invasive temperature monitoring · Neck hyperthermia treatment · Temperature dependent dielectric properties · M-sequence radar technology

1 Introduction to Hyperthermia

1.1 Clinical Motivation

Despite the constant improvement of existing medical capabilities and the emergence of new technologies aimed at improving the quality of diagnosis and efficiency of clinical treatment, oncological diseases still remain one of the most common causes of death. Worldwide, 19.3 million new cancer cases and almost 10.0 million cancer deaths occurred in 2020 [1]. Certainly, early stage cancer detection and essential cancer treatment are of a great interest.

Standard procedures of oncological tumors treatment are surgery, radiotherapy and chemotherapy. Additionally, immunotherapy, hormone therapy and bone marrow transplant, also known as a stem cell transplant, can be used to increase the doses of therapy to treat the cancer [2]. Some patients may need only one type of treatment, but in majority of oncological cases a combination of treatments, such as surgery with chemotherapy and/or radiation therapy are more successful offering higher survival rate together with better life quality. But even combined treatment approaches sometimes are not sufficient for the tumor remission. Another method to enhance the clinical outcome of the oncological treatment is the application of thermal therapies. According to the induced temperature increase, thermal therapies applied in cancer treatment are represented by hyperthermia (HT) and thermal ablation. While ablation assumes high temperature increase (up to 60–70 °C) and is an alternative to the surgical approach, hyperthermia is reached by lower temperature increase (up to 40–45 °C) and can be applied as an adjuvant procedure combined with radiotherapy or chemotherapy for cancer treatment [3]. Main existing options for treatment of oncological diseases are shown in Fig. 1.

Hyperthermia therapy is mainly used as a support therapy during oncological treatment of advanced tumors and can noticeably enhance its effectiveness and improve the clinical outcome without introducing additional toxicity. Most important effects of the hyperthermia application are damage of cancerous cells and shrinkage of the tumor size. Another important positive aspect of HT is related to improved blood flow and respectively oxygenation in the tumor area. Tumor hypoxia develops because of uncontrollable cell proliferation and abnormal tumor blood vessels resulting in reduced transport of oxygen. Since the main mechanism in radiotherapy is the creation of reactive oxygen species (subset of free radicals that kill cancer cells), hypoxic tumors are therefore radiation resistant [4]. A number of chemotherapeutic drugs have been also shown to be less effective when applied to hypoxic tumorous area, which noticeably decreases the outcome of the treatment [5]. Application of heat during hyperthermia increases the tumor perfusion and therefore provides benefits

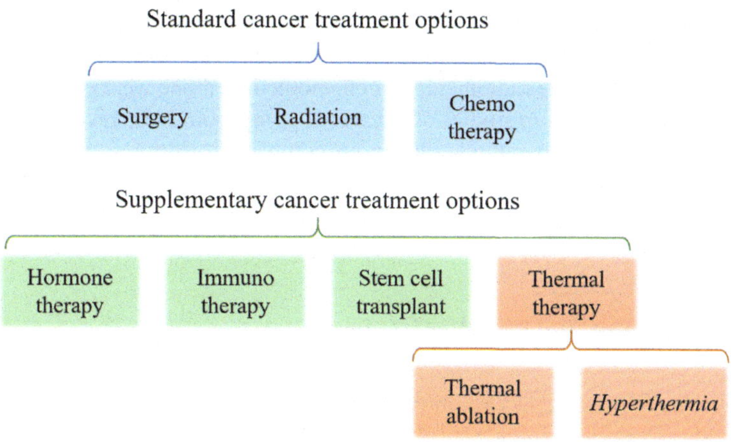

Fig. 1 Main treatment options for oncological diseases

of increased radiosensitivity, higher response to antigens and drugs in combination with chemotherapy, inhibition of the DNA repair in radiation damaged tumor cells and overall increased disease-free survival [6]. Figure 2 is illustrating the advantages of hyperthermia therapy in conjunction with other treatments.

Hyperthermia session duration is around 60 min and it is repeated weekly for a month or longer. Usually in clinical practice HT is applied in combination with radiotherapy or chemotherapy only once a week due to the effect of thermotolerance, when tissues do not respond to the heat for some period of time.

One of the highest positive results of combined therapy was noted in the head and neck area. The study of Datta et al. [8] investigating randomized and nonrandomized clinical studies for various tumors (nasopharynx, oral cavity, oropharynx,

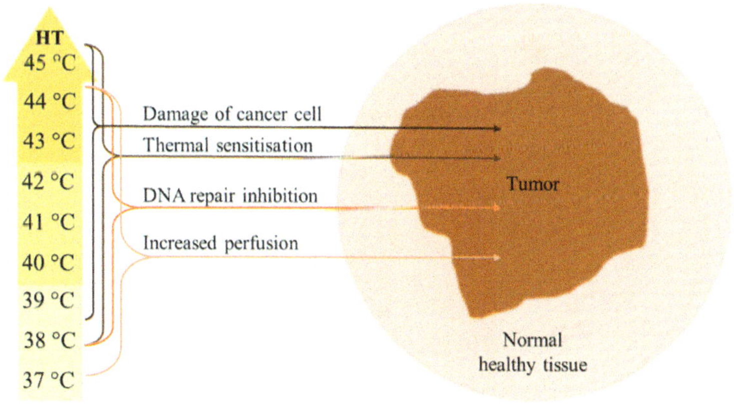

Fig. 2 Benefits of the hyperthermia treatment application [7]

hypopharynx and neck nodes) with radiotherapy (± chemotherapy) versus radiotherapy (± chemotherapy) + HT showed that the addition of HT increased the outcome to 75.3%, while radiotherapy or chemotherapy alone were effective only in 50.3% of cases. Follow-up studies of a research group from India revealed complete recovery in 42.4% of only radiotherapy group compared to 78.6% in the radiotherapy + HT group [9]. Information about the results of many other similar studies, which confirm the already mentioned ones, can be found e.g. on the webpage [10].

1.2 Neck Anatomy and Neck Cancer

The anatomy of the neck has a complex structure with organs and tissues vital for the human life. They are responsible for breathing, speaking, swallowing, regulation of metabolism, support and connection of the brain and cervical spine, blood circulation and lymphatic inflow and outflow from the head [11]. The neck consists of four compartments: vertebral, visceral and two vascular. Vertebral compartment includes cervical vertebra, spinal cord and postural muscles. Visceral compartment contains thyroid, parathyroid, thymus, larynx, pharynx, esophagus and trachea. Each vascular compartment consists of carotid artery, internal jugular vein and the vagus nerve, and is located on both sides of the neck. Organs are supported by muscle and fat tissues. Main structures of the human neck are shown in the Fig. 3.

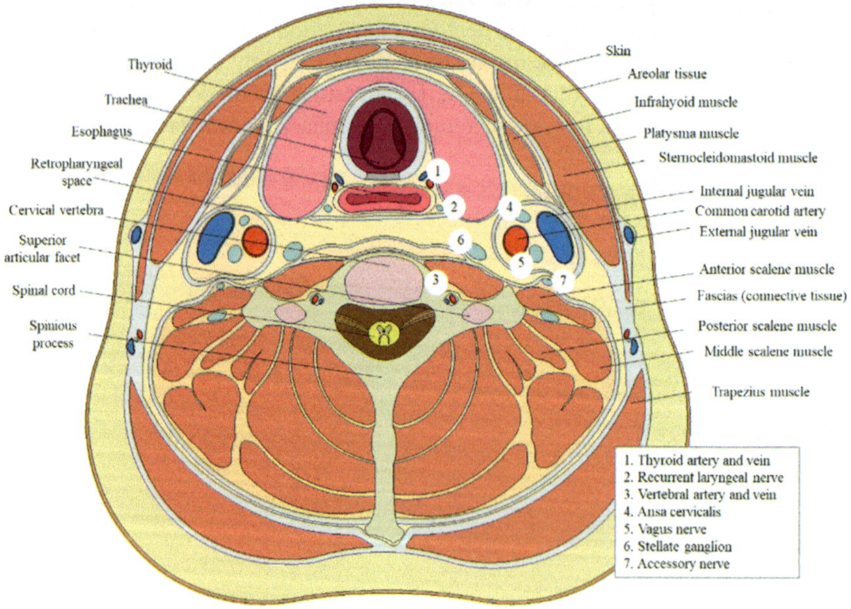

Fig. 3 Anatomical structures of the human neck (drawn according to [12])

Neck related oncological diseases account for more than 10% of all new cancer cases in 2020 with a death rate more than 10.5% above all cancer types [1]. Mainly neck cancers include tumors located in the esophagus, larynx, thyroid gland, trachea and their metastasis in the lymph nodes. Commonly, the tumors of the thyroid and larynx have a shape close to a sphere or an ellipse and a size around 1 to 4 cm depending on the stage, while the esophagus tumors are prolonged and can have a length from 3 up to 7 cm. In case of metastasis in the lymph nodes the tumors have commonly a spherical shape of 1–2 cm in diameter on early stages and 3–6 cm on late ones [13, 14].

1.3 Hyperthermia: State of Art

Thermal therapies and especially hyperthermia, are known and used for medical purposes for a long time. Since the early 80's, investigation and introduction of HT approach in clinical procedures have continuously grown. One of the first groups is the research group from Czech Technical University in Prague (CTU), which in cooperation with Charles University and Institute of Radiation Oncology started the technical development of a hyperthermia system in 1981 and treated cancer patients by combination of radiotherapy and hyperthermia in 1982.

Currently, some of the HT systems already used in clinics are Pyrexar, USA [6], MedLogix, Italy [15] and so called capacitive HT systems in Japan and China [16]. There are also impressive university technical research activities in this area: HT Hypercollar at Rotterdam Medical Center [17], waveguide HT system in Amsterdam Medical Center [18] and research at Chalmers University, Sweden [19].

Hyperthermia systems can be divided regarding the size of the area to be heated into four main modes: local also known as superficial, regional, loco-regional or part body and full body hyperthermia [20]. The mode of application depends on cancer type, size of the tumor and its position inside the human body. Local hyperthermia is indicated for superficial diseases such as chest wall recurrences, superficial malignant melanoma lesions or lymph node metastases of head and neck tumors. This technique is performed by positioning the heating applicators coupled with water bolus on the patient's skin. In regional hyperthermia an array of phase-controlled antennas surrounds the cross section of the target organ contaminated with the tumor. Interference patterns can be shaped according to the location of the cancerous tissue. This technique is especially useful for the treatment of deep located tumors. Part body hyperthermia is based on similar principles. The array of applicators is placed around the certain area of the body and provides heat to the target region for a prolonged time period. In the last mode, which is whole body hyperthermia, the patient is placed in thermal chamber and is exposed to electromagnetic (EM) radiation. The aim of this approach is to raise temperature of the whole body during a specific time interval and improve e.g. the outcome of chemotherapy during treatment of metastatic cancer.

In general, all of these systems require temperature control. The present work is aimed at the development of a system for non-invasive temperature monitoring during regional hyperthermia treatment of neck tumors.

1.4 Temperature Monitoring During Hyperthermia

Despite the progress in the implementation of hyperthermia systems, one of the remaining challenges to be solved in this area of research is temperature monitoring. Accurate temperature control during the hyperthermia procedure is essential for two reasons. First of all, to monitor the achieved temperature distribution inside the tumor and ensure sufficient heat level application to cancerous cells and secondly, to keep the temperature increase outside the tumorous region within the tolerable limits in order to prevent damage of the healthy cells. Currently, the most common ways of temperature measurement are minimally-invasive endoluminal thermometry [21] and invasive implementation of fiber optic catheters [22]. Since both methods are invasive, they are first of all painful for the patient, moreover catheters provide only limited information of temperature values at a single point and must follow the anatomical restrictions on their placement inside the human body. Furthermore, hyperthermia therapy will be applied in repeated sessions, so that invasive methods are not the best option. On the contrary, non-invasive temperature monitoring would be advisable to improve the patient comfort, decrease the possibility of infection as well as to reduce the treatment costs.

For this purpose, various medical technologies are being investigated, e.g. magnetic resonance imaging (MRI) [23], ultrasound technology [24], passive [25] and active microwave based methods [26, 27].

MRI provides high spatial resolution, however it has several limitations. Firstly, the presence of various biomedical implants and devices (e.g. carotid stents) in the body part to be treated could make it challenging for patients to undergo magnetic resonance imaging procedures [28]. In addition, any magnetic parts of the equipment for hyperthermia therapy have to be placed far enough from the magnetic resonance imaging scanner, which requires a larger space and hence results in even higher costs, or they have to be MRI compatible (as e.g. MRCollar [29]). Ultrasound thermometry method offers a good penetration depth and is easily focused in the area to be treated but cannot penetrate through air cavities (e.g. trachea and esophagus lumen, hollow bones) and therefore is not able to image these tissues as well as tissues behind them.

Medical microwave imaging (MWI) has become a rapidly developing topic for investigations by many scientific groups working with imaging, as well as for the clinical partners. Microwave sensing is cost-effective in comparison to other imaging approaches (e.g. MRI, computer tomography (CT)), safe for the patient due to non-ionizing radiation and, therefore, promising for several medical monitoring and imaging tasks. Among others, microwave imaging is used for early stage breast cancer diagnostics [30–32], brain stroke imaging and follow-up [33–35] or lung and skin cancer detection [36, 37], respectively.

We are dealing with active microwave imaging, in particular ultra-wideband radar imaging for non-invasive tissue temperature monitoring during hyperthermia. Measuring the changes in scattering behavior of the treated tissues caused by temperature difference, the aim is to detect and image the changes of the dielectric properties of treated tissue during the therapy. This technology and the methodology for tissue temperature estimation are described in the next section.

1.5 Chapter Sections Description

This chapter focuses on the development of a system for UWB MWI-guided hyperthermia application in the neck area and is organized as follows: Sect. 2 covers the methodology for ultra-wideband microwave imaging of temperature induced changes in tumorous tissue, algorithms for image reconstruction as well as limitations and requirements of the hybrid system for hyperthermia treatment with microwave temperature monitoring. Section 3 is devoted to the electromagnetic modelling of heating and sensing antennas and their suitable arrangement in the hybrid prototype. Section 4 is dedicated to the description of the experimental setup used in this work, materials under test and the design of the measurement scenario. Moreover, it includes the results of first experimental validation of the temperature estimation methodology. The last part of Sect. 4 presents the developed prototype and a heterogeneous anatomically realistic neck phantom.

2 Methodology of Non-invasive Temperature Estimation

2.1 Temperature Dependent Dielectric Properties

The dielectric properties of biological tissues play a key role in microwave imaging. It is well-known, that they are not only frequency dependent, but also temperature dependent.

The temperature dependent dielectric properties are described by the complex relative permittivity $\underline{\varepsilon}$ and the effective conductivity σ:

$$\begin{aligned}\underline{\varepsilon}(\omega,\vartheta) &= \varepsilon'(\omega,\vartheta) - i\varepsilon''(\omega,\vartheta)\\ \sigma(\omega,\vartheta) &= \omega\varepsilon_0\varepsilon''(\omega,\vartheta)\end{aligned} \tag{1}$$

where the real part ε' represents the relative permittivity, the imaginary part ε'' the relative dielectric loss, ε_0 the permittivity of free space, $\omega = 2\pi f$ the angular frequency, f the frequency and ϑ the temperature.

The magnitude of dielectric parameters change during hyperthermia treatment depends on tissues water content. High water content tissues like blood, glands or

Table 1 Dielectric properties of relevant human neck tissues at 2.45 GHz

Tissue type	Relative permittivity	Effective conductivity in S/m
Muscle	52.7	1.7
Fat	5.3	0.1
Cervical vertebra	11.4	0.4
Spinal cord	30.1	1.0
Trachea	39.7	1.4
Trachea lumen	1.0	0.0
Esophagus	62.2	2.2
Esophagus lumen	1.0	0.0
Thyroid gland	57.2	1.9
Skin	38.0	1.5
Blood	58.3	2.5

muscle exhibit larger changes of their dielectric properties as a function of temperature in comparison to low water content tissues e.g. fat or bones. A database of tissue specific temperature dependent dielectric properties is required to calculate the temperature in the area of treatment based on the values of relative permittivity and effective conductivity estimated from imaging.

There are various studies where numerically modelled or experimentally measured values of dielectric properties of biological tissues are reported, e.g. [38–41]. Mostly, these studies include investigations only at body temperature. For example, Table 1 represents the values of neck specific tissues according to the study of Gabriel [40] at 2.45 GHz.

There are a few studies with the focus on the temperature dependency of the dielectric properties of human tissues or tissue mimicking materials in the microwave frequency range. Majority of these studies are limited to liver tissue [42–44]. Temperature dependent studies of lung tissue over a temperature range from 21 to 90 °C and over the frequency range 0.5 to 8.5 GHz are presented in a paper by Bonello et al. [45]. A study of fat tissue mimicking material for UWB measurements can be found in [46]. A large-scale UWB temperature dependent dielectric spectroscopy study of four porcine tissues including blood in the frequency range from 0.5 to 7 GHz and in the temperature range between 30 and 50 °C is published in [47]. From the results of this study we can see that temperature dependency of the relative permittivity of blood is the highest of all investigated tissue types and is approximately 0.20 decrease per degree Celsius between 1 and 4 GHz. Liver and muscle tissues have smaller water content and, therefore, slighter changes of their dielectric properties of around 0.10 per degree Celsius in maximum. As for the fat tissue, it is practically temperature independent due to its low water content and the decrease of relative permittivity is not more than 0.02 per degree Celsius between 1 and 7 GHz. Obviously, the temperature dependency of relative permittivity and conductivity is tissue and frequency dependent. In [47], the temperature dependencies are represented in detail by a second-order polynomial fit of two-pole Cole–Cole model parameters. Depending on the tissue types under investigation for temperature control and their water content, these values can be used to convert estimated dielectric properties to temperature. Furthermore, the information about varying temperature sensitivity at different frequencies can be exploited to define optimal working conditions (bandwidth) in specific scenarios.

2.2 UWB Sensor Technology

Active medical microwave imaging is represented by two techniques, tomographic and radar-based imaging. Both approaches are widely used in different clinical applications (breast cancer detection [48–50], diagnosing brain diseases such as Alzheimer [51] or stroke [52], heart and respiration rate registration [53, 54], etc.). Microwave tomography is based on the solution of the inverse electromagnetic scattering problem by iterative minimization of the difference between measured and simulated field values and provides estimation of tissue dielectric properties at certain frequencies (quantitative imaging). UWB radar imaging algorithms detect and localize scatterers in the resultant qualitative images reconstructing the reflectivity inside the region of interest. Moreover, beside interpretation in the time domain, ultra-wideband radar data can be processed by the Fourier Transform and analyzed in the frequency domain when required by type of the imaging approach.

We are working with radar imaging based on the UWB M-sequence technology developed at Technische Universitaet Ilmenau, Germany [55]. This technology applies pseudo-noise codes (M-sequences) generated by high-speed digital shift registers as stimulation signal. The impulse response of the scenario under test is created by means of cross correlation between received signal and ideal M-sequence. Therefore, the amplitude unit is arbitrary. In contrast to impulse radar, the signal energy is distributed over the continuous stimulus signal, reducing the voltage exposure of the medium under test and, thus, making this technology suitable for medical applications. Furthermore, M-sequence technology is characterized by long-term stability and low jitter. Complex Multiple Input Multiple Output (MIMO) systems are implemented by cascading simple sensor modules consisting of a transmitter port (Tx) and two receiver ports (Rx) [50, 55].

2.3 Flowchart of the MWI for Temperature Monitoring

The methodology of non-invasive temperature monitoring during hyperthermia is based on continuous estimation of the changing tissue dielectric properties in the area of interest during the treatment by means of microwave sensing. The flowchart illustrating this approach is presented in the Fig. 4.

Generally, this methodology can be divided into two parts: preparation stage and ongoing monitoring during HT treatment. Knowledge of the temperature dependency

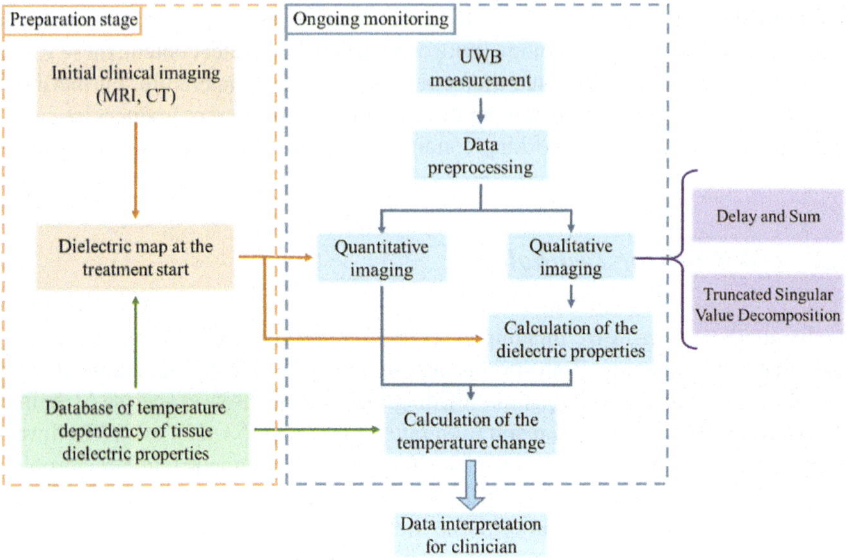

Fig. 4 Flowchart of the tissue temperature monitoring methodology

of relative permittivity and effective conductivity of specific tissues in the body region to be treated by hyperthermia in the relevant microwave frequency range is a keystone of the approach. Assuming that hyperthermia is an adjunctive therapy for cancer treatment, the availability of a CT or MRI scan of the patient is expected before treatment begins. To this end, individual patient specific maps with tissue type segmentation and corresponding dielectric properties are created in the first part. These dielectric tissue maps of the region of interest are used as input (initial guess) for microwave imaging.

The second part is the actual tissue temperature monitoring by means of UWB measurements performed in real-time during the whole hyperthermia procedure. The spatial reconstruction of changing dielectric properties inside the human body is made via imaging algorithms. Depending on the type of algorithm, the dielectric properties are calculated directly (quantitative imaging) or in an intermediate step using the reconstructed reflection intensity (qualitative imaging). Afterwards, the change of the temperature in the area of interest is estimated based on the known temperature dependency of the specific tissues.

2.4 *Microwave Imaging Algorithms*

The methodology for tissue temperature monitoring is not strictly limited to a specific imaging algorithm. Practically, due to the fact that the temperature estimation has to be done in real-time during the hyperthermia treatment, the calculation effort plays an important role. Therefore, we do not consider imaging algorithms which require the iterative solution of the non-linear inverse scattering problem for image reconstruction due to their high computational and respectively time effort. Hence, two imaging algorithms which we investigate and apply to our approach are Delay and Sum (DAS) [30, 56] and Truncated Singular Value Decomposition (TSVD) [57]. The details of the application of these imaging algorithms for our approach are presented in [58].

DAS algorithm is based on the principle of coherent addition of backscattered radar signals, which are collected while illuminating the target with electromagnetic waves. Its main advantages are simplicity, robustness, and short computational time. The DAS beamformer equation can be written as:

$$I_{DAS}(\boldsymbol{r}_0, T) = \sum_{n=1}^{N} s_{h,n}(t_0 + \tau_n(\boldsymbol{r}_0), T), \tag{2}$$

where I is the image, N is the number of channels, T is the hyperthermia treatment time, t is the propagation time with t_0 as the time when the electromagnetic wave takes off the transmitter antenna, τ_n is the channel dependent time of flight and is calculated based on the assumed mean propagation speed of the signal. $s_{h,n}(t, T)$ is the ongoing and patient specific signal of channel n after clutter removal determined by the tissue region to be monitored during the treatment. In case of differential imaging, it is substituted by the differential signal $s_{\Delta\vartheta,n}(t, T) = s_{h,n}(t, T) - s_{h,n}(t, T_0)$ representing the difference between signals received at the hyperthermia treatment time T and at the start of treatment T_0. The signals received at each channel are time-aligned for each focal point $\boldsymbol{r}_0 = (x, y, z)$ and coherently summed. The summed signal is assigned to the reflection intensity of the focal point and this process is repeated for all focal points within the region of interest.

TSVD based microwave imaging is solving the linear inverse problem represented as:

$$A(\boldsymbol{r}_0)x(\boldsymbol{r}_0, T) = S_h(T), \tag{3}$$

where A is the scattering operator, mapping the unknown contrast function x at different treatment time T into the frequency domain data $S_h(T)$, obtained as the Fourier Transform of $s_h(t, T)$. For each channel the scattering operator A is represented by the discretized version of a linear and compact integral operator, whose kernel is relying on the Born approximation and exploiting the reciprocity principle. In order to retrieve the solution from the data, the scattering operator A must be inverted. To this end, its representation in terms of singular functions can be exploited:

$$A = UDV^*, \tag{4}$$

where U represents left singular vectors, V represents right singular vectors, D is a diagonal matrix, where main diagonal is non-zero, with decreasing singular values.

Since the inverse scattering problem is ill-posed, a regularized solution is necessary to retrieve reliable images. Hence, a truncation is considered in the reconstruction formula:

$$I_{TSVD}(\boldsymbol{r}_0, T) = V_C \cdot D_C^{+} \cdot U_C^{*} \cdot S_h(T), \tag{5}$$

where the symbol $^+$ denotes the inverse of the matrix, C represents the regularizing truncation parameter, which is chosen such to meet a good tradeoff between accuracy and stability of the reconstruction.

It is worth to mention that SVD of the scattering operator can be computed off-line at the preparation stage, which is time-efficient and qualifies the algorithm for the real-time medical applications.

2.5 Tumor Temperature Estimation Approach

Since qualitative radar imaging does not directly provide dielectric properties values, the next step is estimation of relative permittivity in the region of interest from the reconstructed images [59]. We presume that the image values $I(\boldsymbol{r}_0, T)$ in the region of treatment are depending on two parts, the dielectric contrast between tumor and surrounding healthy tissue as well as on the accumulation of temperature independent influencing parameters (e.g. radar cross section of the tumor, signal path attenuation, impulse response of radar and antennas, etc.) which are not known exactly, and cannot be quantified separately, even more because the treatment area is located in the near field. Therefore, we summarize them within the parameter $\underline{F}$:

$$I(\boldsymbol{r}_0, T) = \underline{F}(\boldsymbol{r}_0) \cdot \underline{\Gamma}(\boldsymbol{r}_0, \omega, T). \tag{6}$$

Assuming that parameters included in $\underline{F}$ remain constant during hyperthermia treatment, only the dielectric contrast between background and tumor will change [60]. The effect of dielectric contrast in (Eq. 2) and (Eq. 5) can be approximated by means of the complex reflection coefficient $\underline{\Gamma}$. Based on the initial clinical imaging and the database of tissue specific dielectric properties, the permittivities of tumor and surrounding tissues at the beginning of treatment are known. The reflection coefficient and the corresponding image at the starting time T_0 are used as reference, so that during the treatment we relate the ongoing images to the reference one:

$$\frac{I(\boldsymbol{r}_0, T)}{I_{ref}(\boldsymbol{r}_0, T_0)} = \frac{\underline{\Gamma}(\boldsymbol{r}_0, \omega, T)}{\underline{\Gamma}_{ref}(\boldsymbol{r}_0, \omega, T_0)}. \tag{7}$$

In this way, the effect of $\underline{F}(\boldsymbol{r}_0)$ will be eliminated. Simplifying the problem by assuming idealized conditions of specular reflection and plane waves, the changing permittivity $\hat{\underline{\varepsilon}}_t(\boldsymbol{r}_0, \omega, T)$ inside the tumor region can be estimated as

$$\hat{\underline{\varepsilon}}_t(\mathbf{r}_0, \omega, T) = \underline{\varepsilon}_{bg}(\omega)\left(\frac{I_{ref}(\mathbf{r}_0, T_0) - \underline{\Gamma}_{ref}(\mathbf{r}_0, \omega, T_0)I(\mathbf{r}_0, T)}{I_{ref}(\mathbf{r}_0, T_0) + \underline{\Gamma}_{ref}(\mathbf{r}_0, \omega, T_0)I(\mathbf{r}_0, T)}\right)^2 \quad (8)$$

where $\underline{\varepsilon}_{bg}$ is the permittivity of the healthy tissue.

This equation is applied to the area of the tumor in each three-dimensional (3D) image voxel by voxel. In case of DAS based temperature estimation Eq. (8) is applied to estimate the dielectric properties within the radar bandwidth, while for a single frequency reconstruction via TSVD only values at the specified frequency are computed. After permittivity is estimated, temperature values can be calculated as described in Sect. 2.1.

2.6 *Microwave Hyperthermia System*

2.6.1 HT System Components and Their Parameters

Main components of a microwave hyperthermia system are microwave power generator connected via coaxial cable with one or several applicators for emitting electromagnetic power and water bolus which is placed between the applicator and the human body or phantom. For the hyperthermia application in the neck area we use a system developed at CTU Prague [61]. The heating applicator is a waveguide (WG) antenna with strip line horn aperture, with body made from aluminum and sidewalls from plexiglass [62]. The main advantages of the designed heating applicator are ease of application and possibility of modifications, i.e. disconnecting the front plexiglass membrane, adjusting transmission coefficient with the tuning probe. The microwave generator that we use to power these applicators during the experiments is UHF-POWER-GENERATOR PG 70.150.2 (CTU Prague) with working frequency of 434 MHz. The generator consists of two separate power channels with maximum power of 150 watts.

2.6.2 Requirements and Limitations for Hybrid Microwave System Development

The inclusion of UWB sensors for non-invasive temperature monitoring close to the microwave hyperthermia heating system, requires overcoming several challenges. First of all, interference problems caused by similar working frequencies of imaging and heating parts of the system have to be solved. Another challenge is the limited space for positioning of both sensing antennas and heating applicators of quite large dimensions in the relatively small area to be treated in the neck region. To achieve the best possible imaging resolution, appropriate antenna configurations with low interference between heating and sensing parts of the system have to be found.

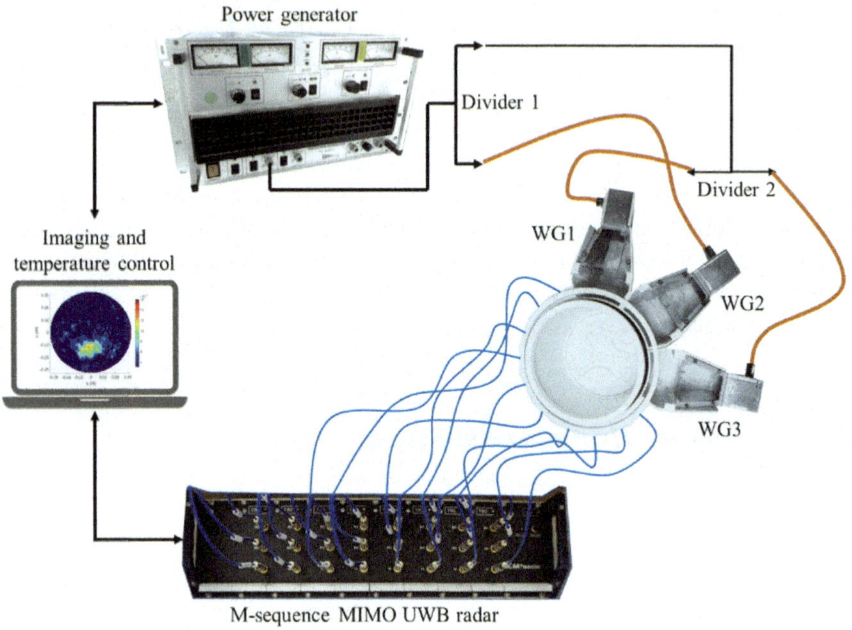

Fig. 5 Laboratory prototype of the hybrid microwave system for neck hyperthermia with non-invasive temperature monitoring via UWB technology

The average adult human neck dimensions are 10–11 cm in height and 10–13 cm in diameter [63]. To provide efficient and accurate heating to the tumorous tissue, we use three hyperthermia waveguide applicators. In this case, the middle applicator should be placed directly in front of the tumor and provide the main amount of heat, while the other two applicators, shifted to the left and to the right, will improve heat focusing in the tumorous region after phase and amplitude optimization procedure. The dimensions of effective aperture of the waveguide are 10 cm by 5 cm. Due to the average neck height as well as the applicator height, it is not possible to place the sensing antennas above or below the hyperthermia applicators. This leads us to approximately 40% of the neck circumference for the hyperthermia applicators, 40% for sensing antennas and the remaining space is left for the adaption of the system to different neck diameters. On this basis, a laboratory prototype was designed. Its schematic is presented in the Fig. 5.

3 Numerical Investigation of the Antenna Array Configurations

We use software platform Sim4Life® for all presented here numerical simulations since the capabilities of this software allow to create an accurate biological model including its dielectric properties, define UWB sources, adapt the mesh and voxelization for each part of the simulation model and calculate the electromagnetic field distribution via Finite-Difference Time-Domain method. The source signal for the hyperthermia part of the system (in our case WG applicators) is chosen as continuous wave at frequency 434 MHz. The UWB source signal of the sensing antennas (UWB bow-ties) is chosen as Gaussian modulated pulse in the bandwidth 1–6 GHz. The mesh is defined for each simulation setup separately and will be specified for each model.

3.1 Influence of Water Bolus Thickness on Reflection Coefficient of the WG Antenna

As mentioned above, the water bolus is one of the crucial parts of the hyperthermia system helping to couple EM energy between applicator and area to be treated in combination with the cooling effect of the patient skin. The water bolus thickness is influencing the applicator parameters, for instance reflection coefficient and distribution of the Specific Absorption Rate (SAR).

At first, we model and validate the influence of the water bolus thickness on the reflection coefficient $|S_{11}|$ in the frequency range 100–800 MHz. The reason to perform this study is to find out the ideal working conditions for the water filled WG applicator with strip-line horn aperture in the hyperthermia system. The simulation setup is presented in Fig. 6. It consists of metallic WG applicator (grey color, set as Perfect Electric Conductor), water bolus of variable thicknesses (blue color, $\varepsilon' = 78$, $\sigma = 0.03$ S/m) and muscle tissue (yellow color, $\varepsilon' = 54$, $\sigma = 0.8$ S/m) with an inserted spherical tumor imitate (red color, $\varepsilon' = 65$, $\sigma = 0.7$ S/m). We performed a set of nine simulations changing the water bolus thickness in the range 0.5–4.5 cm with step of 0.5 cm. The mesh of the whole model consists of 17.3 million cells (MCells). Each part of the model was meshed separately by adaptive procedure. The WG applicator and tumorous tissue have highest priority and small mesh elements (0.3 mm). Water bolus and muscle phantom have lower priority and middle size of elements (0.5 mm).

The results of the simulations of reflection coefficient $|S_{11}|$ for various water bolus thicknesses in the considered frequency band are presented in the Fig. 7. The intersection of red lines is indicating the WG applicator working frequency of 434 MHz and the threshold −10 dB. From the Fig. 7 it is obvious that for all water bolus thicknesses the $|S_{11}|$ parameter is below −10 dB at this frequency, which is acceptable for hyperthermia since the realized WG applicators have the possibility of the tuning by capacitive pins.

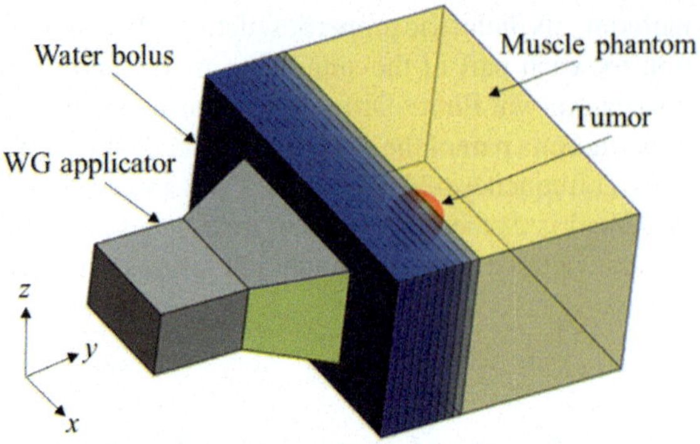

Fig. 6 Simulation setup with WG applicator attached to the water bolus of variable thicknesses

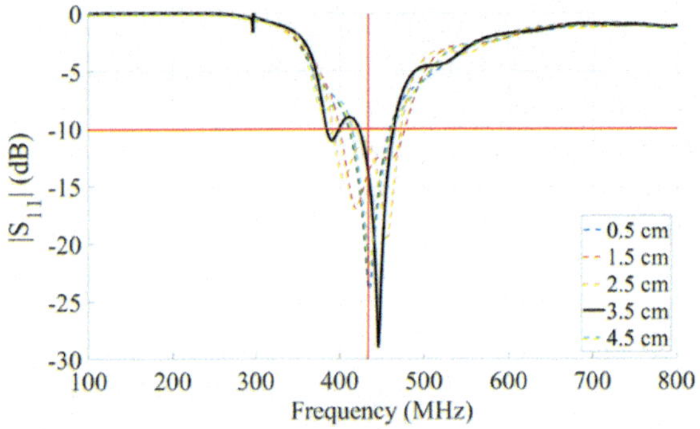

Fig. 7 $|S_{11}|$ parameter for different water bolus thicknesses

3.2 Codependent Antenna Positioning in One Setup

Next, we simulate the hybrid microwave system consisting of WG hyperthermia applicators and dipole sensing antennas. The main goal of this numerical study is to investigate the coexistence possibilities of WG heating applicators and sensing UWB antennas intended for temperature measurement in one setup. We study the influence of the presence of the sensing bow-tie antenna array on the optimized SAR distribution in different setups. First, we modify the microwave hyperthermia system intended for head and neck hyperthermia treatment developed at CTU [61] by reducing the number of WG applicators from four to three. We propose a general numerical setup consisting of a homogeneous cylindrical neck (13 cm in diameter, 15 cm in height, $\varepsilon' = 35$, $\sigma = 0.5$ S/m) in which a spherical target (4 cm in diameter, $\varepsilon' = 65$, $\sigma = 0.7$ S/m) is inserted. The water bolus represented by the 3.5 cm water layer (15 cm in height, $\varepsilon' = 78$, $\sigma = 0.03$ S/m) is placed between the WG applicators and the neck phantom. Three WG applicators with stripline horn apertures filled with distilled water and 12 sensing bow-ties antennas (placed in three rows and four columns) are attached to the model in all considered configurations presented in the following figures. The sensing bow-tie antennas are implemented on the FR-4 substrate (0.08 cm thickness, $\varepsilon' = 4.4$, $\sigma = 0.01$ S/m) with dimensions of 10 mm by 5.6 mm and are fed via printed circuit board (16.0 mm by 17.5 mm) with balun.

The optimization procedure is recommended if the hyperthermia system is complex, this means that the system is composed of two or more hyperthermia applicators. This procedure improves energy delivery to the target and reduces unwanted hotspots by amplitude and phase settings of each applicator. The optimization process is performed for all numerical SAR simulations presented in this section and is divided into two steps:

(1) Numerical simulations of SAR distributions of separated WG applicators.
(2) Optimization of the resulting SAR distribution by settings of amplitude and phase of each applicator. The ideal solution is sought using optimization algorithm Generalized Eigenvalue Approach. The cost function L is defined as follows:

$$L = \frac{\int_{target} w(x, y, z)SAR(x, y, z)dV}{\int_{healthy\_tissue} w(x, y, z)SAR(x, y, z)dV} \tag{9}$$

where SAR is specific absorption rate, $w(x, y, z)$ is a weighting parameter determining the priority and sensitivity of tissue (all tissues are set to 1) and V is the volume. The optimization algorithm is searching for the maximum value of the used cost function L. In other words, the aim is to maximize the SAR in the target while minimizing the SAR in the healthy tissue.

3.2.1 Evaluated Parameters

For the results evaluation of the SAR based simulations, two quantifiers to assess the influence of sensing antennas and WG applicators positioning for all considered scenarios are calculated:

1. *aPA* is the ratio of average power absorbed in the target tissue vs. average power absorbed in the remaining tissue and is defined as follows:

$$aPA = \frac{\overline{P_{target}}}{\overline{P_{healthy\_tissue}}} \tag{10}$$

where $\overline{P_{target}}$ is the mean value of the power absorbed in the target and $\overline{P_{healthytissue}}$ is the mean value of the power absorbed in the healthy tissue. The value of the *aPA* depends on the neck and tumor size but is very useful for comparing results of test scenarios based on same phantom. The *aPA* ratio shows how well the energy is focused to the target.
2. Target coverage (TC_{25}) is a quotient expressing how much of the target is covered by at least 25% iso-SAR calculated from the maximum SAR in the tissue. The TC_{25} is defined:

$$TC_{25} = \frac{V_{target} > 0.25 \cdot \{maxSAR\}}{V_{target}} \cdot 100[\%] \tag{11}$$

where V_{target} is the target volume. TC_{25} quotient values greater than 75% in the target are generally considered in head and neck region as with high potential for successful treatment with sufficient temperature rise in the entire target and TC_{25} values in the range of 25–75% are allowed with temperature monitoring [64].

3.2.2 Initial Scenarios (I–III) – Simulation I

We initially proposed three scenarios how to position the WG applicators and the sensing antenna array (Fig. 8). In the scenario I, the WG applicators are placed next to each other and the sensing antennas are symmetrically placed directly in front of WG applicators. In the scenario II, the WG applicators are at the same position as in the scenario I, but the sensing antennas are moved to the surface of the water bolus and shifted aside from WG applicators 1 and 3. In the scenario III, the WG applicators 1 and 3 are rotated around the center of neck phantom and the sensing antennas are placed in the gaps between the applicators. The mesh quality of the numerical setups has high impact on the accuracy. The mesh for all cases is chosen sensitively with the following number of elements: scenario I $\approx$ 110 MCells, scenario II $\approx$ 120 MCells, and scenario III $\approx$ 140 MCells. For all three arrangements the set of EM simulations and the SAR optimization with presence and without presence of sensing antennas are performed.

The tumor, which is the target, is placed in front of the central WG applicator (WG 2). This applicator supplies most of the power to the tumor (for all scenarios around 80%). The role of the side antennas (WG 1 and WG 3) is to focus and shape the EM field with the aim of increasing SAR in the target and suppressing the hotspots outside it.

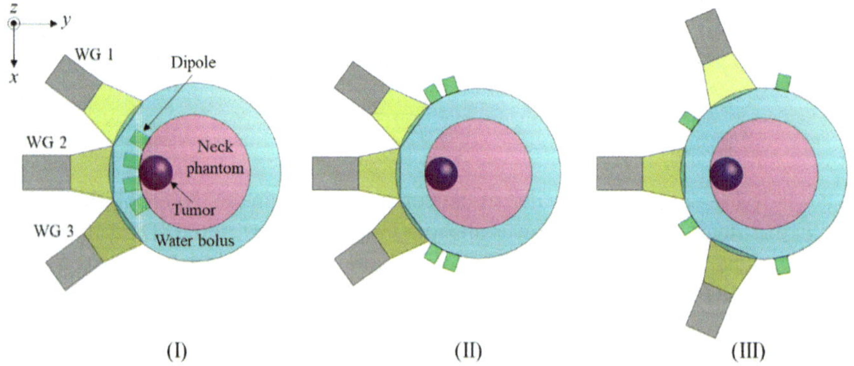

Fig. 8 Initial testing scenarios I–III

Table 2 Resulting parameters for scenarios I–III

Scenario	I		II		III	
Parameter	aPA (−)	TC_{25} (%)	aPA (−)	TC_{25} (%)	aPA (−)	TC_{25} (%)
No antennas	27	92.1	27	92.1	21	52.3
Antennas	21	63.1	26	90.2	21	52.4

The results gained for each scenario are quantified by the aPA ratio and TC_{25} quotient and are presented in the Table 2. In scenarios I and II without sensing antennas, the value of the TC_{25} quotient is 92.1% and aPA ratio is 27. Due to the presence of sensing antennas, the TC_{25} quotient decreases to 63.1% and aPA ratio decreases to 21 in scenario I. When the sensing antennas are moved next to the side WG applicators 1 and 3 (scenario II), the TC_{25} decreases to 90.2% when aPA ratio is 26. In scenario III, the TC_{25} in target reaches low values of around 52% and aPA value is of 21.

The obtained analysis showed that the scenario I is not applicable for hybrid system due to high influence caused by sensing antennas (decrease of TC_{25} by 29%), which are placed directly in front of the applicators. The scenario II is also not applicable because the sensing antennas are located too far from the target (distance 6.5 cm). The scenario III is not applicable because of the low values of TC_{25} for both cases (with/without sensing antennas) as well as the big distance from sensing antennas to the target.

3.2.3 Additional Scenarios (IV–V) – Simulation II

None of the three previously discussed scenarios satisfy both hyperthermia and temperature microwave imaging requirements. Thus, we introduce two additional scenarios (IV and V) which are compromises based on previous scenarios. The additional scenarios IV and V are presented in the Fig. 9. In scenario IV which is based on the scenario II, the water bolus is only in front of the active zone of applicators. So, it is possible to move the sensing antennas directly to the phantom. This change reduces the distance between area of interest and sensing antennas. Scenario V is a modified version of scenario III. The side applicators are rotated around their own axis to be more turned to the target and the water bolus is divided into separated parts for each applicator. The sensing antennas are attached directly to the neck phantom closer to the target than in scenario III.

The simulation proceeds same as for the scenarios I–III. The calculated evaluation parameters for scenarios IV and V are listed in Table 3. From the small TC_{25} changes (91.1% without sensing antennas and 87.2% with presence of sensing antennas) we can deduce that the influence of sensing antennas in scenario IV is very low while the target SAR coverage remains high. The parameters aPA confirm this statement. For scenario V the TC_{25} still reaches low values, namely 25.4% without sensing antennas and 36.1% with antennas, respectively.

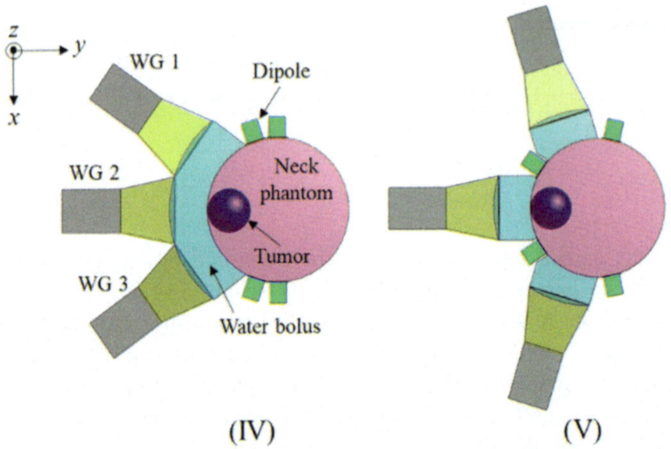

Fig. 9 Additional compromise scenarios IV and V

Table 3 Resulting parameters for scenarios IV and V

Scenario	IV		V	
Parameter	*aPA* (−)	TC_{25} (%)	*aPA* (−)	TC_{25} (%)
No antennas	26	91.1	16	25.4
Antennas	25	87.2	18	36.1

The overview of the outcomes derived from performed simulations is presented in Table 4. Scenarios I, II and IV turned out to be the most suitable for heating due to the high field homogeneity and high energy focus on the target without any significant hotspots (*aPA* is the highest of all tested scenarios) after the optimization procedure. The results of scenarios III and V show that due to stretching of the WG applicators, with the aim of creating a gap for sensing antennas, the field homogeneity decreases and thus the parameter TC_{25} in the target volume will be lower. However, we see the potential in scenario V for improvement since this scenario fits into the requirement of TC_{25} being between 25 and 75% which can be applied with temperature control during HT. From the temperature imaging point of view the scenario I, III and V offers the best situation for good signal acquisition from the heated region. The position of the sensing antennas in scenario IV is not ideal due to the limited antenna positions near the target, but the distance is acceptable. The results of TC_{25} and *aPA* parameters for scenario I show very significant influence by presence of sensing antennas as well as the risk of high power transmission between WG applicators and sensing antennas which can heat the antenna and disturb the data acquisition. This influence can be evaluated as lower for all other tested scenarios. Finally, we can conclude from Table 4 that scenarios IV and V are the appropriate compromises and it is worth pursuing them in further investigations.

Table 4 Overview of the performance of five tested scenarios of hybrid system

	Scenario I	Scenario II	Scenario III	Scenario IV	Scenario V
Heating	+	+	±	+	±
Imaging	+	−	±	±	+
Interference	−	±	±	+	+

3.3 *Sensing Antenna Array Configurations*

After investigating acceptable ways of arranging heating and sensing antennas in general within the hybrid system for MWI-guided HT, in this section we define arrangements of sensing antennas in more detail considering the requirements given by our radar systems.

As mentioned above, the measurement setup for neck hyperthermia temperature monitoring has a cylindrical shape with 10 cm in height. Placing the sensing antennas in quasi direct contact to the skin according to scenarios IV and V from Sect.3.2, the presumed diameter of 13 cm results in approximately 41 cm neck circumference allowing to arrange 16 sensing antennas in a proper way around the neck. As also mentioned above, only 40% of neck circumference is available for sensing antennas placement. Therefore, assuming four sensing antennas can be placed one above the other, 24 sensing antennas arranged in six columns of four antennas can be included in a practical scenario.

As a matter of principle, we specify four basic configurations defining the number of sensing antenna columns between the three heating waveguide applicators, named and numerated as channel configurations ChCnf 0...3, where the digit indicates the number of columns in between. The exact distribution of transmitting and receiving antennas is determined by the fact that our MIMO radar systems consist of single radar units having one transmitter and two receivers. So, in general, we have twice as many receivers as transmitters in each configuration. Figure 10 shows the four configurations where the blue slots represent transmitting antennas, orange indicates receiving ones and the space for hyperthermia applicators is represented in green color.

3.4 *Simulative Validation of the Antenna Configurations*

The imaging capabilities of the four channel configurations are validated numerically in this section based on three tumor imitates which represent three basic tumor types: spherical tumor with 2 cm in diameter (mimics early stage cancer of thyroid gland, larynx or lymph node cancer), spherical tumor of 4 cm in diameter (thyroid cancer in late stages) and cylindrical tumor with 4 cm in height and 2 cm in diameter (esophageal cancer). The dielectric properties correspond to the ones in 3.2, tumors are centered at x = 0 cm, y = −3 cm, z = −5 cm.

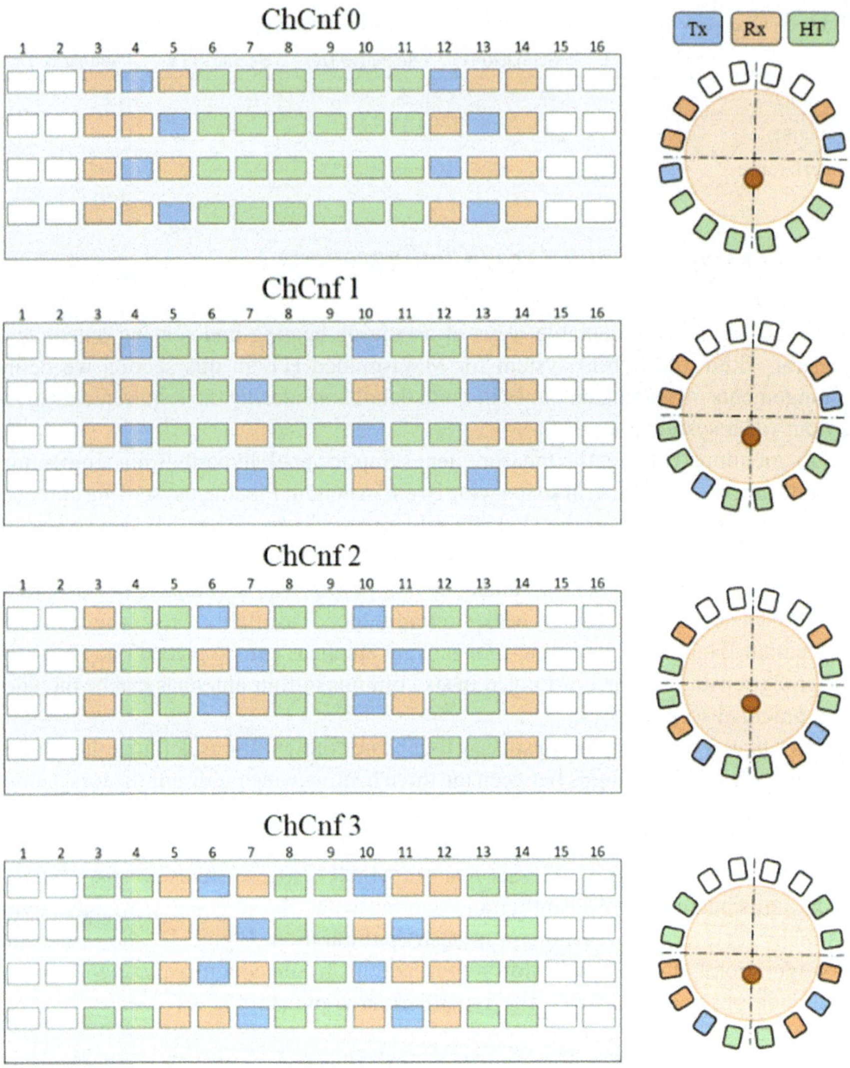

Fig. 10 Antenna placement maps (schematics of the unfolded corpus) of four basic array configurations. Tumor position is shown in brown color

These numerical investigations are also performed with the Sim4Life EM simulation software. After creating the biological model, we define UWB sources (bandwidth 1–6 GHz) to simulate radar signals and then image the recorded signals using Delay and Sum algorithm. In Fig. 11, the UWB images represent reconstructions of horizontal plane at $z = -5$ cm and in Fig. 12 of vertical plane at $x = 0$ cm for each ChCnf and each of three tumor types, respectively.

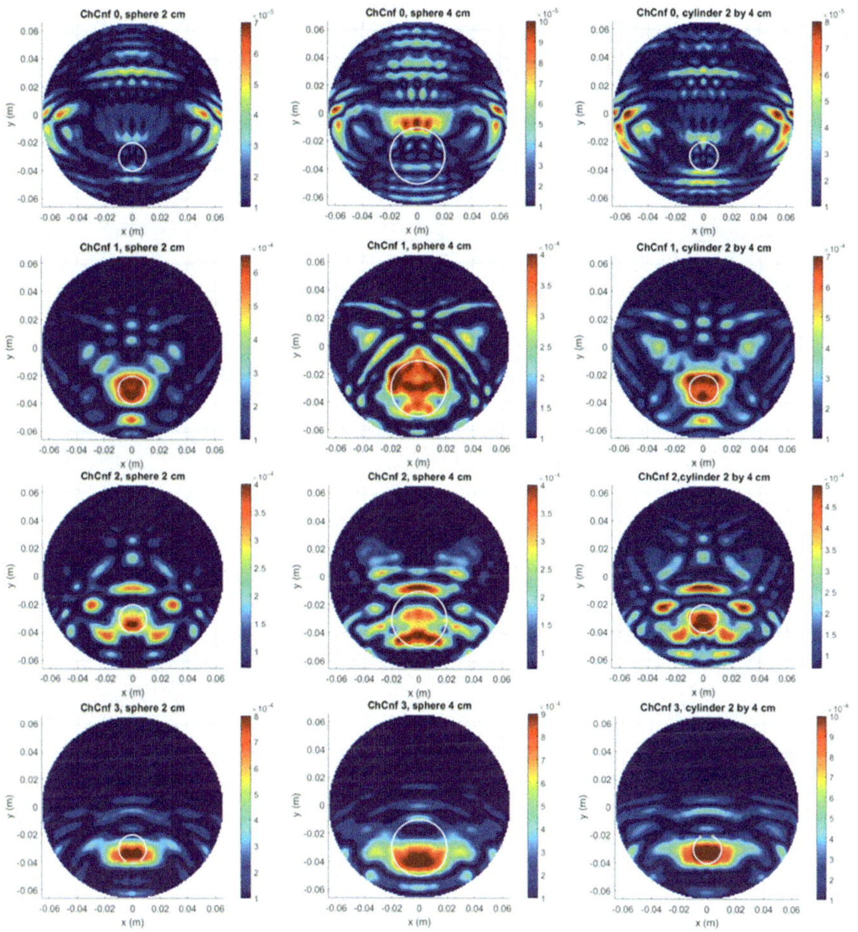

Fig. 11 Horizontal slices of 3D UWB images of the numerical neck phantom with three different tumors applying four investigated antenna arrangements, z = −5 cm, white line indicates shapes and positions of the targets

As we can see from the UWB images based on ChCnf 0 (Figs. 11 and 12 first row), none of the simulated targets is reconstructed and the signal level is almost one order lower than in the other three scenarios. This result is expected, since all sensing antennas are positioned too far from the target area (Fig. 10). Other channel configurations have fully or partly imaged each of three modelled tumors. The most accurate reconstruction results are based on ChCnf 1 since in this version antennas are closer to the tumor and are placed both in front of the target and behind it. In case of the small tumor, image produced by ChCnf 1 is slightly less accurate in comparison to ChCnf 3 due to the small artefact. However, in this configuration can we see the big sphere and the cylinder in their full volume. ChCnf 3, where all sensing antennas are located close to the tumorous region, deals really well with smaller targets but the

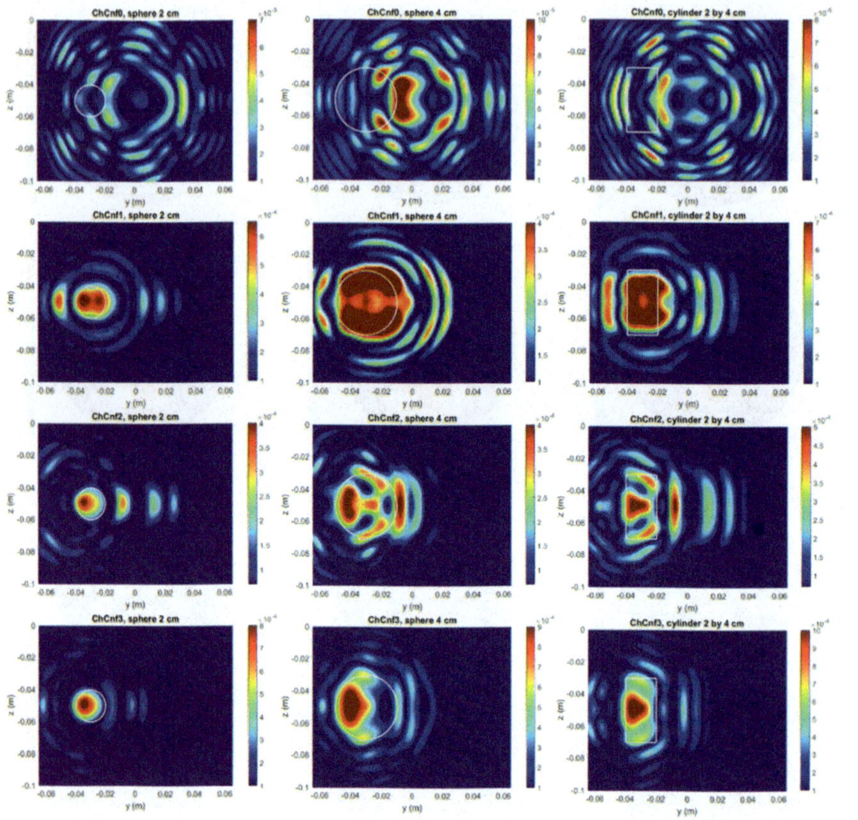

Fig. 12 Vertical slices of 3D UWB images of the numerical neck phantom with three different tumors applying four investigated antenna arrangements, x = 0 cm, white line indicates shapes and positions of the targets

reconstruction of big tumors is only partial (limited to front part of the tumor). The images of ChCnf 2 illustrate correctly that this configuration, having two columns of dipoles close to the target and one column on the side, is an intermediate version of ChCnf 1 and ChCnf 3.

Based on the imaging results, it can be concluded that ChCnf 0 is not suitable for further investigations due to the very limited imaging possibilities. Contrariwise, ChCnf 1 and ChCnf 3 performed well and proved themselves as promising configurations for prototype development. ChCnf 2 representing an intermediate scenario between these versions and has no particular advantages based on these simulation results. Additional simulations with tumor varieties provide supplementary information about possible benefits of the application of each arrangement for real clinical hyperthermia scenarios. In case of early stage neck cancer, while the tumor is small or is located close to the surface of the neck, ChCnf 3 can be advantageous. For late stages or in case of esophageal or tracheal cancer that commonly have prolonged shape and larger dimensions (> 2 cm), ChCnf 1 is more appropriate.

4 Experimental Validation

This section is mainly devoted to the experimental evaluation of the methodical studies and numerical simulations. The experimental setup is introduced including a description of measurement scenario, biological phantom and tissue mimicking materials. Next, qualitative UWB measurements mimicking temperature increase in the tumor during hyperthermia treatment and first quantitative validation of the suggested non-invasive temperature estimation approach are presented. Lastly, a complex anatomically realistic neck phantom as well as a laboratory prototype of the hybrid system developed for UWB MWI-guided hyperthermia in the neck region are presented.

4.1 Measurement Setup

The experimental setup for microwave imaging is based on the UWB MIMO M-sequence radar system (bandwidth 6.5 GHz) presented in [65], which provides 24 transmitters and 48 receivers in maximum. In our application, 8 transmitters and 16 receivers are necessary for each configuration, which results in 128 channels. The bow-tie antennas are differentially fed by passive baluns inserted in the 3D printed corpus with 64 slots (four rings one above the other with 16 slots) allowing to create different antenna arrangements including the considered three configurations ChCnf 1–3. The corpus has a cylindrical inner shape and a 16-sided polygonal outer shape for exact antenna placement.

The neck phantom consists of a cylinder with liquid material mimicking average neck dielectric properties and a tube which imitates the tumor. The tube is inserted in front of the position that is reserved for the central waveguide and, therefore, not equipped with sensing antennas (sides 8 and 9). The antennas are placed in direct contact with a matching layer, which is produced from silicone and carbon mixture, in order to enhance the antenna connection to the cylinder and the wave penetration into the phantom material as well as to unify the antenna crosstalk. The photo of the experimental setup is presented in Fig. 13.

Tissue mimicking materials

For the neck phantom we use a liquid background medium, namely cream. Due to emulsion of fat and water, cream is well suited for mimicking the dielectric properties of average neck tissue in the microwave frequency range. To imitate a realistic decrease of tumor's permittivity during HT treatment, we use water–acetone mixtures with different concentrations. The start of imitated thermal treatment is represented

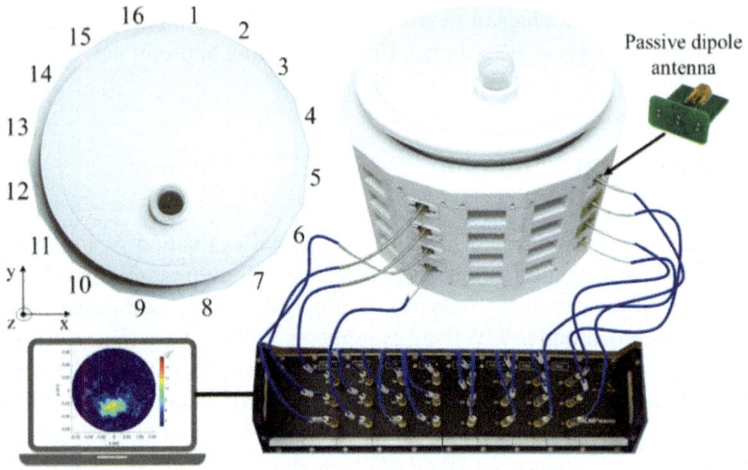

Fig. 13 Measurement setup for microwave UWB imaging including corpus with 64 slots for antennas, matching layer, neck cylinder and tumor mimicking tube

by tumor mimicking material T_0 and heated tumors at five HT stages are indicated as T_1–T_5. The dielectric properties of the tumor imitates are shown in Fig. 14 and their compositions are presented in Table 5.

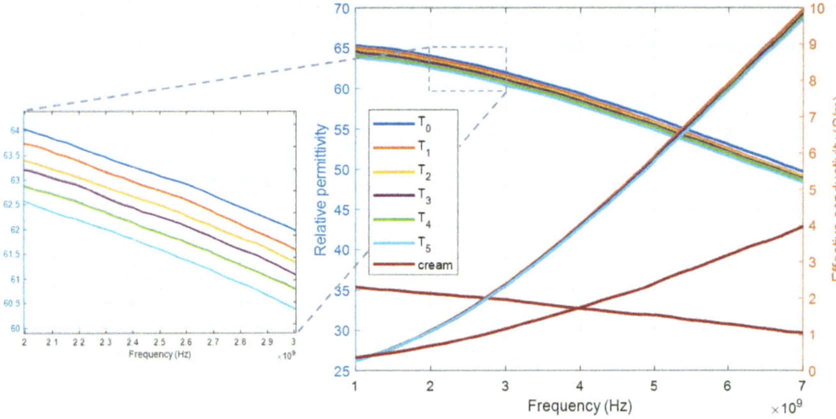

Fig. 14 Relative permittivity and effective conductivity of tumor mimicking materials and cream during imitated hyperthermia treatment

Table 5 Materials under test used in the experiments

Tumor imitate	Composition
T_0	70.0 vol% ddH_2O + 30.0 vol% acetone
T_1	69.7 vol% ddH_2O + 30.3 vol% acetone
T_2	69.4 vol% ddH_2O + 30.6 vol% acetone
T_3	69.1 vol% ddH_2O + 30.9 vol% acetone
T_4	68.8 vol% ddH_2O + 31.2 vol% acetone
T_5	68.5 vol% ddH_2O + 31.5 vol% acetone

4.2 *Experimental Validation of the Antenna Array Configurations*

Differential UWB images (imaging of the differential signal $T_{1...5} - T_0$) for three investigated configurations are generated based on the Delay and Sum beamforming algorithm and presented in Figs. 15, 16 and 17, respectively. The shape and position of the tumor are shown with white line. Color bars represent the intensity of the image in the arbitrary units. We can see that the small changes in tumor's relative permittivity (step of approximately −0.3) were detected and reconstructed. The qualitative correlation between the increasing image values and the increasing change of dielectric properties inside the tumor imitate in comparison to the reference (start of heating) is visible in the images. Please note that the permittivity of the tumor region and, therefore, the dielectric contrast between tumor and healthy surrounding tissue decreases during heating—but the differential UWB signal (i.e. the signal difference between measurements at a certain time during heating $T_{1...5}$ and the start of heating T_0) increases with increasing temperature.

As we can see from the comparison of Figs. 11 and 12 with Figs. 15, 16 and 17, the numerical results seem to be more accurate and error-free. This can be explained by the fact that the simulation is a representation of the ideal experiment: without noise, measurement uncertainties, influence of the presence of plastic phantom cylinder, tube for tumor imitate and matching layer between antennas and the phantom. Despite this, all three investigated antenna arrangements provide basic information about tumor's location in the neck phantom and its shape. In this specific measurement scenario and with application of standard DAS, images reconstructed based on the ChCnf 1 show that target location is not very accurate, moreover, there are artefacts of almost same intensity as in the tumorous region. Contrariwise, images from ChCnf 2 and ChCnf 3 are clean from artefacts and tumor is reconstructed close to its original position in the phantom. However, while in the vertical reconstruction plane the images from these two configurations are similar, in horizontal plane ChCnf 2 has higher precision regarding tumor dimension. Although, this arrangement did not show particularly advantages against ChCnf 1 and ChCnf 3 in numerical investigations, in experimental validation its performance is preferable.

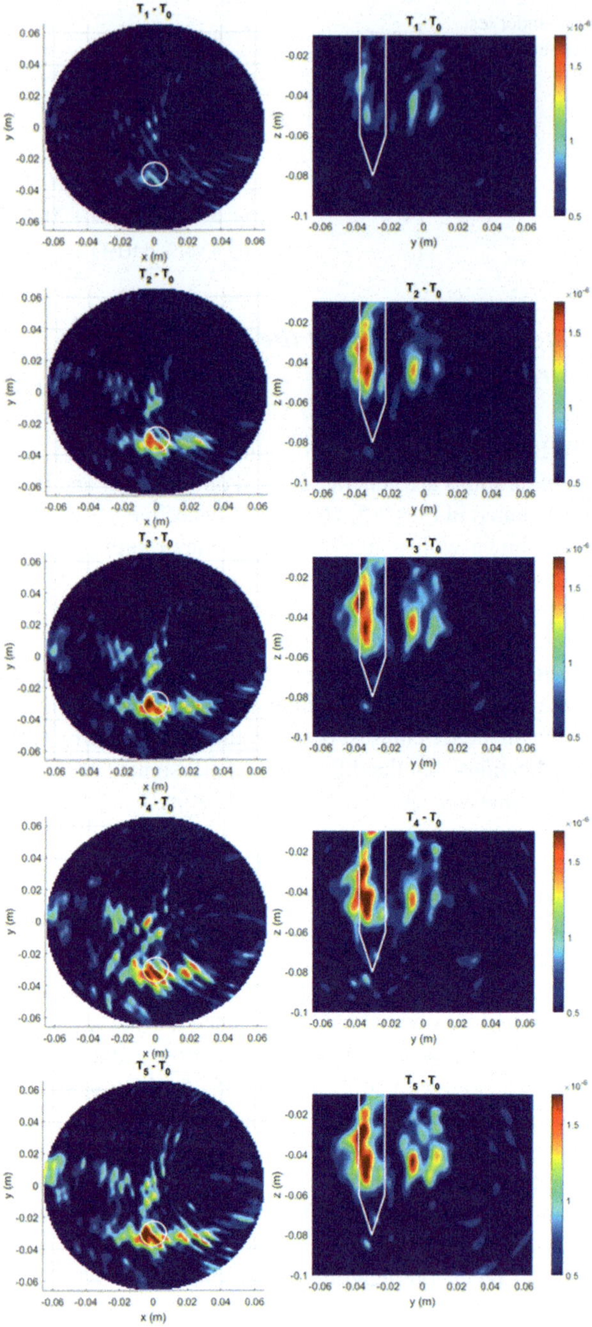

Fig. 15 Differential UWB images based on ChCnf 1

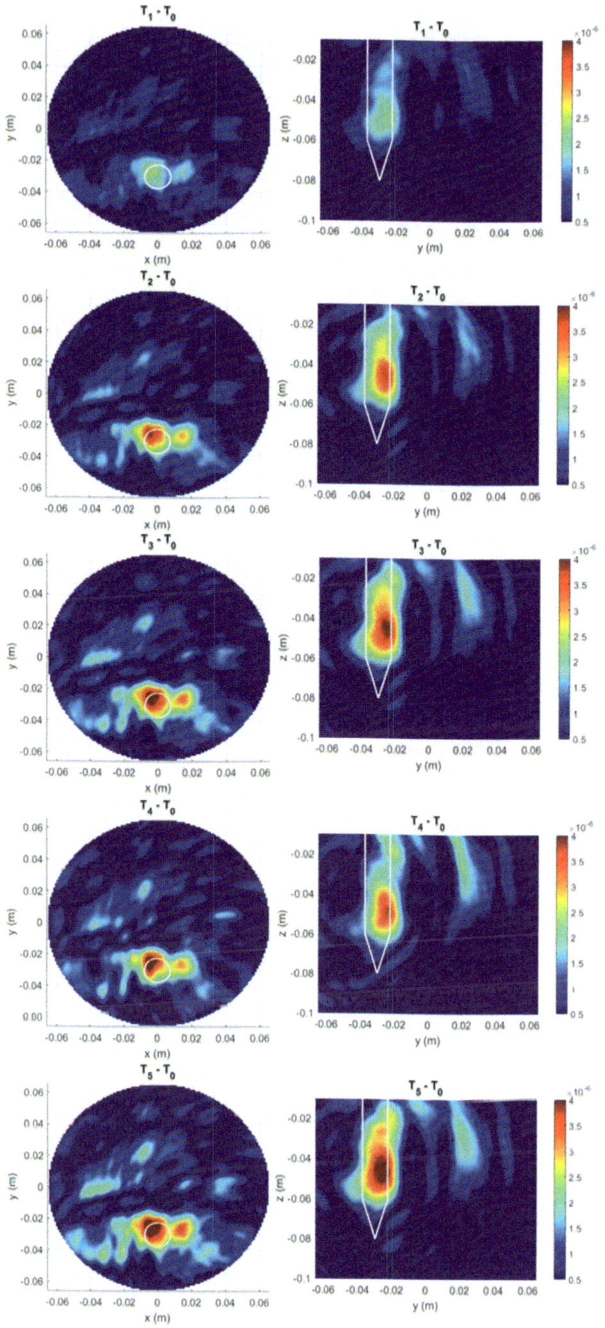

Fig. 16 Differential UWB images based on ChCnf 2

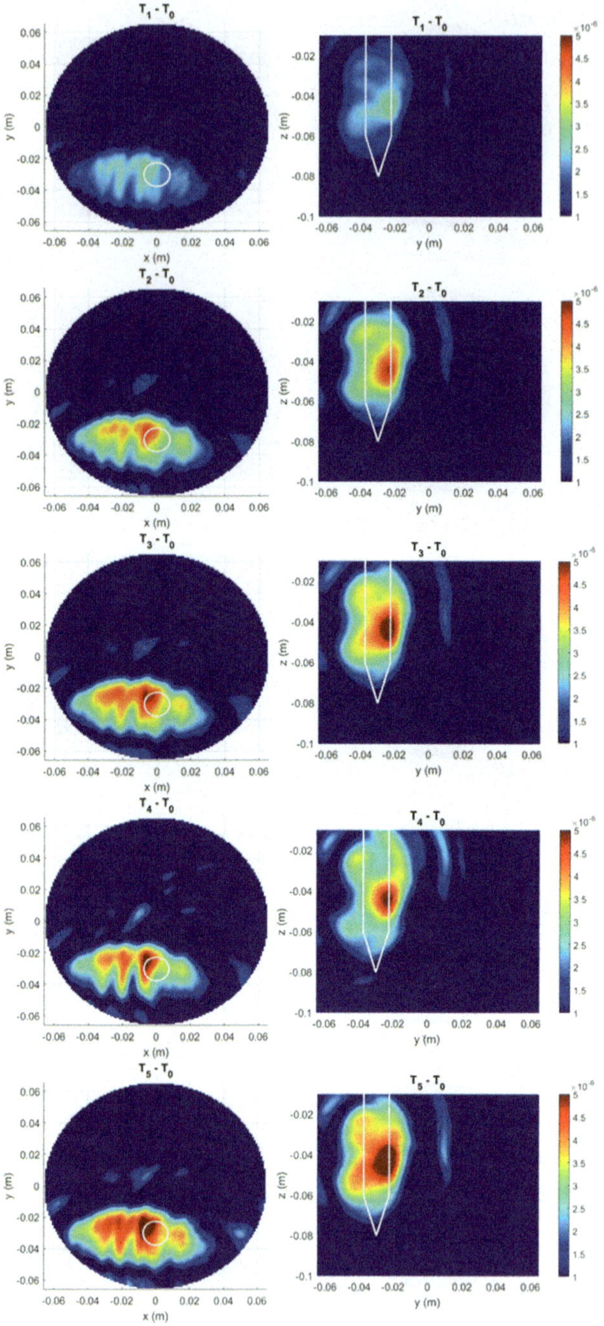

Fig. 17 Differential UWB images based on ChCnf 3

4.3 Experimental Validation of the Non-invasive Temperature Estimation Methodology

This section describes first experimental validation of the temperature estimation approach introduced in Sect. 2.5 and is a summary of the experimental investigations published in [58]. In this experiment we simulate heating of the tumor during hyperthermia treatment and present the quantitative translation of signal amplitudes into changes of the dielectric properties in the tumorous region and respectively, temperature values. Neck phantom and tumor imitates are similar to those described in previous section (cream and water–acetone mixtures). Dielectric properties of the tumor mimicking material at the beginning of HT treatment T_0 are known and the dielectric properties of the tumor imitates simulating increased heating $T_1 \ldots T_4$ are estimated.

After preprocessing (antenna crosstalk removal, subtraction of the effect of plastic tube for tumor imitate [59]), we generate 3D images from each measurement via adapted DAS and TSVD algorithms. In TSVD, the truncation parameter C plays the most important role in the accuracy of further temperature estimation. In this experiment, the optimal parameter is equal to 17 singular values. To monitor the level of the provided heat in the tumor region during hyperthermia, Eq. (8) is applied. The mean values of the estimated and measured via dielectric spectroscopy permittivity decrease inside the tumor imitate are shown in Table 6. The estimated values of relative permittivity can be converted to temperature values based on their temperature dependency. Assuming a temperature sensitivity of 0.1 per 1 °C as described above, all values above 0.8 indicate overheating of the tumorous tissue (8 °C is maximum increase of temperature for HT therapy).

This first validation demonstrate that the estimated permittivity values conform qualitatively with the experiment, i.e. the monotonically ongoing decrease of tumor's permittivity is clearly obvious in both columns of Table 6. The absolute estimation errors are lower than 0.5. This estimation accuracy does not yet live up to practical expectations, but is encouraging for further investigation of this approach. In particular, the estimation algorithm is based on the temperature dependent changes of the reflection coefficient at the boundary between healthy tissue and tumorous tissue to be treated and simplifying assumes that the behavior of the ongoing changing complex reflection coefficients can be linearly described by the ratio of the real-valued UWB

Table 6 Estimated and measured values of mean relative permittivity decrease in the 3D tumor area for DAS and for TSVD

HT stage	Est., DAS, 3 GHz	Est., TSVD, 2 GHz	Measured
$\varepsilon'(T_0) - \varepsilon'(T_1)$	0.63	0.28	0.53
$\varepsilon'(T_0) - \varepsilon'(T_2)$	1.05	0.97	0.92
$\varepsilon'(T_0) - \varepsilon'(T_3)$	1.47	1.38	1.12
$\varepsilon'(T_0) - \varepsilon'(T_4)$	1.83	1.73	1.38

image values. It also assumes that this relationship is frequency independent and valid for the real part as well as for the imaginary part of the permittivity. Obviously, due to different dispersity of tissues and tissue mimicking materials, this assumption does not hold in reality. Furthermore, it has to be noted that the results of [58] presented here were based on a preliminary experimental setup for breast imaging including antenna rotation what will be not possible during hyperthermia of the neck.

4.4 Towards EM Prototype: Design and Realization

After qualitative validation of antenna array arrangements and quantitative validation of the suggested methodology for tissue temperature estimation, a laboratory prototype of the hybrid system for neck hyperthermia with UWB microwave temperature monitoring is designed and assembled.

4.4.1 Design of the Realistic Neck Phantom

The design of the hybrid prototype should include the requirements for both imaging and heating systems as well as meet the criteria of realistic clinical cases of neck cancer. Therefore, a complex anatomical neck phantom is designed and 3D printed. The phantom has cylindrical shape with the average adult neck diameter and height and includes the following structures: thyroid, trachea, esophagus, cervical vertebra, spinal cord, internal carotid arteries, internal jugular veins along with the possibility of skin layer insertion. To provide a higher variety of the possible organ combinations depending on the goal of the experiment, each tissue structure is inserted in a specific holder in the bottom of the phantom and is removable and replaceable. The photo of the printed neck phantom is presented in Fig. 18.

4.4.2 Design of the Laboratory Prototype

Since HT system requires a water bolus between the heating antennas and the phantom/patient's body, individual waveguide holders with a cavity for water bolus are build. Main advantages of this construction are ability of defining individual temperatures of the water bolus for each applicator for more accurate control of the applied hyperthermia and decreased probability of interference between heating and sensing antennas due to absence of water layer in front of the dipole antennas. The photos of the waveguide and the designed holder are presented in Fig. 19.

To provide a possibility to reassemble antenna arrangement depending on the tumor location, size and shape, the corpus side placement is flexible. It enables arbitrary measurement scenarios with appropriate numbers of sensing antennas and heating waveguide applicators.

Fig. 18 Complex heterogeneous neck phantom including: relocatable tumor (1), trachea (2), thyroid gland (3), esophagus (4), two veins (5), two arteries (6), spinal cord (7), vertebra (8) and skin layer (9)

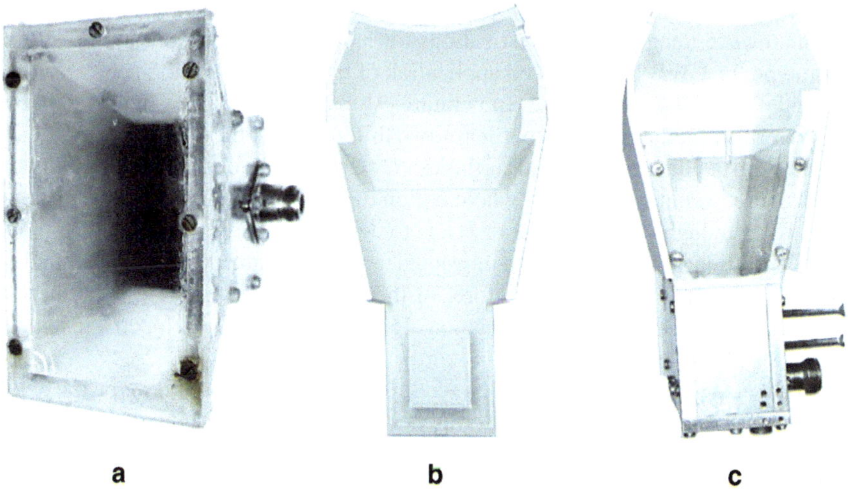

Fig. 19 Photos of the water filled hyperthermia waveguide applicator with strip line horn aperture **a**, 3D printed waveguide holder **b** and assembled applicator in the holder **c**

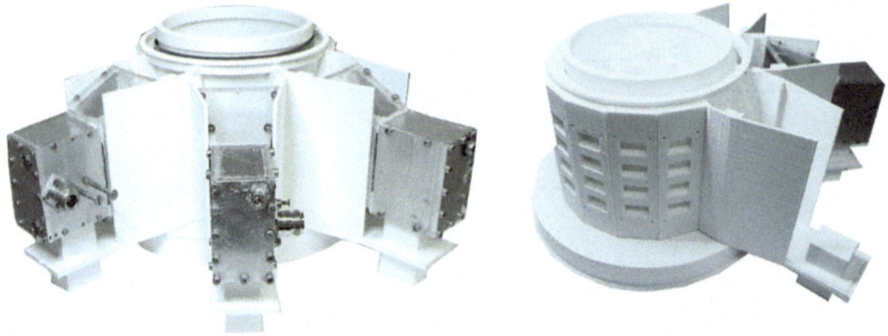

Fig. 20 Photos of the hybrid laboratory prototype for hyperthermia treatment

The laboratory prototype including realistic neck phantom with anatomical structures, corpus with slots for sensing antennas and three waveguide holders with heating applicators is presented in Fig. 20.

5 Conclusions

This chapter presents a methodology for non-invasive temperature estimation based on microwave imaging and investigations towards development of a hybrid system prototype for UWB MWI-guided application of hyperthermia in the neck area.

Initially, we suggest a tissue temperature estimation approach, which is based on the reflection coefficient at the border of two different biological tissues—tumor and healthy surrounding tissue—and furthermore we present its experimental validation. This method is not valid for the whole region of interest, but only for the cancerous area. Even though the main goal of HT is to deliver high level of heat to the tumor itself, detection and localization of unwanted temperature hotspots in the surrounding area is highly needed, too. Therefore, to increase the safety of the hyperthermia procedure, an extension of the estimation algorithm has to be developed and applied prospectively.

In the next step, the clinical design requirements for a hybrid prototype are determined. Two main challenges are interference between heating and sensing antennas as well as limited space for their placement in one setup due to small area of application around the neck. The first issue can be solved by introducing separate water boluses for the HT applicators (as shown in 4.4) and placing heating and sensing antennas so that their polarizations are perpendicular. As for the second point, a good solution presented in this chapter is shifting the two side WG antennas away from the central main WG, so that space for dipoles between the HT applicators is created, providing much better imaging possibilities.

Numerical modeling and simulation provide a good fundament for the construction of appropriate antenna array configurations from both heating and sensing points of view. After the general initial scenarios are tested numerically, they are implemented in our measurement setup and validated experimentally. Finally, a complex heterogeneous neck phantom with anatomical structures is produced and a laboratory prototype of the hybrid system is assembled. After proof of concept based on homogeneous phantoms (tumor imitate and surrounding healthy tissue), the presented system has to be validated based on this realistic neck phantom in the future.

The results of this work revealed that temperature induced changes in the hyperthermia range can be detected and quantitatively interpreted with UWB radar technology, which qualifies the approach for non-invasive temperature monitoring during hyperthermia. In addition, the approach is not limited to neck area and can be adapted for the application in different body regions (e.g. abdominal, breast), since the principles described here hold for any other biological tissues as well. Furthermore, application of Deep Learning for classification of the heated regions during UWB-guided hyperthermia therapy can be promising [see Chap. "Deep Learning Enhanced Medical Microwave Imaging" of this book].

Acknowledgements This work was done in the framework of EMERALD project funded by European Union Horizon 2020 Research and Innovation Programme under the Marie Sklodowska-Curie Actions, Grant Agreement No. 764479.

References

1. H. Sung, J. Ferlay, R.L. Siegel, M. Laversanne, I. Soerjomataram, A. Jemal, F. Bray, Global cancer statistics 2020: GLOBOCAN estimates of incidence and mortality worldwide for 36 cancers in 185 countries. CA. Cancer J. Clin. **71**(3), 209–249 (2021)
2. National Cancer Institute, Types of cancer treatment, in *Nih* (2015), pp. 1–4
3. E.S. Glazer, S.A. Curley, The ongoing history of thermal therapy for cancer. Surg. Oncol. Clin. N. Am. **20**(2) (2011)
4. B.S. Sørensen, M.R. Horsman, Tumor hypoxia: impact on radiation therapy and molecular pathways, in *Frontiers in Oncology*, vol. 10 (Frontiers Media S.A., 21 April 2020), p. 562
5. K. Graham, E. Unger, Overcoming tumor hypoxia as a barrier to radiotherapy, chemotherapy and immunotherapy in cancer treatment. Int. J. Nanomed. **13**, 6049–6058 (2018). Dove Press
6. R. Merten, O. Ott, M. Haderlein, S. Bertz, A. Hartmann, B. Wullich, B. Keck, R. Kühn, C. M. Rödel, C. Weiss, C. Gall, W. Uter, R. Fietkau, Long-term experience of chemoradiotherapy combined with deep regional hyperthermia for organ preservation in high-risk bladder cancer (Ta, Tis, T1, T2) (2019)
7. M.M. Paulides, EM hyperthermia in radiation oncology. 9–13 (Sept 2019)
8. N.R. Datta, B. Pestalozzi, P.A. Clavien, A. Siebenhüner, E. Puric, S. Khan, C. Mamot, O. Riesterer, J. Knuchel, C.S. Reiner, S. Bodis, M. De Oliveira, H. Petrowsky, T. Winder, M. Guckenberger, S. Lang, P. Bauerfeind, B. Bode, O. Tschalèr, D. Marder, T. Roeren, R. Grobholz, B. Seifert, HEATPAC—a phase II randomized study of concurrent thermochemoradiotherapy versus chemoradiotherapy alone in locally advanced pancreatic cancer. Radiat. Oncol. **12**(1), 1–10 (2017)
9. N.G. Huilgol, S. Gupta, C.R. Sridhar, Hyperthermia with radiation in the treatment of locally advanced head and neck cancer: a report of randomized trial. J. Cancer Res. Ther. **6**(4), 492–496 (2010)

10. European Society of Oncology, Publications. [Online]. Available: https://www.esho.info/publications/. Accessed 15 July 2022
11. Z.K. Roesch, P. Tadi, *Anatomy, Head and Neck* (StatPearls Publishing, Neck, 2019)
12. Chapter 25. Overview of the Neck|The Big Picture: Gross Anatomy | AccessMedicine|McGraw Hill Medical." [Online]. Available https://accessmedicine.mhmedical.com/content.aspx?bookid=381&sectionid=40140034. Accessed 15 July 2022
13. National Cancer Institute, A to Z list of cancer types—national cancer institute, in *National Cancer Institute*, 2016. [Online]. Available https://www.cancer.gov/types. Accessed 15 July 2022
14. Head and Neck Cancer: Stages|Cancer.Net. [Online]. Available https://www.cancer.net/cancer-types/head-and-neck-cancer/stages. Accessed 15 July 2022
15. R. Zweije, H.P. Kok, A. Bakker, A. Bel, J. Crezee, Technical and clinical evaluation of the ALBA-4D 70 MHz loco-regional hyperthermia system, in *2018 48th European Microwave Conference, EuMC 2018* (2018), pp. 328–331
16. H. Kato, M. Kuroda, K. Shibuya, S. Kanazawa, Focused deep heating with an inductive type applicator. Therm. Med. **23**(3), 133–143 (2007)
17. M.M. Paulides, G.M. Verduijn, N. Van Holthe, Status quo and directions in deep head and neck hyperthermia. Radiat. Oncol. **11**(1) (2016)
18. H.P. Kok, G. Van Stam, J. Sijbrands, A. Bakker, M.W. Kolff, G. Van Tienhoven, M.C.C.M. Hulshof, J. Crezee, Clinical use of a waveguide hyperthermia system for superficial tumors with deep infiltration, in *European Microwave Week 2017: A Prime Year for a Prime Event, EuMW 2017—Conference Proceedings; 47th European Microwave Conference, EuMC 2017*, vol. 2017 (Jan 2017), pp. 868–871
19. H.D. Trefná, J. Vrba, M. Persson, Physics in medicine and biology to cite this article: Hana Dobšíek Trefná et al. Phys. Med. Biol. Phys. Med. Biol **55**, 2167–2185 (2010)
20. American Cancer Society, Hyperthermia to treat cancer. Am. Cancer Soc. 1–5 (2016)
21. P. Wust, C.H. Cho, B. Hildebrandt, J. Gellermann, Thermal monitoring: Invasive, minimal-invasive and non-invasive approaches. Int. J. Hyperth. **22**(3), 255–262 (2006)
22. E. Schena, D. Tosi, P. Saccomandi, E. Lewis, T. Kim, Fiber optic sensors for temperature monitoring during thermal treatments: an overview. Sensors (Switzerland) **16**(7) (22 July 2016). MDPI AG
23. F. Adibzadeh, K. Sumser, S. Curto, D.T.B. Yeo, A.A. Shishegar, M.M. Paulides, Systematic review of pre-clinical and clinical devices for magnetic resonance-guided radiofrequency hyperthermia. Int. J. Hyperth. **37**(1), 15–27 (01 Jan 2020). Taylor and Francis Ltd
24. D. Mazumder, R.M. Vasu, D. Roy, R. Kanhirodan, A remote temperature sensor for an ultrasound hyperthermia system using the acoustic signal derived from the heating signals. Int. J. Hyperth. **34**(1), 122–131 (2018)
25. E. Groumpas, M. Koutsoupidou, I.S. Karanasiou, C. Papageorgiou, N. Uzunoglu, Real-time passive brain monitoring system using near-field microwave radiometry. IEEE Trans. Biomed. Eng. **67**(1), 158–165 (2020)
26. G. Chen, J. Stang, M. Haynes, E. Leuthardt, M. Moghaddam, Real-time three-dimensional microwave monitoring of interstitial thermal therapy. IEEE Trans. Biomed. Eng. **65**(3), 528–538 (2018)
27. J.C. Bolomey, L. Jofre, Three decades of active microwave imaging achievements, difficulties and future challenges, In *2010 IEEE International Conference Wireless Information Technology System ICWITS 2010*, no. 1 (2010)
28. F.G. Shellock, Magnetic resonance safety update 2002: implants and devices. J. Magn. Reson. Imaging. **16**(5), 485–496 (01 Nov 2002)
29. T. Drizdal, K. Sumser, G.G. Bellizzi, O. Fiser, J. Vrba, G. C.V. Rhoon, D.T.B. Yeo, M.M. Paulides, Simulation guided design of the MRcollar: a MR compatible applicator for deep heating in the head and neck region. Int. J. Hyperth. **38**(1), 382–392 (2021)
30. E.C. Fear, X. Li, S.C. Hagness, M.A. Stuchly, Confocal microwave imaging for breast cancer detection: localization of tumors in three dimensions. IEEE Trans. Biomed. Eng. **49**(8), 812–822 (2002)

31. N.R. Epstein, P.M. Meaney, K.D. Paulsen, 3D parallel-detection microwave tomography for clinical breast imaging. Rev. Sci. Instrum. **85**(12), (Dec 2014)
32. T. Henriksson, M. Klemm, D. Gibbins, J. Leendertz, T. Horseman, A.W. Preece, R. Benjamin, I.J. Craddock, Clinical trials of a multistatic UWB radar for breast imaging, in *LAPC 2011–2011 Loughborough Antennas and Propagation Conference* (2011)
33. P.H. Tournier, F. Hecht, F. Nataf, S. Semenov, M. Bonazzoli, F. Rapetti, V. Dolean, I. El Kanfoud, I. Aliferis, C. Migliaccio, C. Pichot, Microwave tomography for brain stroke imaging, in *2017 IEEE Antennas and Propagation Society International Symposium, Proceedings*, vol. 2017 (Jan 2017), pp. 29–30
34. J. Ljungqvist, S. Candefjord, M. Persson, L. Jönsson, T. Skoglund, M. Elam, Clinical evaluation of a microwave-based device for detection of traumatic intracranial hemorrhage. J. Neurotrauma **34**(13), 2176–2182 (2017)
35. A.T. Mobashsher, A.M. Abbosh, Y. Wang, Microwave system to detect traumatic brain injuries using compact unidirectional antenna and wideband transceiver with verification on realistic head phantom. IEEE Trans. Microw. Theory Tech. **62**(9), 1826–1836 (2014)
36. A. Zamani, S.A. Rezaeieh, A.M. Abbosh, Lung cancer detection using frequency-domain microwave imaging. Electron. Lett. **51**(10), 740–741 (2015)
37. F. Töpfer, S. Dudorov, J. Oberhammer, Millimeter-wave near-field probe designed for high-resolution skin cancer diagnosis. IEEE Trans. Microw. Theory Tech. **63**(6), 2050–2059 (2015)
38. E. Porter, A. La Gioia, S. Salahuddin, S. Decker, A. Shahzad, M.A. Elahi, M. O'Halloran, O. Beyan, Minimum information for dielectric measurements of biological tissues (MINDER): a framework for repeatable and reusable data. Int. J. RF Microw. Comput. Eng. **28**(3), e21201 (2018)
39. S. Salahuddin, L. Farrugia, C.V. Sammut, M. O'Halloran, E. Porter, Dielectric properties of fresh human blood, in *Proceedings of the 2017 19th International Conference on Electromagnetics in Advanced Applications, ICEAA 2017* (2017), pp. 356–359
40. S. Gabriel, R.W. Lau, C. Gabriel, The dielectric properties of biological tissues: II. Measurements in the frequency range 10 Hz to 20 GHz. Phys. Med. Biol. **41**(11), 2251–2269 (1996)
41. A. Martellosio, M. Pasian, M. Bozzi, L. Perregrini, A. Mazzanti, F. Svelto, P.E. Summers, G. Renne, L. Preda, M. Bellomi, Dielectric properties characterization from 0.5 to 50 GHz of breast cancer tissues. IEEE Trans. Microw. Theory Tech. **65**(3), 998–1011 (2017)
42. V. Lopresto, R. Pinto, G.A. Lovisolo, M. Cavagnaro, Changes in the dielectric properties of ex vivo bovine liver during microwave thermal ablation at 2.45 GHz. Phys. Med. Biol. **57**(8), 2309–2327 (2012)
43. A.P. O'Rourke, M. Lazebnik, J.M. Bertram, M.C. Converse, S.C. Hagness, J.G. Webster, D.M. Mahvi, Dielectric properties of human normal, malignant and cirrhotic liver tissue: in vivo and ex vivo measurements from 0.5 to 20 GHz using a precision open-ended coaxial probe. Phys. Med. Biol. **52**(15), 4707–4719 (2007)
44. M. Lazebnik, M.C. Converse, J.H. Booske, S.C. Hagness, Ultrawideband temperature-dependent dielectric properties of animal liver tissue in the microwave frequency range. Phys. Med. Biol. **51**(7), 1941–1955 (2006)
45. J. Bonello, A. Elahi, L. Farrguia, C.V. Sammut, M. O'Hollaran, E. Porter, R. Balduino, Temperature-Dependent Dielectric Properties of Lung, in *2018 2nd URSI Atlantic Radio Science Meeting, AT-RASC 2018* (2018)
46. A. Prokhorova, S. Ley, O. Fiser, J. Vrba, J. Sachs, M. Helbig, Temperature dependent dielectric properties of tissue mimicking phantom material in the microwave frequency range, in *14th European Conference Antennas Propagation, EuCAP 2020* (2020), pp. 1–5
47. S. Ley, S. Schilling, O. Fiser, J. Vrba, J. Sachs, M. Helbig, Ultra-wideband temperature dependent dielectric spectroscopy of porcine tissue and blood in the microwave frequency range. Sensors (Switzerland) **19**(7), 1707 (2019)
48. D. Byrne, M. Sarafianou, I.J. Craddock, Compound radar approach for breast imaging. IEEE Trans. Biomed. Eng. **64**(1), 40–51 (2017)

49. E.J. Bond, B.D. Van Veen, S.C. Hagness, X. Li, An overview of ultra-wideband microwave imaging via space-time beamforming for early-stage breast-cancer detection. IEEE Antennas Propag. Mag. **47**(1), 19–34 (2005)
50. M. Helbig, B. Faenger, S. Ley, I. Hilger, Multistatic M-sequence UWB radar system for microwave breast imaging, in *2021 IEEE conference on antenna measurements and applications, CAMA 2021* (2021), pp. 540–545
51. I. Saied, T. Arslan, S. Chandran, C. Smith, T. Spires-Jones, S. Pal, Non-invasive RF technique for detecting different stages of Alzheimer's disease and imaging beta-amyloid plaques and tau tangles in the brain. IEEE Trans. Med. Imaging **39**(12), 4060–4070 (2020)
52. D.O. Rodriguez-Duarte, J.A. Tobon Vasquez, R. Scapaticci, G. Turvani, M. Cavagnaro, M.R. Casu, L. Crocco, F. Vipiana, Experimental validation of a microwave system for brain stroke 3-d imaging. Diagnostics **11**(7) (2021)
53. J. Sachs, M. Helbig, R. Herrmann, M. Kmec, K. Schilling, E. Zaikov, Remote vital sign detection for rescue, security, and medical care by ultra-wideband pseudo-noise radar. Ad Hoc Networks **13**(PART A), 42–53 (Feb 2014)
54. H. Xu, M.P. Ebrahim, K. Hasan, F. Heydari, P. Howley, M.R. Yuce, Accurate heart rate and respiration rate detection based on a higher-order harmonics peak selection method using radar non-contact sensors. Sensors **22**(1), 83 (2022)
55. J. Sachs, Handbook of ultra-wideband short-range sensing: theory, sensors, applications. (Wiley-VCH 2012)
56. M. Helbig, K. Dahlke, I. Hilger, M. Kmec, J. Sachs, Design and test of an imaging system for UWB breast cancer detection. Frequenz **66**(11–12), 387–394 (2012)
57. R. Scapaticci, V. Lopresto, R. Pinto, M. Cavagnaro, L. Crocco, Monitoring thermal ablation via microwave tomography: an Ex vivo experimental assessment. Diagnostics **8**(4), 81 (2018)
58. A. Prokhorova, S. Ley, A.Y. Ruiz, R. Scapaticci, L. Crocco, M. Helbig, Preliminary investigations of microwave imaging algorithms for tissue temperature estimation during hyperthermia treatment, in *2021 International Conference on Electromagnetics in Advanced Applications, ICEAA 2021* (2021), pp. 79–84
59. A. Prokhorova, S. Ley, M. Helbig, Quantitative interpretation of UWB radar images for non-invasive tissue temperature estimation during hyperthermia. Diagnostics **11**(5), 818 (2021)
60. S. Ley, O. Fiser, I. Merunka, J. Vrba, J. Sachs, M. Helbig, Preliminary investigations for non-invasive temperature change detection in thermotherapy by means of UWB microwave radar, in *Conference Proceedings Annual International Conference IEEE Engineering Medicine Biology Society*, vol. 2018 (July 2018), pp. 5386–5389
61. O. Fišer, Microwave hyperthermia for treatment of head and neck tumors controlled by non-invasive temperature monitoring based on UWB radar doctoral thesis (2018)
62. O. Fiser, I. Merunka, J. Vrba, Waveguide applicator system for head and neck hyperthermia treatment. J. Electr. Eng. Technol. **11**(6), 1744–1753 (2016)
63. A.N. Vasavada, J. Danaraj, G.P. Siegmund, Head and neck anthropometry, vertebral geometry and neck strength in height-matched men and women. J. Biomech. **41**(1), 114–121 (2008)
64. M.M. Paulides, G.M. Verduijn, N. Van Holthe, Status quo and directions in deep head and neck hyperthermia. Radiat. Oncol. **11**(1), 1–14 (2016)
65. M. Helbig, J.H. Koch, S. Ley, R. Herrmann, M. Kmec, K. Schilling, J. Sachs, Development and test of a massive MIMO system for fast medical UWB imaging, in *Proceedings of the 2017 19th International Conference on Electromagnetics in Advanced Applications, ICEAA 2017* (2017), pp. 1331–1334

An Initial Assessment of a Microwave Imaging System to Monitor Microwave Ablation Treatments

Mengchu Wang, Marta Cavagnaro, Rosa Scapaticci, Sandra Costanzo, and Lorenzo Crocco

Abstract This chapter presents design guidelines and an initial experimental assessment of a microwave imaging system to monitor liver ablation treatments. The hypothesis at the basis of using microwave imaging for real-time monitoring of thermal ablation procedures is related to the change of the tissue dielectric properties during the treatment. Firstly, the optimal working conditions in terms of operating frequency and dielectric properties of the coupling medium in which the MWI antennas are embedded are determined. In this respect, a general approach based on the use of transmission lines is used. Secondly, a compact slot-loaded antipodal Vivaldi antenna is designed within the proposed working conditions, realized, and experimentally assessed. Following a previous in-silico assessment of a possible experimental arrangement, a simple yet representative experimental set-up for the validation of the imaging system is considered. The set-up foresees 8 antennas

M. Wang (✉) · M. Cavagnaro
Department of Information Engineering, Electronics, and Telecommunications (DIET), University of Rome "La Sapienza", 00184 Rome, Italy
e-mail: wang.m@irea.cnr.it

M. Cavagnaro
e-mail: marta.cavagnaro@uniroma1.it

M. Wang · M. Cavagnaro · R. Scapaticci · S. Costanzo · L. Crocco
National Research Council of Italy Institute for Electromagnetic Sensing of the Environment (CNR-IREA), 80124 Napoli, Italy
e-mail: scapaticci.r@irea.cnr.it

S. Costanzo
e-mail: sandra.costanzo@unical.it

L. Crocco
e-mail: crocco.l@irea.cnr.it

S. Costanzo
Department of Computer Engineering, Modeling, Electronics and Systems (DEIS), University of Calabria, 87036 Rende, Italy

Inter-University National Research Center on Interactions Between Electromagnetic Fields and Biosystems (ICEmB), 16145 Genoa, Italy

F. Vipiana and L. Crocco (eds.), *Electromagnetic Imaging for a Novel Generation of Medical Devices*, Lecture Notes in Bioengineering,
https://doi.org/10.1007/978-3-031-28666-7_11

inserted inside a tank filled with the coupling material; in front of the antenna array, a 3-D printed ellipsoidal phantom filled with tissue-mimicking liquid represents the thermally ablated zone. The proposed system is experimentally assessed with one antenna mechanically moved in a linear motion and measuring the signal in absence and in the presence of the 3-D printed phantom. The retrieved images inside the region of interest show that the designed system is able to detect the position of the ablated zone. The experimental results show the feasibility of the developed system for liver ablation monitoring that might be useful for new researchers to understand the challenges involved.

Keywords Microwave imaging · Liver ablation · Vivaldi antenna · Coupling medium · Dielectric properties measurement · Imaging algorithms

1 Introduction

Liver cancer is a malignancy that occurs in liver cells. In 2020, it was the sixth most common cancer, and the third most deadly cancer worldwide, with an increasing yearly fatality rate [1]. So far, the golden standard treatment for liver cancer is surgery [2]. However, the prognosis for liver cancer is patient-dependent: in practice, only 5–15% of the patients are suitable for surgical removal [3]. Accordingly, other treatments have been developed, such as thermal ablation, embolization therapy, radiation therapy, target drug therapy, chemotherapy, and immunotherapy [2]. Until now, radiofrequency ablation, cryoablation, and more recently, microwave ablation are the main thermal ablation techniques that have been adopted on the clinical side [4]. In particular, microwave thermal ablation is minimally invasive, low-cost, and rapid. Accordingly, thermal ablation can significantly reduce patients' morbidity and the mortality rate as compared to other treatment procedures [5].

Microwave thermal ablation destroys cancer cells by using electromagnetic (EM) energy at microwave frequencies to increase the temperature in the biological tissues. To achieve thermal ablation, temperatures as high as 60 °C are needed. Accordingly, in some systems, close to the radiating antenna in the center of the ablated zone, temperatures above 100 °C are reached, leading to carbonization of the tissue [6]. The ablated zone usually shows an ellipsoidal shape which is also called the coagulation zone, corresponding to the irreversibly damaged biological tissue. There is a thin layer of hyperemic red rim around the coagulation zone called the transition zone, which contains viable cells [7]. Only the coagulation zone is considered to be "successfully ablated". A coagulation zone that includes the whole tumor plus a safety margin, which in the liver tumor is usually about 5 mm, corresponds to a positive conclusion of the procedure, as insufficient ablated tumorous cells will lead to cancer recurrence [8].

It is well understood that the continuous monitoring of the evolution of the ablation zone during ablation procedures remains difficult. Computed tomography (CT) has the potential to provide real-time imaging during thermal ablation [9]. However, the exploited ionizing radiation has an intrinsic cancer risk towards both the patient and the medical staff, thus making it unsuitable [10]. Magnetic resonance imaging (MRI) currently shows well-validated techniques for near real-time temperature monitoring for cryoablation [11]. However, its use as a tool to monitor the microwave thermal ablation process is not straightforward due to compatibility issues with the magnetic field caused by the current flowing on the microwave antenna (presence of rotating forces on the antenna, presence of noise in the images). Additionally, its high cost and bulky size would increase too much the cost and flexibility of the procedure. Finally, Ultrasound (US) has attractive features such as being portable, cost-effective, and capable of real-time imaging. However, it is difficult to monitor the ablation process because the US sensor is blinded by the hyper-echogenic cloud caused by water vaporization [12].

The diversity of outcomes and their unpredictability has been the main hurdle in the development of thermal ablation treatments. In this respect, the lack of an accurate and objective real-time imaging system for temperature monitoring purposes is one of the most important limitations. Because of this drawback, the effectiveness of the treatment mostly relies on the clinician's expertise and skills [13].

Following the above, the development of real-time monitoring systems is a fundamental issue that researchers must confront to enable the deployment of thermal ablation to reach its full potential. A new technology that can overcome all the above issues would be needed. To this end, a novel imaging modality has been recently proposed, namely microwave imaging (MWI) [14]. MWI is portable, low-cost, uses non-ionizing radiation, and it is capable of real-time imaging. The technique reconstructs the map of the dielectric properties of an unknown target from the knowledge of the scattered EM field. Accordingly, a MWI system is made of an array of antennas which surround the region to be investigated, emit the probing EM field, and receive the scattered one. Then, appropriate techniques are used to elaborate the received field and derive the map of the dielectric properties of the investigated area. Up to now, a considerable amount of research work was dedicated to microwave imaging for biomedical applications, such as breast cancer detection [15], brain stroke monitoring [16], thermal therapy monitoring [17], and osteoporosis diagnosis [18]. In these applications, MWI detects the presence of the target (e.g., the tumor) due to the contrast of the permittivity and the conductivity, which is observed between different tissues according to their kind and pathological conditions (e.g., the healthy and malignant tissues) [19]. The hypothesis at the basis of using MWI for real-time monitoring of thermal ablation procedures is related to the change that tissue dielectric properties undergo during the treatment [20]. Indeed, during microwave thermal ablation, the water molecules in the ablation zone dramatically decrease due to the heating of the tissues. This results in a change of the dielectric properties values in the ablated tissue, as compared to the untreated one [20], which can be detected by a microwave imaging system.

An initial feasibility study of MWI for real-time monitoring of thermal ablation procedures of the liver was performed in [21]. The study had some limitations, as the used antennas and configuration were not optimized for the application at hand. Additionally, measurements were performed only before and after a microwave thermal ablation procedure, while the actual goal is to perform continuously monitoring during the the treatment. However, the promising outcomes motivated the research activity presented in this chapter. Here, the guidelines for the design of a MWI system for real-time monitoring of thermal ablation are set and used to design the system. In this respect, it is useful to recall that the MWI system gets the information on the evolution of the thermally ablated area by measuring the scattering parameters of the antennas in the MWI system and elaborating them with MWI techniques.

The organization of the chapter is as follows: in Sect. 2, the optimal MWI system design parameters are discussed. In particular, the system working frequency and the dielectric properties of a coupling medium in with the MWI antennas are embedded are analyzed. The design of the antenna for the MWI system is reported in Sect. 3. Finally, the experimental validation of the MWI system is shown in Sect. 4. The conclusion and future remarks are discussed in Sect. 5.

2 Identification of the Microwave Imaging System Design Parameters

2.1 *Study of Abdomen Tissue Properties*

To design the microwave imaging system for liver thermal ablation monitoring, the optimal parameters of the system should be identified. In particular, at first the optimum frequency band of the imaging system should be defined within the microwave region of the spectrum, i.e., within about 100 MHz to a few GHz [22]. Additionally, since human body tissues are quite different from air, to reduce reflections of the incident EM field at the air-skin boundary, MWI antennas are usually embedded into a material (coupling medium). Accordingly, the dielectric properties of the coupling medium should be optimized as well. To this end, the interaction between the EM field and the abdomen region of the human body has to be studied.

The portion of the abdominal region of interest for the applications at hand mainly consists of four types of tissues, i.e., skin, fat, muscle, and liver. The dielectric properties of the mentioned biological tissues in the frequency band of interest are mainly described by the polarization of the water molecules inside the tissue [23], represented by a single-pole Cole–Cole model [24]:

$$\varepsilon = \varepsilon'(\omega) - j\varepsilon''(\omega) = \varepsilon_\infty + \frac{\Delta\varepsilon}{1 + (j\omega\tau)^{1-\alpha}} + \frac{\sigma_i}{j\omega\varepsilon_0}, \tag{1}$$

where ε' is the real part of the dielectric permittivity, ε'' is the imaginary part of the dielectric permittivity, ε_∞ is the dielectric permittivity at infinite frequency, $\Delta\varepsilon$ is the magnitude of the dispersion, ω is the angular frequency, $\omega = 2\pi f$, f is the frequency, τ is the relaxation time, α is the distribution parameter, σ_i is the static ionic conductivity, and ε_0 is the permittivity of free space.

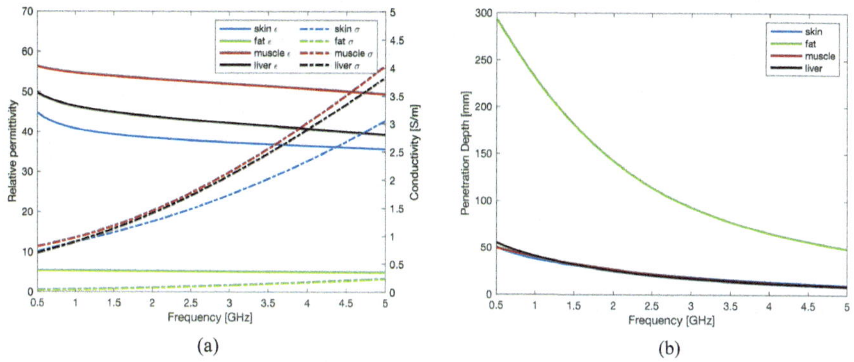

Fig. 1 **a** Human abdomen tissues' properties; **b** Tissue penetration depth

The dielectric properties of the four mentioned tissues in the frequency range of interest are shown in Fig. 1a, while Fig. 1b reports the corresponding EM wave penetration depth [25]. It is found that the penetration depth in lossy tissues like skin, muscle, and liver is much shorter as compared to that in the fat. In general, penetration depth decreases inside the tissue as frequency increases. When the working frequency raises above 5 GHz, the penetration depth in the skin, muscle, and liver tissue reduces to below 10 mm. This is one of the main reasons why the majority of MWI systems operate below 6 GHz, although higher frequencies would entail higher spatial resolution.

To investigate the best MWI set-up in terms of frequency band, number of antennas, array configuration, and so on, as well as to perform numerical studies on the MWI system, the accurate knowledge of the changes of the dielectric properties of tissues as a function of the temperature and frequency is required.

The dielectric properties of the liver, before and after the MWA procedure, are shown in Fig. 2 [26]. From these results, it can be seen that post-ablation permittivity is reduced as compared to these pre-ablation one due to the loss of water.

2.2 *A Simple Model of the Abdomen*

In an MWI system, the coupling medium properties and the antennas' frequency band have to be chosen in such a way that the largest possible portion of the radiated electromagnetic power can enter the abdomen and provide a meaningful backscattered signal after interacting with the liver. For the sake of generality, it is important to translate the complex anatomy structure into a simple model.

As it was already discussed in [27], the region of interest can be represented (in a first-order approximation) by a semi-infinite multi-layer planar slab, in this case composed of skin, fat, muscle, and liver. Similarly, the EM field propagating within the coupling medium and impinging on the layered model can be modeled as a

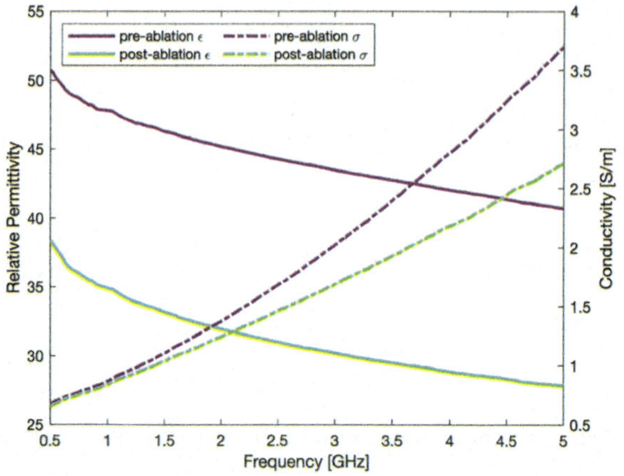

Fig. 2 Pre- and post-ablation liver tissue dielectric properties

plane wave traveling inside the coupling medium in the direction orthogonal to the interface between the medium and the skin. Based on average statistics [28, 29], the thickness of the skin (l_s) was taken equal to 2.3 mm, that of fat $\left(l_f\right)$ 12.2 mm, and that of muscle (l_m) 20.2 mm. The last layer, mimicking the liver, is considered to be semi-infinite in-depth, being the target of the analysis. The above-described layered structure exposed to the plane wave with orthogonal incidence can be conveniently studied through the formalism of transmission lines [30], as sketched in Fig. 3.

Following the transmission line equivalence, the characteristic impedance of the line segment corresponding to the n-th tissue is given as [30]:

$$Z_n = \sqrt{\mu_0/\varepsilon_0\varepsilon_n}, \tag{2}$$

where ε_n is the complex relative permittivity of the n-th tissue, as given in Eq. (1).

In the following, Z_{mm}, Z_s, Z_f, Z_m, and Z_l denote the characteristic impedances of coupling medium, skin, fat, muscle, and liver, respectively. According to the impedance transfer equation, the equivalent impedance at interface CC′ (Fig. 3) between fat and muscle is [30]:

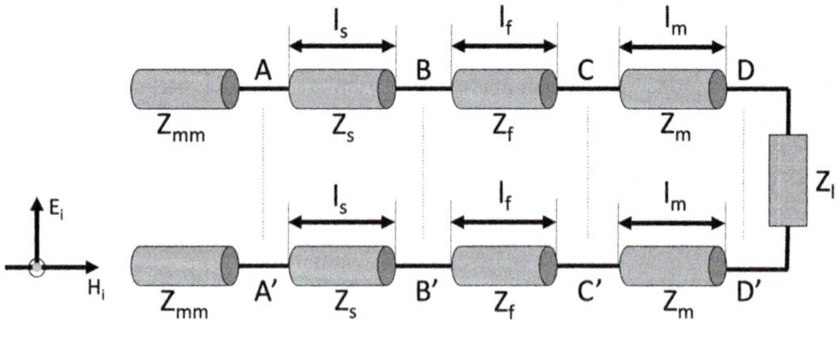

Fig. 3 Transmission line model of a plane wave orthogonally impinging on a planar layered structure

$$Z_{CC'} = Z_m \cdot \frac{Z_{DD'} + jZ_m \cdot tank_m l_m}{Z_m + jZ_{DD'} \cdot tank_m l_m}, \tag{3}$$

where

$$Z_{DD'} = Z_l, \tag{4}$$

represents the impedance of the liver and k_m the complex wavenumber in the lossy tissue. The equivalent impedance at interface BB' between skin and fat is calculated with an equation readily obtained from Eq. (3), simply substituting Z_f to Z_m, l_f to l_m, k_f to k_m and $Z_{CC'}$ to $Z_{DD'}$. Similarly, the equivalent impedance at interface AA' between coupling medium and the skin is:

$$Z_{AA'} = Z_s \cdot \frac{Z_{BB'} + jZ_s \cdot tank_s l_s}{Z_s + jZ_{BB'} \cdot tank_s l_s}, \tag{5}$$

From the knowledge of $Z_{AA'}$ the reflection coefficient of the electric field at the interface between the coupling medium and skin can be finally evaluated as:

$$\Gamma_{AA'} = \frac{Z_{AA'} - Z_{mm}}{Z_{AA'} + Z_{mm}}, \tag{6}$$

while the transmission coefficient for the power at the interface with the coupling medium is given as:

$$T_{AA'} = 1 - |\Gamma_{AA'}|^2 \tag{7}$$

Based on the above formulas, the transmission coefficient at the interface between a generic lossless medium, with relative permittivity varying between 1 and 80 is calculated in a frequency range between 500 MHz and 5 GHz. Figure 4a shows the calculated transmission coefficient as a function of the coupling medium permittivity and frequency. The plot shows that similar to what happens in brain stroke monitoring applications [31], a 'forbidden transmission band' occurs from 2 GHz to 3.5 GHz, where the transmission coefficient is lower than 0.5. Such an effect arises because the layered structure is made by a low permittivity layer (fat), enclosed between two higher permittivity layers (skin and muscle, respectively), behaving like a waveguide at some frequencies [32]. Accordingly, between 2 GHz and 3.5 GHz less power is delivered to the target as compared to other portions of the frequency spectrum. The power transmission increases again above 3.5 GHz. However, the penetration depth of the EM wave severely decreases at higher frequencies (see Fig. 1b), thus making it difficult if not impossible, to accurately measure useful signals.

Following the above considerations, Fig. 4a suggests that the best choice would be a frequency band of 500 MHz–2 GHz and a coupling medium with a permittivity value close to 23. In fact, although lower permittivity values show better transmission at low frequencies, the 2–3.5 GHz stopband shows almost no power transmission to

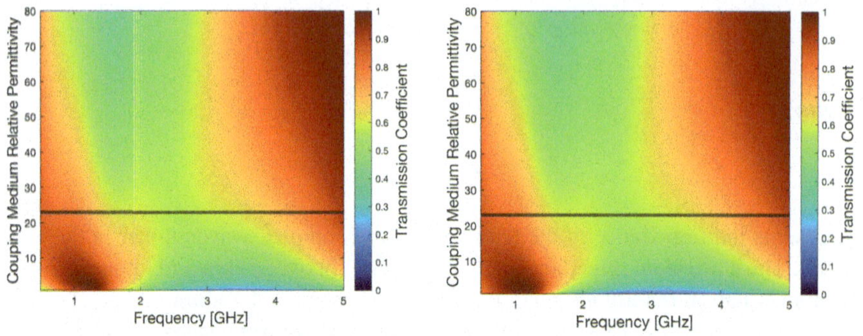

(a) (b)

Fig. 4 Power transmission coefficient for different coupling medium permittivities; **a** lossless coupling medium, **b** coupling medium with a conductivity of 0.07 S/m

the target. On the contrary, higher permittivity values show degraded performances at the low frequencies.

A permittivity value of 23 allows for an improvement in the achievable spatial resolution of a factor of about 5, as compared to the case of free space, while not suffering of the degradation of transmission which is observed for higher permittivity.

To verify if the unavoidable presence of losses may have an effect on the previous outcomes, the transmission coefficient is recomputed by considering a coupling medium with losses. In particular, the losses associated with a realistic medium, obtained with a water–oil emulsion and showing a permittivity of about 23 at 915 MHz [33], are considered. Accordingly, the 1-D transmission analysis is repeated considering a coupling medium with relative permittivity between 1 and 80 and a conductivity of 0.07 S/m. As it can be seen from Fig. 4b, the above-reported findings are confirmed.

2.3 Practical Realization of the Coupling Medium

After defining the theoretical dielectric properties of the coupling medium, its realization is the next step. So far, various coupling mediums have been used for MWI systems, such as saline water [34], triton X-100/water mixture [35], glycerin/water mixture [36], vegetable oil [37], oil/water emulsion [33], solid urethane rubber with graphite powder [38], etc.

The coupling medium for the microwave imaging system to monitor liver ablation should have the following features: almost constant dielectric properties along the frequency of interest (0.5–2 GHz), dielectric properties stable in time, non-toxicity, low cost. Moreover, the medium should be easy to make. The existing media that have been used for other MWI systems have different limitations with respect to the considered application. The permittivity of the saline water is higher than the

Table 1 Coupling medium recipe

Components	Ratio by weight (%)
Distilled water	37.80
Sunflower oil	57.84
Guar gum	0.42
Dishwashing detergent	3.94

target value 23; the conductivity of the triton X-100 or glycerin water mixture is relatively high; the permittivity of the vegetable oil is lower than the target value; oil/water emulsions could achieve the desired dielectric properties, but the water and oil in the medium tend to separate very quickly. The solid urethane rubber with graphite powder could achieve the desired dielectric properties, however, its preparation is complex. Regarding the mentioned difficulties, a new coupling medium is proposed as a mixture of water, oil, dishwashing detergent, and guar gum. The recipe is improved from the oil/water emulsion in [33] by adding guar gum as a thickening agent to increase the fluid viscosity. This solves the separation issue between phases of the mixture (water and oil), and avoids the need for continuous circulation. However, the drawback of the proposed mixture is related to the opacity of the achieved medium. The components leading the coupling medium to achieve the target permittivity of 23, are given in Table 1.

The preparation procedure of the coupling medium works as follows:

- Fill a container with distilled water, then smoothly sprinkle the guar gum powder on the surface of the water. Gently stir the mixture with a whisk until the guar gum powder is dissolved in the water, taking care of avoiding air bubbles during this process. If any lump remains undissolved, use a sieve to remove it.
- Add dishwashing detergent into the solution and gently mix it.
- Add the sunflower oil into the mixture and stir it until all ingredients are well incorporated.

The dielectric properties of the coupling medium were measured with the open-ended coaxial probe technique, using the Keysight high-temperature probe (Keysight 85070E) connected to a vector network analyzer (P5002A Keysight Streamline, 9 kHz to 9 GHz). The measurement uncertainty of the system is equal to 5%. 201 frequency points were measured from 500 MHz to 5 GHz. To verify whether the coupling medium's dielectric properties are stable in time, the medium was measured three times in a one-week observation time. The first measurement (day 0) was performed soon after the preparation of the coupling medium, the second measurement 24 h after the first measurement (day 1), and the third measurement 7 days after the first measurement (day 7). Figure 5a, b demonstrate the measured properties. It is found that, in the frequency band of interest, the relative permittivity of the medium is close to the target value $\varepsilon_r = 23$ on day 0. It is worth also to mention that the coupling medium remains stable during the one-week observation time.

The mean values and standard derivation of the coupling medium dielectric properties along the frequency band of 500 MHz–5 GHz are reported in Table 2.

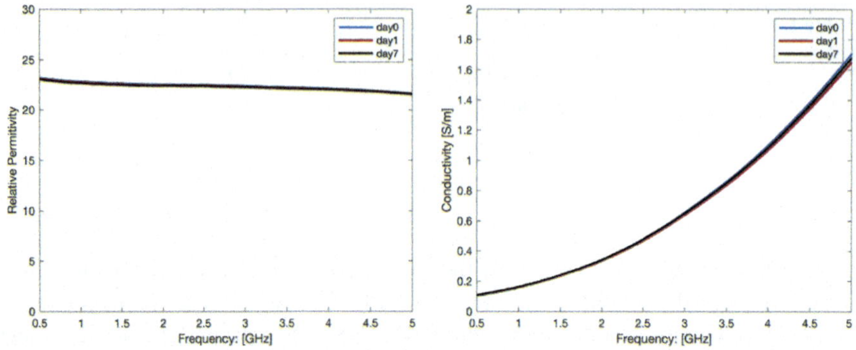

(a) (b)

Fig. 5 Measured dielectric properties of the coupling medium: **a** permittivity; **b** conductivity

Table 2 Dielectric parameters of the coupling medium

Relative permittivity ε_r		Conductivity σ [S/m]	
Mean	STD	Mean	STD
22.36	1.55	0.67	6.37×10^{-3}

3 Antennas for the Microwave Imaging System

3.1 Antenna Design

In an MWI system, the antenna plays a crucial role, as it establishes the amount of electromagnetic power transmitted to the investigated region, and it represents the sensing element which receives the scattered field [39]. It is well understood that microwave imaging resolution can be increased either by using high frequencies, or by using multiple antennas arranged in an array [40]. However, the design of antennas for biomedical applications is not a trivial task. These antennas usually have to work in close proximity to human body tissues, wherein the electromagnetic wave attenuates severely as the frequency increases [41]. Besides, as discussed in the previous section, even at lower frequencies some forbidden bands can occur. This limits the portion of the electromagnetic spectrum that can be effectively used. Finally, due to the restricted body area allowed for measurements [42], it is crucial to design an antenna with compact dimensions, while preserving its performances at lower frequencies. Summarizing the reported considerations, biomedical MWI systems demand the antennas to be wideband, efficient in terms of power transmission, and compact.

The challenge of antenna miniaturization was early discussed by Hansen et al. [43]. It is well understood that the reduction of antenna physical size always comes together with a reduction of its bandwidth and Q factor. Several different antenna miniaturization techniques have been investigated to scale down the antenna without degradation of its performances. Among these, etching slots on the antenna's radiator proved to be an efficient way. This method could elongate the surface current path and hence increase the antenna's electrical length without increasing the overall dimension [44]. Promising studies such as [45, 46] have shown that the slots on the radiator could enhance the antipodal Vivaldi antenna's lowest bandwidth without increasing its physical dimension. Furthermore, adding parasitic elements at the aperture side of the antenna showed improvements in the antenna radiation and enhancement of its bandwidth at low frequencies [47].

With reference to the optimal working conditions described in the previous section (coupling medium properties $\varepsilon_r = 23$ and working frequency band 0.5–2 GHz), three possible antennas were designed. The first antenna was an antipodal Vivaldi antenna using RT/Duroid 6010LM as substrate [27]. Then, to improve antenna radiation when working inside the coupling medium closely to the abdomen region, an antipodal Vivaldi antenna with a cone-shaped ceramic lens was proposed [48]. The presence of the cone-shaped ceramic lens improved the antenna's radiation inside the phantom, without degrading the antenna's matching or increasing the antenna's dimensions. However, the drawback of the antenna in [48] is related to its fabrication difficulty. Therefore, to further reduce the dimension of the antenna, a slot-loaded antipodal Vivaldi antenna was designed, aiming at the miniaturization of antenna dimension, while still maintaining its working bandwidth at lower frequencies.

The slot-loaded antipodal Vivaldi antenna's geometry is shown in Fig. 6. To reduce the antenna dimensions with respect to those proposed in [27], the following miniaturization techniques were applied: firstly, the antenna's width was trimmed by 20 mm (dimension Wa in Fig. 6). This step reduced the antenna's aperture in front of the human abdomen region from 60 mm down to 40 mm. However, the reduction of the antenna's physical dimension also reduced the antenna's electrical length, hence, the antenna's lowest working frequency was shifted towards higher frequencies. To maintain the antenna's lowest working frequency as close as possible to 500 MHz, a slot was etched in the middle of the radiator and the ground plane to elongate the current path. The slot width, length, and position were not chosen arbitrarily. These parameters were selected after analyzing the influence of the different geometrical characteristics on the antenna's matching. In particular, simulations were performed with the parameter sweep function of the CST-MW Studio software (Dassault Systèmes, Vélizy-Villacoublay, France) on the following parameters: R_1, R_2, D_1, D_s, θ, L_s, W_p, W_s (see Fig. 6). The resulting antenna's dimensions are listed in Table 3. In this design, due to realization availabilities, the chosen substrate is Arlon AR1000, with a relative permittivity $\varepsilon_r = 9.8$, a thickness of 1.575 mm (0.062″), and a conductive copper layer thickness equal to 0.035 mm.

3.2 *Experimental Validation of the Antenna*

This section shows the experimental realization and characterization of the antenna described in the previous section. The antenna's performance is characterized both with a single antenna as well as in the array configuration.

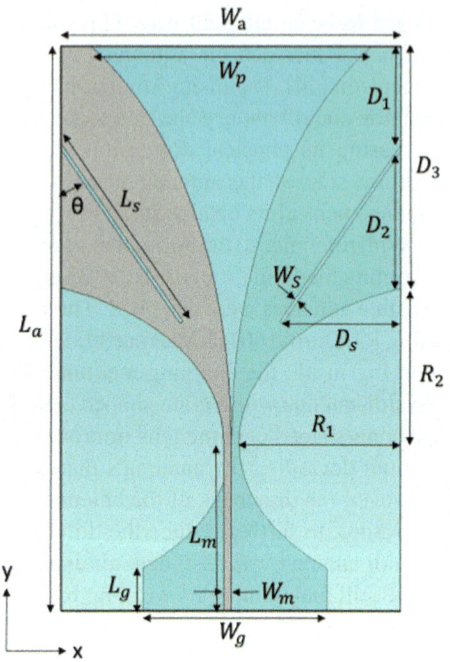

Fig. 6 Slot-loaded antipodal Vivaldi antenna geometry (gray: metallic layer, blue: antenna substrate)

Table 3 Dimensions of the slot-loaded antipodal Vivaldi antenna

Parameters	Value [mm]	Parameters	Value [mm]
W_a	40	L_a	65
W_p	32.9	L_g	5
W_g	21.7	L_m	19.5
W_m	0.9	L_s	24.9
W_s	0.4	D_1	11.2
R_1	19.1	D_2	15.8
R_2	17.8	D_3	27.7
θ	35°	D_s	14

The fabricated antenna is shown in Fig. 7a. The antenna's matching is measured by connecting one antenna to a vector network analyzer (P5002A Keysight Streamline, 9 kHz to 9 GHz) and placing the antenna in a container filled with the coupling medium. The coupling medium was made on the same day of the experiment. As discussed in [27], the antenna experiences a mismatching issue when it is placed inside the coupling medium and fed by an SMA connector. Similar to the simulation result, the same mismatching issue is observed during the antenna measurement. This issue is fixed by applying epoxy resin on the connector pin, in order to isolate it from the coupling medium, as shown in Fig. 7b.

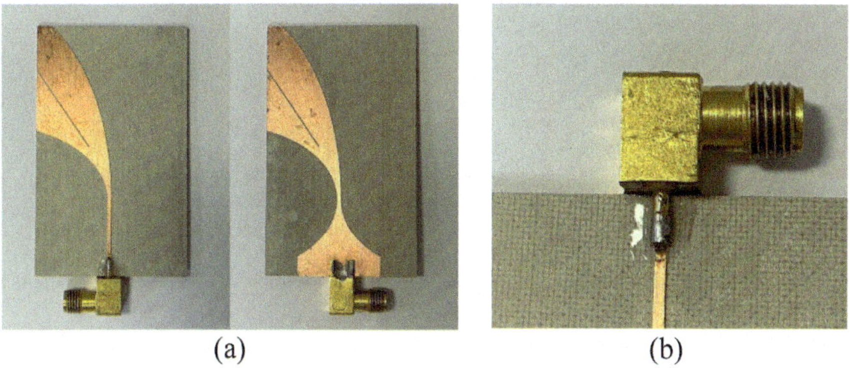

Fig. 7 **a** Fabricated antenna sample; **b** SMA connector pin covered by epoxy resin (realization performed into ERMIAS laboratory at university of Calabria)

The comparison between simulations and measurements results is reported in Fig. 8, where a quite good agreement is achieved. It is worth mentioning that the antenna is simulated with the coaxial connector covered by the epoxy resin and in the coupling medium whose dielectric properties are those measured on day 0. From the measured S-parameter, a matching from 600 MHz to 3 GHz is obtained, with the possibility of working even at higher frequencies.

Given the scope of the antenna, i.e., to be used close to the human abdomen, the near field distribution is of interest. The E-field distribution of the antenna placed inside the coupling medium is shown in Fig. 9. It can be seen that the E-field amplitude decreases as frequency increases. This is due to the fact that the coupling medium has greater losses at higher frequencies.

In order to verify antenna's performances when located in the array configuration, the antenna's matching was evaluated placing two antennas close to each other (see

Fig. 8 Simulated and measured S-parameter of one antenna in the coupling medium

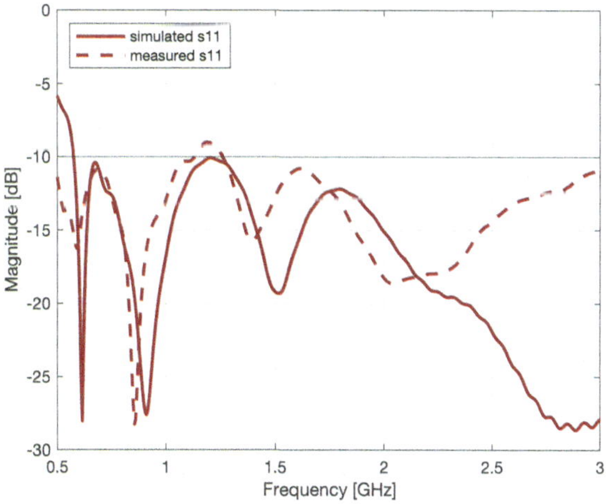

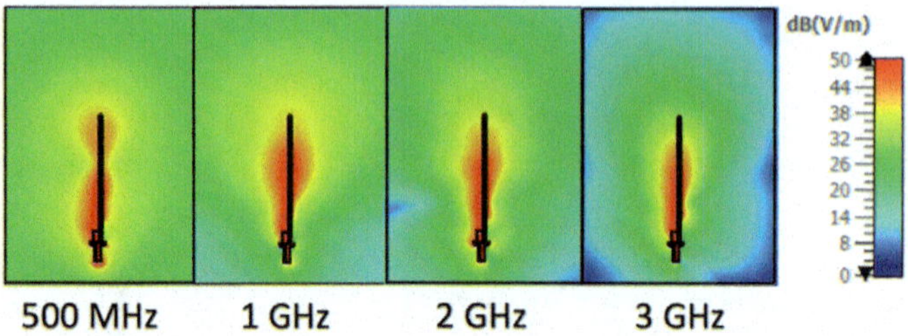

Fig. 9 Slot-loaded antipodal Vivaldi antenna's E-field distribution in the coupling medium at different frequencies

Fig. 10a). The distance between each antenna was 23 mm, which is larger than a quarter wavelength at 1 GHz in the coupling medium. The experimental set-up is shown in Fig. 10b.

The simulated and measurement result are shown in Fig. 11. In the figure only S_{12} is reported since the system is reciprocal, and as a consequence S_{21} is exactly the same. It is found that the simulation result well agrees with the measurement one. The lower experimental values of S_{12} with respect to the numerical ones are most likely related to higher losses in the realized coupling material. Additionally, it can be seen that the antenna's matching at low frequency, especially around 600 MHz, is preserved even with the presence of another antenna, thus assessing the good performance of the designed slot-loaded antipodal Vivaldi antenna against mutual

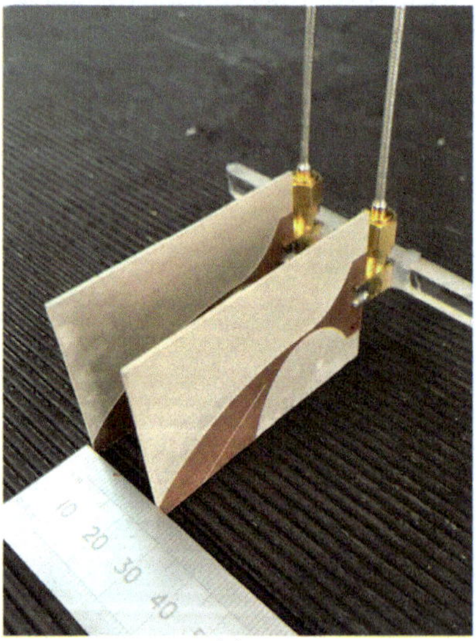

(a)

(b)

Fig. 10 **a** Antennas in array configuration; **b** antennas' measurement experimental set-up (ERMIAS laboratory at university of Calabria)

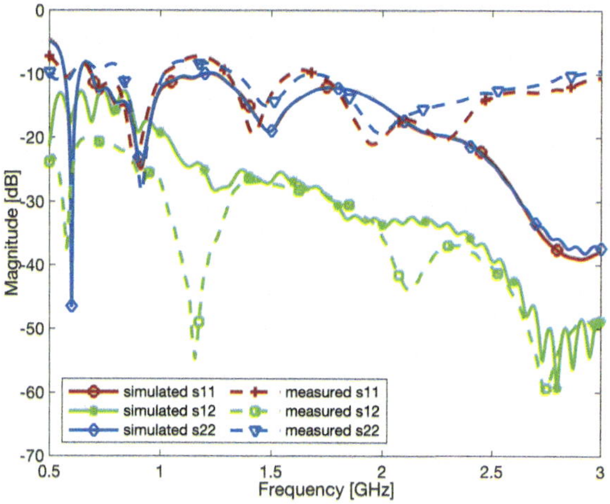

Fig. 11 Simulation and measurement result of two antennas in an array configuration in the coupling medium

coupling. This achievement is partly due to the antenna design and partly to the presence of the lossy background medium.

4 Experimental Validation of the Microwave Imaging System

4.1 MWI Experimental Set-Up

In [49] an MWI device to monitor liver thermal ablation treatment was proposed. The MWI system was made by 8 antennas located on a straight line, with a distance of 23 mm between two adjacent radiators. The performances of the system were verified in-silico, showing its ability to reveal the position and the dimension of the ablation zone, as well as different ablation stages.

Following the satisfying results in [49], the experimental validation of the MWI system should be the next step. For the sake of simplicity, instead of using 8 antennas as in the MWI system in [49], in this study only one antenna is placed in the coupling medium and moved linearly in front of an ablation liver phantom in 8 different positions, in step of 23 mm.

Figure 12a shows the top view (the projection in the y–z plane) of the experimental set-up: a container made of acrylonitrile butadiene styrene (ABS) and with a size of $210 \times 210 \times 210$ mm^3, (width $\times$ length $\times$ height) is filled with the coupling medium designed in Sect. 2.3; the antenna is inserted into the container, completely immersed into the coupling medium, and moved linearly in front of the phantom; the phantom is located 35 mm away from the antennas tip. In Fig. 12a, the region of interest (ROI) for the imaging system, which has a dimension of $160 \times 60 \times 60$ mm^3 (y–x–z), is marked.

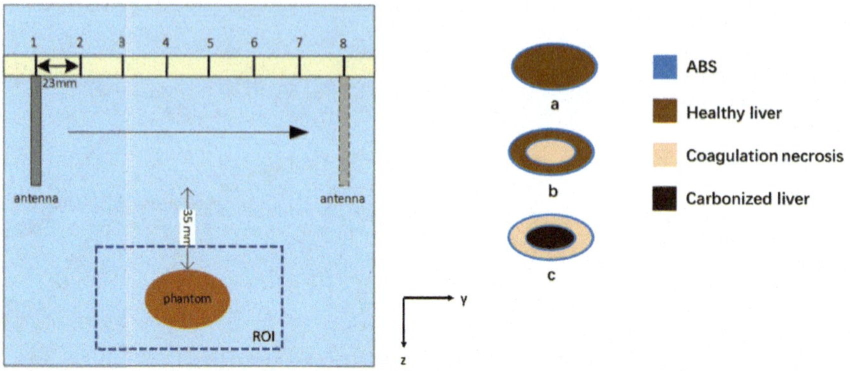

(a) (b)

Fig. 12 **a** Top view of the set-up configuration; **b** different phantoms considered in the experimental set-up

As the ablation zone typically exhibits an ellipsoidal shape, a simple ellipsoidal structure is designed to mimic the thermally ablated area evolution during the ablation treatment [50]. The dimensions of the phantom are translated from the experimental results in [50], while the dielectric properties of the pre-ablated liver and the coagulated region are taken from the data in [26]. The dielectric properties of the carbonized liver are taken from [20], which are obtained at a single frequency ($\varepsilon_r = 8.33$, $\sigma = 0.39$ S/m, @ 2.45 GHz). In more detail, the actual phantom is fabricated with 3D-printed ABS structures ($\varepsilon_r = 3$, $\sigma = 4 \times 10^{-3}$ S/m) whose thickness is 1.5 mm. The phantom consists of two nested ellipsoids, the outer one having axes with dimensions of $60 \times 40 \times 40$ mm^3 (y–x–z) and the inner one with dimensions of $34 \times 13 \times 13$ mm^3 (y–x–z). The origin of the z-axis is positioned at the tip of the antennas.

By filling the two ellipsoidal regions with different materials, it was possible to simulate three different stages of the treatment, namely (Fig. 12b):

(a) Pre-ablation treatment scenario: both ellipsoids are filled with the untreated liver tissue mimicking material (phantom **a** in Fig. 12b);
(b) Ongoing ablation treatment scenario: the outer ellipsoid is filled with the untreated liver and the inner one with coagulated tissue mimicking material (phantom **b** in Fig. 12b);
(c) Post-ablation treatment scenario: the outer ellipsoid is filled with the coagulated tissue, and the inner one is filled with the carbonized tissue mimicking material (phantom **c** in Fig. 12b).

The phantom structure is shown in Fig. 13a. More details on the panthom can be found in the first chapter of this book. In practice, the phantom is connected to a rack through three pipelines for filling the ellipsoids with the different tissue-mimicking materials and hang above the tank. The experimental implementation of the MWI system is shown in Fig. 13b.

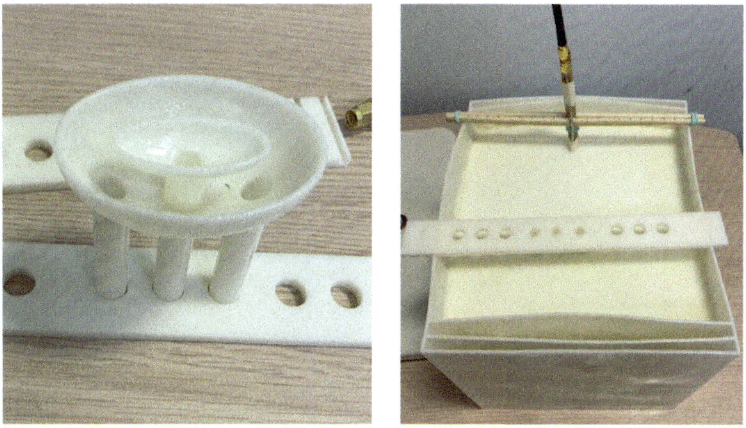

Fig. 13 **a** Phantom inner structure; **b** Experimental implementation of the MWI system

As well evidenced by Fig. 13b, the opacity of the coupling material does not allow verification of the actual position of the antenna. This introduces uncertainties in the measurements, which were not present in the simulated system. On the contrary, the consistency of the material did not impede the smooth movement of the antenna.

To fill the different phantoms shown in Fig. 12b, three different tissue-mimicking materials representing the healthy liver, coagulation necrosis, and carbonized liver were fabricated and characterized. The tissue-mimicking materials are made of a Triton-X/water/NaCl mixture whose properties can be tuned by adjusting the Triton-X to water ratio. The NaCl in the recipe was used to increase the mixture conductivities. The dielectric properties measurement was performed with the same set-up as the one for coupling medium measurement, presented in Sect. 2.3. The phantom materials recipes are shown in Table 4.

The measured dielectric properties of tissue-mimicking materials are shown in Fig. 14. In particular, each tissue-mimicking material was measured three times, and the average results of the three measurements are reported. In the figure, the reference dielectric properties are reported also. In particular, the healthy liver and the

Table 4 Tissue mimicking material recipe

	The ratio in mass (g)		
	NaCl	Triton X-100 (%)	Distilled water (%)
Healthy liver	0.39%	32.64	66.97
Coagulation necrosis	0	49.96	50.04
Carbonized liver	0	85.89	14.11

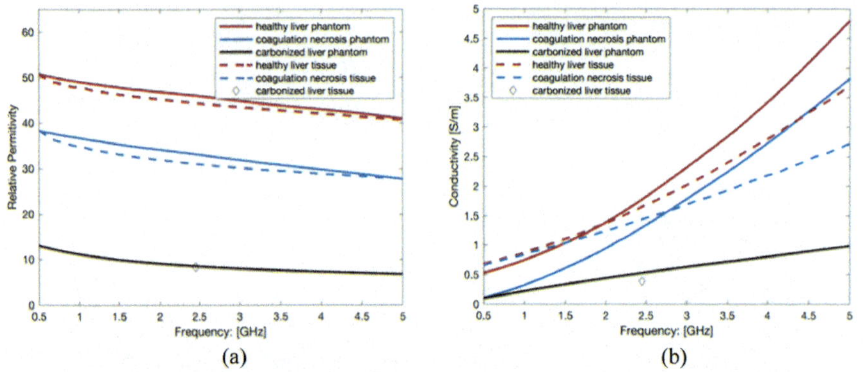

Fig. 14 Dielectric properties of tissue-mimicking material

coagulation necrosis properties are from [26]. The carbonized liver tissue properties, obtained at 2.45 GHz, are from the literature [20].

To further understand the difference between the phantom dielectric properties and the corresponding tissue (i.e., healthy liver [26], coagulation necrosis [26], and carbonized liver [20]), their properties difference along the frequency band 500 MHz–5 GHz were evaluated. The mean values and standard derivations of the difference are reported in Table 5. Since the literature [20] provided the carbonized liver properties at a single frequency (2.45 GHz), the measured value at this frequency is compared only. It is found that the permittivity difference between the phantom and the corresponding tissue is within 5%. However, the conductivity difference is up to 36%; regrettably, due to the limitation of the materials used to make the phantoms, it is not possible to decrease this difference by modifying the ratio of the ingredients. Besides, the conductivity of the phantom could only be increased by adding more NaCl into the recipe.

Table 5 Dielectric difference between the phantom material and the corresponded tissue

Tissue mimicking phantom	Relative permittivity difference		Conductivity difference	
	Mean	STD	Mean	STD
Healthy liver	0.0267	0.0084	0.1531	0.1535
Coagulation necrosis	0.0466	0.0211	0.304	0.2059
Carbonized liver	0.0214	N/A	0.3567	N/A

4.2 Image Reconstruction

In terms of the algorithm to be used in the MWI system for liver ablation monitoring, the study in [51] compared the performances of two approaches, namely the truncated singular value decomposition (TSVD), exploiting the distorted Born approximation (DBA) [22], and the linear sampling method (LSM). Due to the limitation in aspect imposed by the problem geometry (the abdomen can be probed only from one side and only for a limited portion), better performances were expected from TSVD as compared to LSM. As a matter of fact, LSM only provides a satisfying result when the target is illuminated from all directions. The outcome of that study verified this point, demonstrating that the TSVD algorithm could provide more accurate imaging results as compared to the LSM, by using the same measurement configuration to image the evolution of the ablation zone.

In this work, the image reconstruction is carried out through a differential MWI approach similar to the one used in [21, 49]. In such an approach, the imaging task is faced by exploiting the DBA [22]. This approach is suitable for the imaging of small size targets, with low dielectric properties contrast as compared to the background medium. In DBA, the total field induced by each antenna in the ROI $\left(\underline{E}_{tot}\right)$ can be approximated with the incident field $\left(\underline{E}_{inc}\right)$, say $\underline{E}_{tot} \cong \underline{E}_{inc}$. By doing so, the underlying imaging task is reduced to a linear ill-posed inverse problem [52]. Such a simplification has the remarkable consequence of reducing the computational complexity, hence enabling performing real-time imaging for the thermal ablation treatment. On the other hand, this approach can only provide information on the morphology (i.e., presence, position, extent) of the region wherein the contrast variation is occurring since the quantitative estimation of the tissue properties is only achieved when the enforced approximation is completely fulfilled.

It is worth mentioning that if the goal of the imaging task is to obtain an estimate of the permittivity in the treated tissue, the iterated (non-real-time) version of the DBA as in [53] could be appropriate. However, the goal of this study is to develop a MWI device for monitoring the treatment procedure, which lasts no more than 10 min. A suitable algorithm would be one that can determine the position and size of the treated tissue in a timely manner.

Same as the in-silico analysis in [49], to form the MWI images, the input datum of the imaging procedure was given by the differential scattering parameters calculated as the difference between the scattering parameters in two different scenarios. From all the considered scenarios, five differential data-sets are generated, namely: a-s0, b-s0, c-s0, b-a, c-b. The data sets a-s0, b-s0, and c-s0 account for the changes that occurred with respect to the reference scenario, and therefore the goal was to image the extent of the outer ellipsoid. In the meanwhile, the b-a data-set is representative of a scenario in which the changes occur only in the inner ellipsoid, wherein the tissue ablated in the initial part of the treatment is located. Finally, c-b account for the difference between the intermediate and final stages of the treatment, and the goal is to image the external ellipsoid, and possibly the transition from necrosis to carbonization occurring in the inner part.

The scattering parameters were recorded at 8 different positions (see Fig. 12) for the following scenarios: ROI with phantoms (a, b, c) and ROI without phantom (s0). For each position, the S-parameters were recorded 5 times without the phantom, and 5 times with the presence of the phantom. The S-parameters results were averaged before the data processing.

The levels of the differential signals calculated from the measured scattering parameters for all 5 scenarios are reported in Fig. 15. It is found that the signal level for all scenarios is similar: in the range from −30 to −90 dB. Such a signal level is above the dynamic range of a commercial VNA. It is also worth noting that the differential signal level from 0.5 to 2 GHz is above −50 dB, which is higher than that from 2 to 3 GHz. This phenomenon is in agreement with the conclusion of the optimal MWI system working conditions: 0.5–2 GHz is the most suitable frequency band for this MWI system. Therefore, the considered bandwidth for the data processing is from 0.7 to 1.6 GHz.

The algorithm used for processing the experimental data is the same as the one in [49]. It is worth noting here that even though TSVD should in principle be able to give quantitative information, in this case it can only provide qualitative knowledge of the imaged area, mainly because of the aspect-limited input data (the abdomen is probed from one side only). However, the qualitative information, i.e. the location where the dielectric properties changed, is enough to estimate the extension of the ablated area.

The reconstructions of the five scenarios achieved from the elaboration of the experimental differential data (a-s0, b-s0, c-s0, b-a, c-b) are reported in Fig. 16 (corresponding to y–z plane in Fig. 12). In particular, in the figure the contrast amplitude normalized to the maximum value is shown. As above reported, the data sets a-s0, b-s0, c-s0 account for the changes that occurred with respect to the reference scenario (s0). The b-a data set is representative of a scenario in which the changes occur only in the inner ellipsoid, c-b accounted for the difference between the intermediate and final stages of the treatment. In Fig. 16, a rectangular mask in the region of interest ranging from −41 mm to 41 mm along the y-axis and from −21.7 mm to 25.9 mm along the z-axis, is used to highlight the target area. The red contour in the figure indicates the ground truth of the outer ellipsoid, and the black contour in the figure indicates the ground truth of the inner ellipsoid. It is found that the reconstruction of the position and dimension of the target overlaps with the ground truth.

The results show that the mono-static MWI system is capable of detecting the differential signal caused by the changing of dielectric properties in the region of interest, even though the variation of the inner ellipsoid is hard to identify. Thus, while proving the concept, the mono-static MWI system is unable to reveal the ablation stages during the treatment.

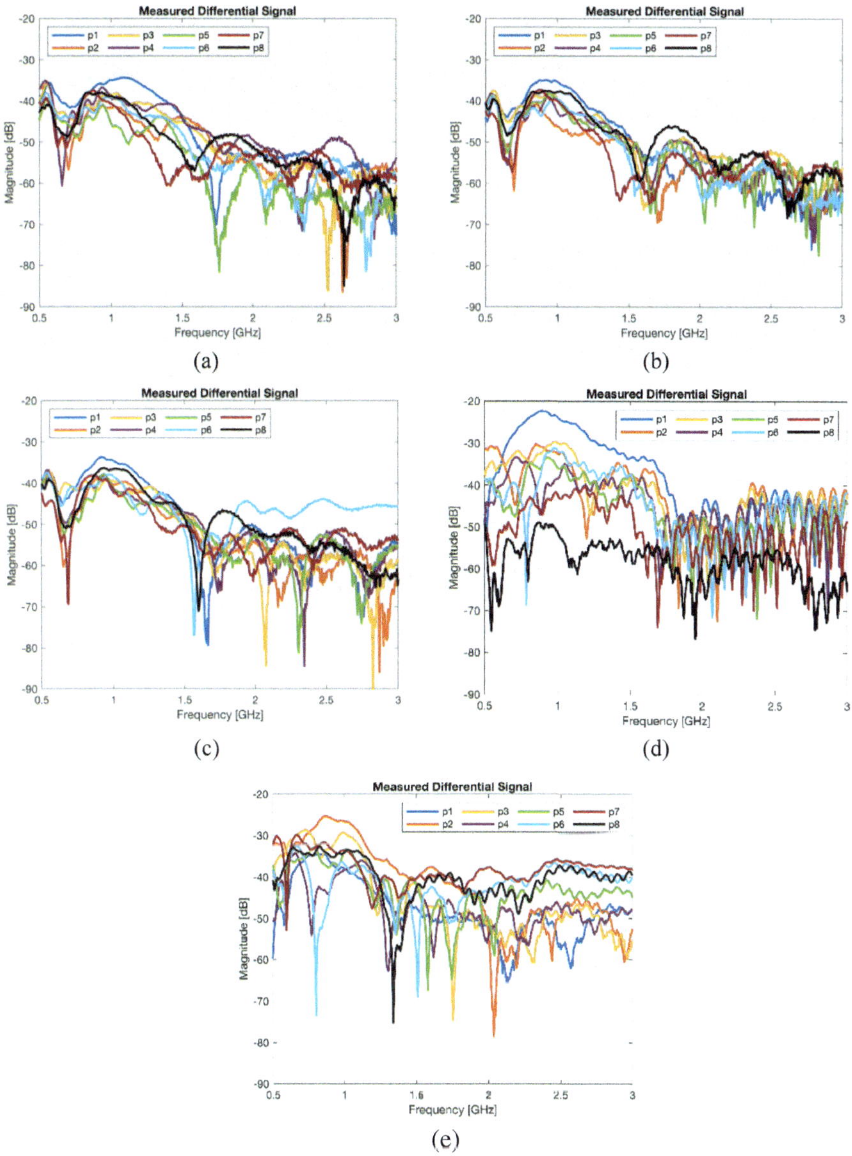

Fig. 15 Measured differential signal of scenario: **a** 'a-s0'; **b** 'b-s0'; **c** 'c-s0'; **d** 'b-a'; **e** 'c-b' at all positions. P1 to p8 represents the 8 different positions of the antenna inside the coupling medium

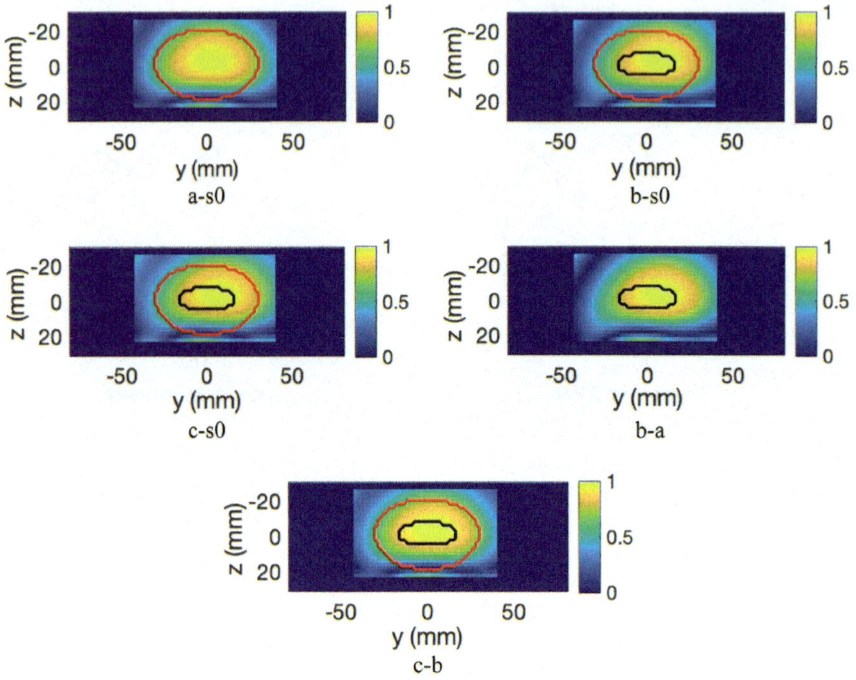

Fig. 16 Reconstructions of target from five differential scenarios

5 Conclusions

The aim of this chapter is to show the guidelines for the development of an experimental MWI system for real-time monitoring of thermal ablation procedures. Over the years, real-time monitoring of liver thermal ablation treatments has been considered a challenging task because conventional modalities such as X-ray, CT, PET/CT, MRI, and Ultrasound all have their constraints. Therefore, the successful execution of thermal ablation treatment still heavily relies on the clinician's experience. Microwave imaging is a modality with the advantages of being non-invasive, based on non-ionizing radiation, portable, cost-effective, and capable of real-time monitoring. It has been enormously investigated in recent decades, and demonstrates promising results for biomedical applications. However, to the existing knowledge, it is the first time that a microwave imaging system has been conceived for real-time monitoring of liver tumor ablation.

The design guidelines of a MWI system for liver thermal ablation monitoring were given. The goal of the guideline is to precisely determine the useful frequency band and coupling medium properties that can be utilized for the MWI system following a rational approach towards the system design.

Aiming for the optimal performance of the MWI system, the apparatuses were specially conceived within the guidelines. In this respect, a slot-loaded antipodal Vivaldi antenna was designed. The uncomplicated slot-loaded design is easy to fabricate and demonstrates a good performance towards reducing the mutual coupling effect between neighboring antennas. Finally, the antenna was fabricated and its performance characterized in the coupling medium. The antenna's simulation and measurement results demonstrated a good agreement with each other.

An initial experimental assessment was performed to validate the feasibility of the MWI system. To this end, a simple multi-view mono-static MWI system consisting of one antenna moving linearly in front of the phantom was realized. The 2D imaging results show that it is possible to localize the phantom position and the dimension in a satisfactory way, in agreement with previous in-silico analysis. It is worth noting here that the resolution expected from MWI is about $\lambda/4$ (a quarter of wavelength), i.e. about 2 cm for the frequencies and coupling material dielectric properties used in this application. This dimension is close to the dimensions of the ellipsoid (3.4 $\times$ 1.3 cm) in the experiment, showing that the foreseen imaging can be actually achieved.

While the performed experiment took advantage of a simple shape of the experimental set-up, it is worth noticing here that the ellipsoidal shape of the thermally ablated area corresponds to experimental findings in real clinical scenario. Additionally, in the actual scenario, the location of the tumor and positioning of the microwave antenna are well known, and can be used to help microwave imaging processing thus balancing the additional noise and uncertainties linked to the real environment.

Finally, since the more time-demanding computational task is the one performed by TSVD, which can be performed before the beginning of the procedure, the proposed approach can be exploited in real-time, to achieve the monitoring during the thermal ablation procedure.

Future work could include the experimental assessment of a multi-view multi-static MWI system with 8 antennas placed in front the phantom in array configuration. Moreover, studies on the measurement configuration, such as the numbers of antennas and their placement could be done looking for the optimal results, as the final goal of the study is to implement the MWI system is a clinical setting. Therefore, an investigation devoted to monitoring ex-vivo thermal liver ablation treatment with the proposed MWI system should be performed.

Acknowledgements This work was supported by the EMERALD project funded from the European Union's Horizon 2020 research and innovation program under the Marie Skłodowska-Curie grant agreement No. 764479. The Authors would like to thank Soroush Abedi, Nadine Joachimowicz, and Hélène Roussel for providing the 3D-printed phantom of the experimental validation.

References

1. H. Sung, J. Ferlay, R.L. Siegel, M. Laversanne, I. Soerjomataram, A. Jemal, F. Bray, Global cancer statistics 2020: GLOBOCAN estimates of incidence and mortality worldwide for 36 cancers in 185 countries. CA: Cancer J. Clin. **71**(3), 209–249 (2021)
2. C.Y. Liu, K.F. Chen, P.J. Chen, Treatment of liver cancer. Cold Spring Harb. Perspect. Med. **5**(9), a021535 (2015)
3. P. Srivatanakul, H. Sriplung, S. Deerasamee, Epidemiology of liver cancer: an overview. Asian Pac. J. Cancer Prev. **5**(2), 118–125 (2004)
4. S.A. Wells, J.L. Hinshaw, M.G. Lubner, T.J. Ziemlewicz, C.L. Brace, F.T. Lee, Liver ablation: best practice. Radiol. Clin. **53**(5), 933–971 (2015)
5. M.F. Meloni, J. Chiang, P.F. Laeseke, C.F. Dietrich, A. Sannino, M. Solbiati, E. Nocerino, C.L. Brace, F.T. Lee Jr, Microwave ablation in primary and secondary liver tumors: technical and clinical approaches. Int. J. Hyperth. **33**(1), 15–24 (2017)
6. V. Lopresto, R. Pinto, M. Cavagnaro, Experimental characterisation of the thermal lesion induced by microwave ablation. Int. J. Hyperth. **30**(2), 110–118 (2014)
7. S. Mulier, Y. Ni, L. Frich, F. Burdio, A.L. Denys, J.F. De Wispelaere, B. Dupas, N. Habib, M. Hoey, M.C. Jansen, M. Lacrosse, L. Michel, Experimental and clinical radiofrequency ablation: proposal for standardized description of coagulation size and geometry. Ann. Surg. Oncol. **14**(4), 1381–1396 (2007)
8. E.K. Abdalla, J.N. Vauthey, L.M. Ellis, V. Ellis, R. Pollock, K.R. Broglio, K. Hess, S.A. Curley, Recurrence and outcomes following hepatic resection, radiofrequency ablation, and combined resection/ablation for colorectal liver metastase. Ann. Surg. **239**(6), 818 (2004)
9. P. Bruners, G.D. Pandeya, E. Levit, E. Roesch, T. Penzkofer, P. Isfort, B. Schmidt, M.J. Greuter, M. Oudkerk, T. Schmitz-Rode, C.K. Kuhl, A.H. Mahnken, CT-based temperature monitoring during hepatic RF ablation: feasibility in an animal model. Int. J. Hyperth. **28**(1), 55–61 (2012)
10. L. Shi, S. Tashiro, Estimation of the effects of medical diagnostic radiation exposure based on DNA damage. J. Radiat. Res. **59**(suppl_2), ii121–ii129 (2018)
11. B. Quesson, J.A. de Zwart, C.T. Moonen, Magnetic resonance temperature imaging for guidance of thermotherapy. J. Magn. Reson. Imaging: Official J. Int. Soc. Magn. Reson. Med. **12**(4), 525–533 (2000)
12. M. Ahmed, L. Solbiati, C.L. Brace, D.J. Breen, M.R. Callstrom, J.W. Charboneau, M.H. Chen, B.I. Choi, T. De Baère, G.D. Dodd III, D.E. Dupuy, S.N. Goldberg, Image-guided tumor ablation: standardization of terminology and reporting criteria—a 10-year update. Radiology **273**(1), 241–260 (2014)
13. D.A. North, R.T. Groeschl, D. Sindram, J.B. Martinie, D.A. Iannitti, M. Bloomston, C. Schmidt, W.S. Rilling, T.C. Gamblin, R.C. Martin, Microwave ablation for hepatic malignancies: a call for standard reporting and outcomes. Am. J. Surg. **208**(2), 284–294 (2014)
14. R.C. Conceição, J.J. Mohr, M. O'Halloran (eds.), *An introduction to microwave imaging for breast cancer detection* (Springer International Publishing, Basel, Switzerland, 2016)
15. N.K. Nikolova, Microwave imaging for breast cancer. IEEE Microwave Mag. **12**(7), 78–94 (2011)
16. R. Scapaticci, L. Di Donato, I. Catapano, L. Crocco, A feasibility study on microwave imaging for brain stroke monitoring. Prog. Electromagnet. Res. B **40**, 305–324 (2012)
17. M. Haynes, J. Stang, M. Moghaddam, Real-time microwave imaging of differential temperature for thermal therapy monitoring. IEEE Trans. Biomed. Eng. **61**(6), 1787–1797 (2014)
18. B. Amin, A. Shahzad, L. Crocco, M. Wang, M. O'Halloran, A. González-Suárez, M.A. Elahi, A feasibility study on microwave imaging of bone for osteoporosis monitoring. Med. Biol. Eng. Comput. **59**(4), 925–936 (2021)
19. M. Lazebnik, D. Popovic, L. McCartney, C.B. Watkins, M.J. Lindstrom, J. Harter, S. Sewall, T. Ogilvie, A. Magliocco, T.M. Breslin, W. Temple, A large-scale study of the ultrawideband microwave dielectric properties of normal, benign and malignant breast tissues obtained from cancer surgeries. Phys. Med. Biol. **52**(20), 6093 (2007)

20. V. Lopresto, R. Pinto, G.A. Lovisolo, M. Cavagnaro, Changes in the dielectric properties of ex vivo bovine liver during microwave thermal ablation at 2.45 GHz. Phys. Med. Biol. **57**(8), 2309 (2012)
21. R. Scapaticci, V. Lopresto, R. Pinto, M. Cavagnaro, L. Crocco, Monitoring thermal ablation via microwave tomography: an ex vivo experimental assessment. Diagnostics **8**(4), 81 (2018)
22. M. Pastorino, *Microwave Imaging*, vol. 208. (Wiley, 2010)
23. A. Peyman, C. Gabriel, Cole–Cole parameters for the dielectric properties of porcine tissues as a function of age at microwave frequencies. Phys. Med. Biol. **55**(15), N413 (2010)
24. S. Gabriel, R.W. Lau, C. Gabriel, The dielectric properties of biological tissues: III. Parametric models for the dielectric spectrum of tissues. Phys. Med. Biol. **41**(11), 2271 (1996)
25. D. Andreuccetti, R. Fossi, C. Petrucci, An internet resource for the calculation of the dielectric properties of body tissues in the frequency range 10 Hz–100 GHz. IFAC-CNR, Florence (Italy), Based on data published by C. Gabriel et al. [Online]. Available http://niremf.ifac.cnr.it/tissprop/ (1997)
26. M. Wang, L. Farina, L. Crocco, M. Cavagnaro, Changes in dielectric properties following a microwave thermal ablation procedure, in 2021 *International Conference on Electromagnetics in Advanced Applications*, 9–13 August 2021. (Honolulu, Hawaii, USA)
27. M. Wang, L. Crocco, M. Cavagnaro, On the design of a microwave imaging system to monitor thermal ablation of liver tumors. IEEE J. Electromagnet. RF Microwaves Med. Biol. (2021)
28. I.G. Zubal, C.R. Harrell, E.O. Smith, Z. Rattner, G. Gindi, P.B. Hoffer, Computerized three-dimensional segmented human anatomy. Med. Phys. **21**(2), 299–302 (1994)
29. O. Akkus, A. Oguz, M. Uzunlulu, M. Kizilgul, Evaluation of skin and subcutaneous adipose tissue thickness for optimal insulin injection. J. Diabetes Metab. **3**(216), 2 (2012)
30. C.A. Balanis, *Advanced Engineering Electromagnetics*. (Wiley, 1999)
31. R. Scapaticci, M. Bjelogrlic, J.A.T. Vasquez, F. Vipiana, M. Mattes, L. Crocco, Microwave technology for brain imaging and monitoring: physical foundations, potential and limitations, in *Emerging Electromagnetic Technologies for Brain Diseases Diagnostics, Monitoring and Therapy*, ed. by L. Crocco, I. Karanasiou, M.L. James, R.C. Conceição (Springer, 2018), pp. 7–35
32. J. VanderKruk, C. Steelman, A. Endres, H. Vereecken, Dispersion inversion of electromagnetic pulse propagation with in freezing and thawing soil waveguides. Geophys. Res. Lett. **36**(18) (2009)
33. J. Stang, M. Haynes, P. Carson, M. Moghaddam, A preclinical system prototype for focused microwave thermal therapy of the breast. IEEE Trans. Biomed. Eng. **59**(9), 2431–2438 (2012)
34. P.M. Meaney, D. Goodwin, A.H. Golnabi, T. Zhou, M. Pallone, S.D. Geimer, G. Burke, K.D. Paulsen, Clinical microwave tomographic imaging of the calcaneus: a first-in-human case study of two subjects. IEEE Trans. Biomed. Eng. **59**, 3304–3313 (2012)
35. R. Scapaticci, J. Tobon, G. Bellizzi, F. Vipiana, L. Crocco, Design and numerical characterization of a low-complexity microwave device for brain stroke monitoring. IEEE Trans. Antennas Propag. **66**(12), 7328–7338 (2018)
36. M. Meaney, T. Zhou, D. Goodwin, A. Golnabi, E.A. Attardo, K.D. Paulsen, Bone dielectric property variation as a function of mineralization at microwave frequencies. J. Biomed. Imaging (2012)
37. X. Li, S.K. Davis, S.C. Hagness, D.W. Van der Weide, B.D. Van Veen, Microwave imaging via space-time beamforming: experimental investigation of tumor detection in multilayer breast phantoms. IEEE Trans. Microw. Theory Tech. **52**(8), 1856–1865 (2004)
38. D.O. Rodriguez-Duarte, J.A. Tobon Vasquez, R. Scapaticci, G. Turvani, M. Cavagnaro, M.R. Casu, L. Crocco, F. Vipiana, Experimental validation of a microwave system for brain stroke 3-D imaging. Diagnostics **11**(7), 1232 (2021)
39. B. Borja, J.A. Tirado, H. Jardon, An overview of UWB antennas for microwave imaging systems for cancer detection purposes. Prog. Electromagnet. Res. B **80** (2018)
40. O.M. Bucci, L. Crocco, R. Scapaticci, G. Bellizzi, On the design of phased arrays for medical applications. Proc. IEEE **104**(3), 633–648 (2016)

41. A. Yakovlev, S. Kim, A. Poon, Implantable biomedical devices: wireless powering and communication. IEEE Commun. Mag. **50**(4), 152–159 (2012)
42. G. Ruvio, M.J. Ammann, A miniaturized antenna for UWB-based breast imaging, in *2009 3rd European Conference on Antennas and Propagation.* (IEEE, Mar 2009), pp. 1864–1867
43. R.C. Hansen, Fundamental limitations in antennas. Proc. IEEE **69**(2), 170–182 (1981)
44. P. Fei, Y. Jiao, W. Hu, F. Zhang, A miniaturized antipodal Vivaldi antenna with improved radiation characteristics. IEEE Antennas Wirel. Propag. Lett. **10**, 127–130 (2011)
45. Q. Ying, W. Dou, Simulation of two compact antipodal Vivaldi antennas with radiation characteristics enhancement, in *2013 Proceedings of the International Symposium on Antennas and Propagation*, vol. 1. (IEEE, Oct 2013), pp. 523–526
46. X. Fang, M. Ramzan, Q. Wang, D. Plettemeier, Compact antipodal Vivaldi antennas for body area communication, in *Advances in Body Area Networks I*. (Springer, Cham, 2019), pp. 357–369
47. J. Ao, J. Huang, W. Wu, N. Yuan, A miniaturized Vivaldi antenna by loading with parasitic patch and lumped resistor. AEU-Int. J. Electron. Commun. **81**, 158–162 (2017)
48. M. Wang, L. Crocco, M. Cavagnaro, Antipodal Vivaldi antenna with ceramic cone lens for biomedical microwave imaging systems, in *2021 15th European Conference on Antennas and Propagation (EuCAP)*, 22–26 March 2021. Virtual Conference pp. 1–5
49. M. Wang, R. Scapaticci, M. Cavagnaro, L. Crocco, Towards a microwave imaging system for continuous monitoring of liver tumor ablation: design and in silico validation of an experimental set-up. Diagnostics **11**(5), 866 (2021)
50. V. Lopresto, L. Strigari, L. Farina, S. Minosse, R. Pinto, D. D'Alessio, B. Cassano, M. Cavagnaro, CT-based investigation of the contraction of ex vivo tissue undergoing microwave thermal ablation. Phys. Med. Biol. **63**, 055019 (2018)
51. R. Scapaticci, G.G. Bellizzi, M. Cavagnaro, V. Lopresto, L. Crocco, Exploiting microwave imaging methods for real-time monitoring of thermal ablation. Int. J. Antennas Propag. (2017)
52. M. Bertero, P. Boccacci, *Introduction to Inverse Problems in Imaging*. (CRC Press, Boca Raton, FL, USA, 2020)
53. G. Chen, J. Stang, M. Haynes, E. Leuthardt, M. Moghaddam, Real-time three-dimensional microwave monitoring of interstitial thermal therapy. IEEE Trans. Biomed. Eng. **65**(3), 528–538 (2017)

The manufacturer's authorised representative in the EU is Springer Nature Customer Service Centre GmbH, Europaplatz 3, 69115 Heidelberg, Germany. If you have any concerns regarding our products, please contact ProductSafety@springernature.com

Printed and bound by CPI Group (UK) Ltd, Croydon, CR0 4YY
01/10/2026
02225902-0010